Anti-Idiotypes, Receptors, and
Molecular Mimicry

D. Scott Linthicum   Nadir R. Farid
Editors

Anti-Idiotypes, Receptors, and Molecular Mimicry

With 92 Illustrations

Springer Science+Business
Media, LLC

D. Scott Linthicum
Department of Pathology
  and Laboratory Medicine
University of Texas
  Health Science Center
Houston, TX 77225, USA

Nadir R. Farid
Department of Medicine
Memorial University
St. Johns, Newfoundland
Canada, A1B 3V6

Library of Congress Cataloging-in-Publication Data
Anti-idiotypes, receptors, and molecular mimicry.
   Proceedings of a pre-congress symposium held
in Quebec, June 28–July 1, 1986, in conjunction with
the Sixth International Congress of Immunology held
in Toronto.
   Bibliography: p.
   Includes index.
   1. Immunoglobulin idiotypes—Congresses.   2. Cell
receptors—Congresses.   I. Linthicum, D. Scott.
II. Farid, Nadir R.
QR186.7.A55   1988   616.07'9   87-14814

Cover photo illustrates the interaction of the hapten (neuroleptic drug haloperidol with striped balls) with a three dimensional model of the Fv region of a monoclonal idiotypic antibody. The model was generated on an IBM PC/AT using a molecular modeling program called MOLDISP, which was written by Michael B. Bolger.

Typeset by David E. Seham Associates, Inc., Metuchen, New Jersey.

9 8 7 6 5 4 3 2 1

ISBN 978-1-4612-8325-6   ISBN 978-1-4612-3734-1 (eBook)
DOI 10.1007/ 978-1-4612-3734-1

Preface

The pre-congress satellite symposium "Anti-Idiotypes as Probes for the Study of Receptors" was held at the Chateau Montebello, in Quebec, Canada, June 28 through July 1, 1986, in conjunction with the Sixth International Congress of Immunology in Toronto, Ontario, Canada. More than fifty scientists participated in a very intensive exchange of new ideas and discussion of diverse systems. The symposium consisted of formal presentations, poster sessions, and even a limerick contest (if memory serves us correctly it was Glen Gaulton who was voted as most creative for his "penetrating" version of an anti-idiotypic limerick).

This volume represents the written results of the participants. We had hoped that this diverse groups of scientists, each individually interested in different kinds of receptors, but collectively interested in the anti-idiotypic molecular mimicry approach, would not be hesitant in an exhaustive mutual sharing of ideas, concepts, and strategies. Our experiences at Montebello proved positive and the contributions to this volume bear witness to this. A number of collaborations and friendships were formed at Montebello and we anticipate more to follow in the years to come. We would like to express our sincere thanks for the financial support we obtained from our sponsors (see acknowledgments). We also express our deepest thanks to our colleagues who contributed to this effort and wish each of them great success in their future endeavors.

D. Scott Linthicum
Nadir R. Farid

Acknowledgments

The organizers and all the participants wish to express their thanks to the sponsors listed below who provided financial support for this symposium:

CBS Scientific Company, Inc., Del Mar, CA 92014, USA
Hoffman La Roche Inc., Nutley, NJ 07110, USA
Hybritech Inc., San Diego, CA 92121, USA
Monsanto Company, St. Louis, MO 63167, USA
Springer-Verlag/Publishers New York Inc., New York, NY 10010, USA
Cooper Animal Health, Melbourne, Wellcome Australia Ltd.
Boehringer Mannheim GmbH, Werk Tutzing, Federal Republic of Germany
The NutraSweet Company, Skokie, IL 60076, USA

Contents

Contributors

BLAIR ARDMAN
School of Medicine, Cancer Research Center, Tufts University, Boston,
MA 02111, USA

LOREN A. BABIUK
Department of Veterinary Microbiology, Western College of Veterinary
Medicine, University of Saskatchewan, Saskatoon, Saskatchewan,
Canada S7N OW2

J. EDWIN BLALOCK
Department of Physiology and Biophysics, University of Alabama at
Birmingham, Birmingham, AL 35297, USA

MICHAEL B. BOLGER
Division of Medicinal Chemistry, School of Pharmacy, University of
Southern California, Los Angeles, CA 90033, USA

CONSTANTIN BONA
Department of Microbiology, Mount Sinai Medical School, New York,
NY, 10029 USA

KENNETH L. BOST
Department of Physiology and Biophysics, University of Alabama at
Birmingham, Birmingham, AL 35297, USA

RAFAEL BRUCK
Department of Hormone Research, The Weizmann Institute of Science,
Rehovot 76100, Israel

SUSAN BURDETTE
Tupper Research Institute, New England Medical Center Hospital, Boston,
MA 02111, USA

PATRICK R. CARNEGIE
School of Agriculture, La Trobe University, Bundoora, Victoria 33083,
Australia

W. LOUIS CLEVELAND
Department of Microbiology, Roosevelt Hospital, New York,
NY 10032, USA

IRUN R. COHEN
Department of Cell Biology, The Weizmann Institute of Science,
Rehovot 76100, Israel

JEAN-YVES COURAUD
Section de Pharmacologie et d'Immunologie, Département de Biologie,
CEN/Saclay, 91191 Gif-sur-Yvette Cedex, France

ANGEL L. DE BLAS
Department of Neurobiology and Behavior, State University of New York, at
Stony Brook, Stony Brook, NY 11794, USA

ALLEN B. EDMUNDSON
Department of Biology, University of Utah, Salt Lake City,
UT 84112, USA

DANA ELIAS
Department of Hormone Research, The Weizmann Institute of Science,
Rehovot 76100, Israel

NADIR R. FARID
Thyroid Research Laboratory, Health Sciences Center, St. John's,
Newfoundland A1B 3V6, Canada

GLEN N. GAULTON
Department of Pathology, Division of Immunobiology, University of
Pennsylvania School of Medicine, Philadelphia, PA 19104, USA

JAY A. GLASEL
Department of Biochemistry, University of Connecticut Health Center,
Farmington, CT 06032, USA

MARK I. GREEN
Department of Pathology, Division of Immunobiology, University of
Pennsylvania School of Medicine, Philadelphia, PA 19104, USA

CHARLES J. HOMCY
Cardiac Unit, Massachusetts General Hospital, Boston, MA 02114, USA

JYH-HSIUNG HUANG
Department of Molecular Immunology, Roswell Park Memorial Institute,
Buffalo, NY 14263, USA

MARGUERITE M.B. KAY
Research and Medical Services and Division of Geriatric Medicine,
Department of Medicine, Texas A & M University, Olin E. Teague Veterans
Center, Temple, TX 76501, USA

HEINZ KOHLER
Department of Molecular Immunology, Roswell Park Memorial Institute,
Buffalo, NY 14263, USA

D. SCOTT LINTHICUM
Department of Pathology and Laboratory Medicine, University of Texas
Health Science Center at Houston, Houston, TX 77025, USA

RUTH MARON
Department of Cell Biology, The Weizmann Institute of Science,
Rehovot 76100, Israel

PHYLLIS L. OSHEROFF
Department of Molecular Genetics, Roche Research Center, Hoffman-La
Roche, Inc., Nutley, NJ 07110, USA

DONGEUN PARK
Department of Neurobiology and Behavior, State University of New York at
Stony Brook, Stony Brook, NY 11794, USA

ROBERT L. RAISON
Clinical Immunology Research Unit, Sydney University, Sydney NSW 1006,
Australia

LAKSHMI SANGAMESWARAN
Department of Neurobiology and Behavior, State University of New York at
Stony Brook, Stony Brook, NY 11794, USA

DAVID S. SAWUTZ
Cardiac Unit, Massachusetts General Hospital, Boston, MA 02114, USA

YORAM SHECHTER
Department of Hormone Research, The Weizmann Institute of Science,
Rehovot 76100, Israel

MARK A. SHERMAN
Division of Medicinal Chemistry, School of Pharmacy, University of
Southern California, Los Angeles, CA 90033, USA

LAWRENCE R. SMITH
Department of Physiology and Biophysics, University of Alabama at
Birmingham, Birmingham, AL 35297, USA

YASMIN THANAVALA
Department of Molecular Immunology, Roswell Park Memorial Institute,
Buffalo, NY 14263, USA

SYLVIA VAN DRUNEN
LITTEL-VAN DEN HURK
Department of Veterinary Microbiology, Western College of Veterinary
Medicine, University of Saskatchewan, Saskatoon, Saskatchewan,
Canada S7N OW2

JAVIER VITORICA
Department of Neurobiology and Behavior, State University of New York at Stony Brook, Stony Brook, NY 11794, USA

RONALD WARD
Department of Molecular Immunology, Roswell Park Memorial Institute, Buffalo, NY 14263, USA

CECELIA A. WHETSTONE
National Animal Disease Center, USDA Agriculture Research Service, Ames, IA 50010, USA

1
Idiotypes, Paratopes, and Molecular Mimicry

NADIR R. FARID AND D. SCOTT LINTHICUM

Our understanding of the molecular nature of antibody molecules, their interactions with antigens, and even their interactions with other antibody molecules has become a very fast-paced and complex field. Unfortunately, a major stumbling block in this arena is the complicated immunological jargon which has developed over the past several years. Moreover, our knowledge of the precise physical-chemical nature of antigen-antibody interactions is somewhat limited by the sheer size and complexity of the molecules involved. In the past decade the structures of a number of antibody molecules have been studied extensively by X-ray crystallographic techniques (for a review, see Ref. 1), and recent computer modeling techniques to predict molecular structures of antibodies have been validated (2). Although we wish to understand the structural basis for unique biological effects of antibodies, for the most part we are confined to the usage of operational definitions. Immunologists are notorious for developing new and bizarre definitions, pathways, and nomenclature. The collection of work in this volume is probably no exception to this rule, but in this first chapter we wish to outline and clarify several key points which we hope will act as a common thread in the studies presented in this volume and those to follow.

The notion that the hypervariable regions of antibodies can be antigenic in their own right, and therefore the subject of an immune response, was proposed simultaneously by Oudin and Michael (3) and Kunkel *et al.* (4). The term "idiotype" is used to signify the collective antigenic determinants (epitopes) found on a given antibody molecule; these determinants are "idiosyncratic" for the molecule, hence the term. The idiotype may be shared, as a whole or in part, with other antibody molecules raised in response to a specific antigen within an individual or group of individuals. The individual antigenic determinants on an antibody molecule which make up the collective idiotype are termed "idiotopes." For the most part, idiotopes (5) can be localized to the hypervariable regions of heavy chains, light chains, or a combination of the two; there are some exceptions to this rule but we will not consider these at this time. The area of the antibody intimately involved in antigen binding, often a "pocket" or "cleft" formed by the heavy and light chains, is called the "paratope." This region is a "negative molecular mold" of the antigen and hence is often called the "internal image." Unfortunately, this concept constrains us to the idea that only "pockets" or "clefts" are involved in antigen binding and "internal images" are reflections of "pockets." We now know that protruding areas of antibodies can be involved in antigen binding and may represent an "external image" of the antigen which is bound (2). Consequently, one must be cautious in assuming that paratopes are always represented by a binding pocket or cleft.

In general, many investigators will use the term "anti-idiotype antibodies" to designate an antibody mixture or antisera which are directed against all of the idiotypic and paratypic determinants of a particular antibody molecule or mixture of antibodies. Thus, the anti-idiotype antisera that one generates using a given idiotype (Ab_1) are likely to contain (i) antibodies directed to the paratope, (ii) antibodies directed to paratope-related areas, and (iii) antibodies directed to all other idiotopes. The first two types of antibodies are often competitive with antigen for the binding site on Ab_1; the third type of antibody in most cases is not affected by antigen binding (for more detailed discussion of this, see Chapter 12). Only the anti-paratope antibody, however, will represent a true molecular mimic of the antigen. Furthermore, this antibody population will be able to react with Ab_1 molecules from a variety of different species. Immunization with anti-paratypic antibodies will give rise to a population of antibodies functionally equivalent to the original Ab_1.

The possibility that antibodies directed against the paratope could display molecular mimicry of the original antigen was first recognized by Lindemann (6), who dubbed them "homobodies." Jerne incorporated idiotypes and internal images in his network formulation of the immune system. He suggested (7) that the first antibody Ab_1 to an antigen epitope has, in addition to its specific paratope, specific idiotopes (see Fig 1.1). The Ab_1 with its specific paratope and idiotopes is now the subject of an immune response by a population of Ab_2 molecules. Some Ab_2 are solely anti-paratypic ($Ab_{2\beta}$) and some are solely anti-idiotypic ($Ab_{2\alpha}$); we should also extend this hypothesis to include some Ab_2 which may possess both characteristics simultaneously, denoted as $Ab_{2\delta}$ (to signify their "dual" nature). The next response of antibodies, Ab_3, may include the original Ab_1 population, since it is reactive to $Ab_{2\alpha}$ and $Ab_{2\delta}$, as well as other Ab_3 molecules which are part of an ever expanding population of anti-idiotypic molecules.

Sege and Peterson (8) were the first to grasp the fact that antibodies directed against small ligands could act as receptors for these ligands in a fashion similar to their real receptor. Consequently, they demonstrated that these anti-ligand antibodies could act as a mold and thereby produce a second set of antibodies (anti-idiotype) which was able to mimic the

Expanding Idiotypes and Anti-idiotypes

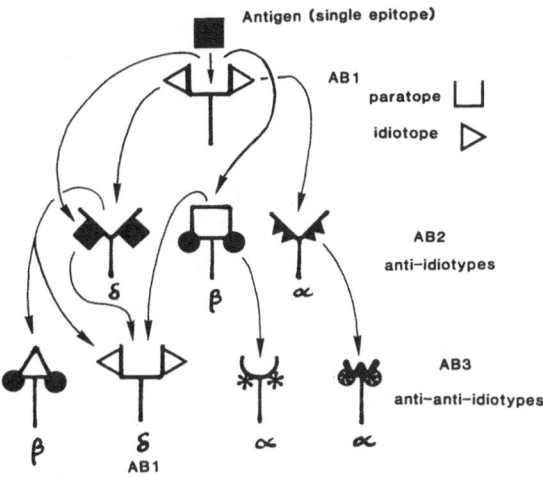

FIGURE 1.1. The idiotype cascade of anti-idiotypes and subsequent anti-anti-idio-types generates a vast array of specificities. In this scheme, antibody (Ab₁) reacts to the antigen epitope and induces three types of anti-idiotypes (Ab₂); Ab₂ᵦ, which has both anti-paratypic and anti-idiotypic specificity, Ab₂ᵦ, which is anti-paratypic, and Ab₂ₐ, which is anti-idiotypic only. The population of anti-anti-idiotype anti-bodies (Ab₃) is made up of new antibodies, plus the original Ab₁ which is induced by Ab₂ᵦ and Ab₂ᵦ.

original ligand (see Fig 1.2). This simple and pragmatic approach has been exploited repeatedly, and many successful studies have validated the original hypothesis. The contributors to this volume have all employed this basic concept in one fashion or another.

The natural assumption that the basis of the molecular mimicry displayed by anti-idiotypes is determined by a complex folding of hypervariable re-gions of the light chain and the heavy chain would suggest that isolated light and heavy chains of the antibodies would not interact with the idio-type. It may be too early to make this assumption, and we may find that the anti-paratypic activity is encoded by specific H- or L-chain regions exposed on the surface of the molecule. It seems unlikely that deep hy-drophobic pockets would be involved. Analysis of shared amino acid se-quences between antigen and anti-paratypic antibodies may reveal striking similarities, and computer-generated models to localize these determinants may reveal the regions potentially responsible for the molecular mimicry (9). These approaches are currently being pursued (see Chapters 9 and 19).

One must consider, however, that molecular mimicry may involve the use of analogues rather than homologues. Analogous molecular contact

Molecular Mimicry

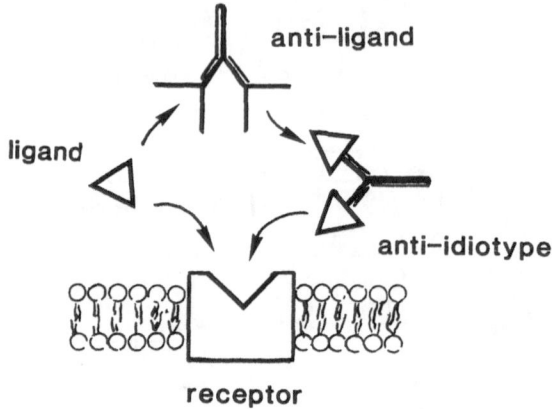

FIGURE 1.2. The working model of anti-idiotype molecular mimicry involves antibodies directed against the ligand which act like receptor and anti-idiotype antibodies against anti-ligand which mimic the original ligand. Both ligand and anti-idiotype antibodies compete for the receptor.

points are likely to be an important mechanism. Similarities in electrostatic interactions may account for some molecular mimicry (10). Such a mechanism could account for anti-paratopes which mimic carbohydrate residues (11). When considering small antigens, such as neurotransmitters and simple sugars, it is difficult to visualize the paratope being constricted to some obligatory residues. The anti-paratope may mimic the antigen because of its behavior in terms of contact points and molecular shape rather than strict homologous molecular composition.

In this volume different authors will present and discuss the ramifications of the use of anti-idiotypic antibodies as probes for hormone receptors, viruses, neurogenic compounds, and small messengers. While most of the work has relied upon the immunization of animals with a collection of exogenous polyclonal idiotypes, some studies have been able to utilize monoclonal reagents of defined specificity. Some of the structural basis of this molecular mimicry will be established in the future, and we will then be able to take advantage of this remarkable system for further study of immunoregulation, vaccines, and molecular engineering of analogues.

Acknowledgments. The authors are supported by grants NIH/NINCDS NS00974, NIH/NINCDS NS 22448, and National MS Society RG 1761-A-4 to D.S.L. and by grants from the Medical Research Council of Canada and the Faculty of Medicine, Memorial University of Newfoundland (Research and Development Fund) to N.R.F.

References

1. Amit, A.G., R.A. Mariuzza, S.E.V. Phillips, and R.J. Poljak. 1986. Three-dimensional structure of an antigen-antibody complex at 2.8 Å resolution. Nature (London) **233**:747.
2. Chothia, C. et al. 1986. The predicted structure of immunoglobulin D1.3 and its comparison with the crystal structure. Nature (London) **233**:755.
3. Oudin, Y., and M. Michael. 1963. Une nouvelle forme d'allotypie des globulins du sérum de lapin apparemment liée à la fonction et à la spécificité antecorps. C.R. Acad. Sci. **257**:805.
4. Kunkel, H.G., M. Mannik, and R.C. Williams. 1963. Individual antigenic specificities of isolated antibodies. Science **140**:1218.
5. Jerne, N.K. 1971. The somatic generation of immune recognition. Eur. J. Immunol. **1**:1.
6. Lindemann, G. 1973. Speculations on idiotypes and homobodies. Ann. Immunol. (Inst. Pasteur) **124**:171.
7. Jerne, N.K. 1974. Towards a network theory of the immune system. Ann. Immunol. (Inst. Pasteur). **125C**:373.
8. Sege K. and Peterson P.A. 1978. Use of anti-idiotypic antibodies as cell surface probes. Proc. Natl. Acad. Sci. USA **75**:2443.
9. Bruck, C., M.S. Co, M. Slaoui, G.N. Gaulton, T. Smith, B.N. Fields, J.I. Mullins, , and M.I. Greene. 1986. Nucleic acid sequence of an internal image-bearing monoclonal anti-idiotype and its comparison to the sequence of the external antigen. Proc. Natl. Acad. Sci. USA **83**:6578.
10. Erlanger, F. 1985. Anti-idiotypic antibodies: what do they recognize? Immunol. Today **6**:10.
11. Roitt, I.M., Y.M. Thanavala, D.K. Male, and F.C. Hay. 1985. Antibodies as surrogate antigens: structural considerations. Immunol. Today **6**:265.

2
Functional Heterogeneity of Anti-Idiotype Antibodies

CONSTANTIN A. BONA

Idiotypes are phenotypic markers of variable-region genes coding the specificity of the antigen receptor of B- or T-lymphocyte clones. There are two major categories of idiotypic determinants. Some individual idiotypes can be expressed on the receptor of a "unique" clone of a single individual in a given species. This individual idiotype is a result of a somatic mutational event which takes place in a "unique" clone subsequent to recombinational events between different gene segments encoding the variable region of heavy and light chains of the immunoglobulin receptor of B cells or variable regions of α and β chains of T-cell receptor. The second category of idiotypes are cross-reactive idiotypes expressed on the antigen receptor of different clones from different individuals of the same species or even of different species. The clones bearing cross-reactive idiotypes may recognize the same antigenic determinant or different antigenic determinants. Obviously, these idiotypes are markers of V-germline genes encoded by conserved DNA segments among individuals of a given species or various species. We defined among cross-reactive idiotypes a special category of idiotypes, termed as regulatory idiotypes, which play an important role in the idiotype-determined regulatory processes (1). The regulatory idiotypes are defined by four criteria:

1. They are autoimmunogenic, an intrinsic quality required to mediate regulatory functions in physiological conditions.
2. They are markers of V-germline genes.
3. They can be shared by antibodies with various specificities.
4. They are recognized and expanded by regulatory T cells (2).

The idiotopes are defined and recognized by clones bearing a receptor which is conventionally defined as anti-idiotype (anti-Id) or Ab₂.

There is a perfect symmetry of anti-idiotype clones in B- and T-cell compartments of the immune system. Indeed, in the B-cell compartments there are clones which are capable of producing antibodies which recognize the idiotypes of the immunoglobulin receptor of other B clones as well as antibodies which recognize antigen receptors of T-cell clones (these are also called "anti-clonotypic" antibodies). Similarly, in the T-cell compartment there are T-cell clones able to recognize the idiotopes of immunoglobulin molecules or the idiotopes of T cells (reviewed in Ref. 3). T cells specific for idiotopes have been identified in both helper and suppressor subsets (4–7).

These data clearly indicate that lymphocyte receptors recognize and are recognized by complementary clones. The portion of the receptor which recognizes an antigenic determinant (epitope) is termed the paratope whereas the determinant which is recognized is the idiotope. It should be pointed out that the distinction between paratope and idiotope is rather semantic since in certain cases the idiotope is identical to the combining site if antigen can inhibit or displace the binding of the anti-Id antibody. The simultaneous presence of complementary clones within the immune system led Jerne to formulate the idiotypic network theory (8), which viewed the immune system in a steady state composed of lymphocyte clones bearing complementary receptors.

This equilibrium is upset by foreign antigens. It therefore appears that the major effect of anti-idiotypes in the steady state is suppression. However, depending on the amount of anti-idiotypes, class of anti-idiotypes, age, and route of antigen administration, anti-Id antibodies can expand the clones bearing corresponding idiotypes. These fundamental observations open a new field for the utilization of anti-idiotypes as biological and therapeutical reagents.

Heterogeneity of Clones Producing Anti-Idiotypic Antibodies

Because the idiotypic determinants are borne by globular proteins, the cloned populations recognizing them are by definition heterogeneous. This is best illustrated by isoelectrofocusing analysis of cloned products of BALB/c mice producing syngeneic antibodies specific for an IdX borne by J558 monoclonal protein specific for 1-3 dextran. Schuller et al. (9) analyzed the IEF pattern of anti-J558IdX antibodies produced by several individual BALB/c mice. From this study they concluded that each BALB/c mouse has 6–8 clones synthesizing anti-idiotypic (anti-Id) antibodies and at least 100 different clones exist in the BALB/c strain. This conclusion is supported by our molecular studies carried out on syngeneic monoclonal antibodies specific for A48 idiotopes expressed on β2-6 functosan antibodies. Study of V_H gene familes used by six syngeneic monoclonal anti-Id antibodies showed that four used V_H genes from the V_H J558 family and two from the QPC 52 family (10). These two examples clearly showed that the population of clones producing anti-Id antibodies is extremely heterogeneous.

Functional Heterogeneity of Anti-Idiotypic Antibodies

From the immunochemical standpoint, anti-Id antibodies are also heterogeneous. They can recognize idiotopes either associated with the combining site or with the framework segments of the variable region. Whereas the binding of Ab_2 to idiotopes associated with the combining site is inhibited by the antigen (11), the binding of Ab_2 to idiotopes located on the framework is antigen noninhibitable. This was clearly demonstrated in an elegant study carried out by Mudgett et al. (12). These authors studied the effect of 3-O-α-N-acetylgalactosaminosyl-N-acetylgaloctosamine on the binding of rabbit anti-Id antibodies to homogeneous anti-polysaccharide antibodies prepared from the serum of a a rabbit immunized with Streptococcus group C. The binding of some anti-Id sera has been inhibited by antigen, whereas that of others has not. Similarly, some anti-Id antibodies passed over a column made with homogeneous anti-polysaccharide antibodies were eluted by hapten whereas others eluted with only ammomium thiocyanate. Interestingly, the binding of anti-Id antibodies eluted with hapten was inhibited by antigen whereas those eluted with ammomium thiocyanate were not inhibited by antigen. These experiments clearly showed that there are two types of anti-idiotypic antibodies, one recognizing idiotopes associated with the combining site and another recognizing idiotopes outside of the combining site.

Based on the immunochemical and functional properties of anti-Id antibodies, they can be classified into three groups: $Ab_{2\alpha}$ anti-idiotypes specific for combining site- or framework-associated idiotopes, $Ab_{2\beta}$ anti-idiotypes carrying the internal image of the antigens, and $Ab_{2\gamma}$ (epibody) anti-idiotype antibodies able to bind to idiotypes as well as to other antigenic determinants. (Fig. 2.1).

$Ab_{2\alpha}$ represents a subset of anti-Id antibodies which can recognize idiotypic determinants associated with either conbining site or framework segments of the variable region of the immunoglobulin receptor. The binding of anti-Id antibodies to the receptor can either stimulate or suppress B-cell clones, depending on the concentration of Ab_2. Generally, it has been observed that high amounts of anti-idiotypes cause suppression whereas minimal amounts induce the expansion of clones bearing the corresponding idiotopes. Suppressive or stimulating effects also depend on the age of the animals. Anti-idiotypic antibodies transferred via the placenta to embryo, via colostrum to newborn, or perenterally injected in young animals generally cause a long-lasting idiotype suppression. However, minimal amounts of anti-Id antibodies specific for the regulatory idiotopes can cause the stimulation of clones as clearly demonstrated by Hiernaux et al. (13) and Rubinstein et al. (14) in the A48Id system. Ab_2 specific for regulatory idiotopes can stimulate clones producing antibodies with various specificities. We called this property of Ab_2 specific for reg-

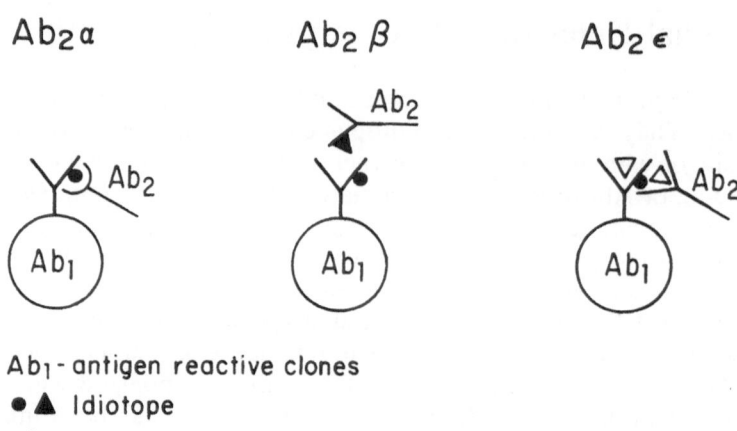

FIGURE 2.1. Binding properties of the major anti-idiotypes Ab$_2$ to idiotype Ab$_1$.

ulatory idiotopes "cross-regulation." We define the cross-regulatory process as the humoral and cellular mechanism which, through the self-recognition of regulatory idiotopes, controls discrete populations of clones which may express diverse antigen-binding specificities (15).

Ab$_{2\beta}$ are anti-Id antibodies carrying the internal image of the antigen. The concept of internal image occurs as a logical statistical necessity of the network theory. Since each paratope recognizes an epitope on molecules other than immunoglobulins, the idiotopes of each paratope should cross-react or mimic the epitopes contained in the antigen dictionary. Because the antigen repertoire of a given individual is reflected in its idiotypic repertoire, for each epitope of foreign and self antigen there may exist an idiotope which is the internal image. It therefore appears that the idiotopes are links between the immune system itself and the universe of antigens.

In principle, the idiotope of Ab$_{2\beta}$ can mimic any antigen borne by bacteria, viruses, and parasites, by cellular antigens of plants and animals, and by biologically active ligands. Obviously these kinds of antibodies can have numerous practical applications as reagents to isolate the cell receptors for various ligands, as vaccines since Ab$_{2\beta}$ can stimulate the clones specific for microbial antigens in lieu of antigen, and therapeutic agents for cancer or autoimmune diseases. The major interest of investigators is to distinguish Ab$_{2\beta}$ from Ab$_{2\alpha}$. This is particularly important since the frequency of Ab$_{2\beta}$ is particularly low.

Indeed, Rubinstein et al. (16) showed that among 12 monoclonal antibodies specific for A48Id, only one was able to stimulate in lieu of the antigen in a humoral antibody response. It should be pointed out that we do not possess a unique criterion to define "à la carte" Ab$_{2\beta}$. Several criteria can be taken into consideration, and these are discussed in the following sections.

Immunochemical Criteria

Immunochemical properties such as inhibition of binding of $Ab_{2\beta}$ to $Ab_{2\alpha}$ by antigen may be useful to distinguish $AB_{2\beta}$ from $Ab_{2\alpha}$. Since the binding of $Ab_{2\beta}$ to Ab_1 mimics the binding of antigen to Ab_1, the antigen inhibition appears to be a good criterion. However, since the $Ab_{2\alpha}$ also can bind to an idiotope associated to a paratope, this criterion by itself does not suffice to define unequivocally an $Ab_{2\beta}$. Location of the idiotype on only the variable region of heavy or light chains appears an attractive criterion since we would envision that only a few contact residues exposed on the surface of globular domains of antibody molecules will be involved in the molecular mimicry phenomenon. Recently, Moran et al. (17) have studied the ability of a monoclonal antibody specific for the idiotype of an antibody against PR8 influenza virus hemagglutinin to vaccinate the mice against various strains of influenza virus type A. By Western blotting it was shown that this anti-Id antibody reacts with only the heavy chain, clearly indicating that it recognizes an idiotope located on V_H. However, an increased anti-HA response was observed only in animals primed with anti-Id and challenged with virus. No increase of anti-HA response was observed in animals primed and challenged with anti-Id antibodies. This suggests that this monoclonal antibody specific for an idiotope solely located on V_H functions as an $Ab_{2\alpha}$.

Structural Criteria

There is no doubt that the structural criterion, namely, the sharing of some sequences of amino acid residues by the variable-region idiotope of antibody molecules and that of the epitope of antigen molecules, could be the most precise in identifying the antibody carrying an internal image. Sequence studies of antibody and antigen in two different systems support this idea. Thus, Mazza et al. (18) found that an anti-Id antibody against a GAT idiotype showed the presence in the segment of a Tyr-Tyr-Glu sequence and that the antibodies produced by animals injected with this Ab_2 are specific for the GT determinant of GAT synthetic antigen. Bruck et al. (19) found a tetrapeptide (Tyr-Ser-Gly-Ser) shared by the neutralization epitope of the hemagglutinin of reovirus and the CDR2 of a monoclonal anti-Id antibody against the antibody specific for viral glycoprotein. This criterion certainly is very important in the case of protein antigens. However, it can not be used in the case of anti-idiotypes carrying the internal image of the polysaccharide antigens or other biologically active substances.

Functional Criteria

For practical reasons the functional criterion, namely, the ability of anti-Id antibodies to elicit an antibody response, certainly remains the best criterion to distinguish $Ab_{2\beta}$ from $Ab_{2\alpha}$.

Basically, an anti-Id antibody can stimulate the expression of four different types of clones: $Id^+ Ag^+$, $Id^+ Ag^-$, $Id^- Ag^+$, and $Id^- Ag^-$ (true Ab_3).

Theoretically, an $Ab_{2\alpha}$ stimulates only the clones bearing corresponding idiotypes and, therefore, can stimulate true Ab_1 ($Id^+ Ag^+$), parallel sets ($Id^+ Ag^-$), and true Ab_3 ($Id^- Ag^-$).

By contrast, an $Ab_{2\beta}$ which stimulates clonal expression by its ability to mimic the antigen can stimulate $Id^+ Ag^+$ as well as $Id^- Ag^+$ clones. It appears, therefore, that the ability of an anti-Id antibody to stimulate clones specific for an antigen which shares the initial idiotype represents the best practical criterion to distinguish $Ab_{2\alpha}$ from $Ab_{2\beta}$.

An additional criterion can be envisioned for anti-Id antibodies carrying the internal image able to induce the expression of T cells. It is well known that T cells cannot recognize and be triggered by soluble antigen. However, there are several examples clearly demonstrating that soluble antibodies (or antibodies coupled with Sepharose) specific for idiotypic determinants and allotypic determinants of the V chain or for antigens associated with TCR (i.e., T_3 or L_3T_4) can induce the expansion of T-cell clones. These anti-Id antibodies may be antibodies of $Ab_{2\alpha}$ type which by virtue of their direct interaction with TCR can induce the activation of T cells. An $Ab_2\beta$ mimicking the antigen should be recognized by T cells like the antigen in association with class I or class II antigens. I propose this criterion as a working hypothesis to distinguish $Ab_{2\alpha}$ from $Ab_{2\beta}$ in the case of T cells. An internal image antigen might be expected to mimic not only the physiological function of the antigen but also its requirements to be recognized and to activate the lymphocyte clones.

Epibodies ($Ab_{2\epsilon}$)

In our studies on the genetics of anti-Id antibodies specific for the major cross reactive idiotype of human macroglobulins exhibiting rheumatoid activity, we have described an anti-idiotype antibody that reacts with the antigen and with the idiotype of an antibody towards the same antigen. We called this multispecific antibody an epibody (20). Recently, another type of epibody was identified which exhibits the ability to bind to a cross-reactive idiotype as well as to its own idiotope (autobody) (21). Initially, epibodies have been described only in the case of autoantibodies. Chen et al. (22) demonstrated that the epibody which binds to CDR1 of the $V_K III$ light chain which contributes to the idiotope of human rheumatoid factor bearing Waldx recognizes a similar sequence (val-ser-ser-ser) borne by Fc γ, the antigen of the rheumatoid factors. Recent studies have shown that an epibody can represent a link of connectivity between various mininetworks specific for foreign antigens and autoantigens, a kind of "superantibody" (23). The existence of an epibody indicates that the idiotype interactions between clones can be mirrored in the binding properties of a single paratope which can recognize both the antigen and the idiotopes. Epibodies specific for idiotopes exhibiting low affinity for antigens may be advantageous since they could increase the stability of the immune complexes or enhance the polymerization of antibodies bound to foreign or self antigens. Both mechanisms can favor the cloning of immune complexes.

Taken together, the data discussed above clearly show that anti-Id antibodies represent heterogeneous populations of antibodies with respect to their clonal origin as well as their immunochemical and functional properties.

Acknowledgment. This study was supported by research grant PCN 110578 from the National Science Foundation.

References

1. Bona, C.A., E. Heber-Katz, and W.E. Paul. 1981. Idiotype-anti-idiotype regulation I. Immunization with a levan-binding myeloma protein leads to the appearance of auto-anti (anti-idiotype) antibodies and to the activation of silent clones. J. Exp. Med. **153**:951.
2. Bonilla, F.A., and C.A. Bona. 1985. Idiotypic immunoregulation: interclonal connections based on a special category of idiotopes, p. 229. *In* M. Reichlin and J.D. Capra (eds.), Idiotypes. Academic Press, Inc., New York.
3. Bona, C. 1987. Regulatory idiotypes. John Wiley & Sons, Inc., New York.
4. Gleason, K., and H. Kohler. 1982. Regulatory idiotopes. T helper cells recognize a shared V_H idiotope on phosphorylcholine-specific antibodies. J. Exp. Med. **156**:539.
5. Rubinstein, L.J., C.B. Victor-Kobrin, and C.A. Bona. 1984. The function of idiotypes and anti-idiotypes on the development of the immune repertoire. Dev. Comp. Immunol. Suppl. **3**:109.
6. Bona, C., R. Hooghe, P-A. Cazenave, C. LeGuern, and W.E. Paul. 1979. Cellular basis of regulation of expression of idiotypes. II. Immunity to anti-MOPC460 idiotype antibodies increases the level of anti-trinitrophenyl antibodies bearing 460 idiotypes. J. Exp. Med. **149**:815.
7. Sherr, D.H., S.T. Ju, and M.E. Dorf. 1981. Hapten-specific T cell responses to 4-hydroxy-3-nitrophenyl acetyl XII. Fine specificity of antiidiotypic suppressor T cells (Ts2). J. Exp. Med. **154**:1382.
8. Jerne, N.K. 1974. Towards a network theory of the immune response. Ann. Immunol. (Inst. Pasteur). **125C**:373.
9. Schuller, W., E. Weiler, and H. Kolb. 1977. Characterization of syngeneic

anti-idiotypic antibody against the idiotype of BALB/c myeloma protein J558. Eur. J. Immunol. **7**:649.

10. Bona, C., and T. Moran. 1985. Idiotype vaccines. Ann. Immunol. (Inst. Pasteur). **136C**:299.

11. Brient, B.W., and A. Nisonoff. 1970. Quantitative investigation of idiotypic antibodies. IV. Inhibition by specific haptens of the reaction of anti-hapten antibody with its idiotypic antibody. J. Exp. Med. **132**:951.

12. Mudgett, M., J.E. Coligan, and T.J. Kindt. 1978. Isolation and characterization of distinct antibody populations from antisera directed against idiotypes of rabbit homogeneous antibodies. J. Immunol. **120**:293.

13. Hiernaux, J., and C. Bona. 1981. Immune network, p. 269. *In* C. Bona and P.A. Cazenave (eds.), Lymphocytic regulation by antibodies. John Wiley & Sons, Inc., New York.

14. Rubinstein, L.J., B. Goldberg, J. Hiernaux, K.E. Stein, and C.A. Bona. 1983. Idiotype-anti-idiotype regulation V. The requirement for immunization with antigen or monoclonal anti-idiotypic antibodies for the activation of β 2→6 and β 2→1 polyfructosan reactive clones in BALB/c mice treated at birth with minute amounts of anti-A48 idiotype antibodies. J. Exp. Med. **158**:1129.

15. Victor-Kobrin, C., F.A. Bonilla, B. Bellon, and C.A. Bona. 1985. Immunochemical and molecular characterization of regulatory idiotopes expressed by monoclonal antibodies exhibiting or lacking β 2-6 fructosan binding activity. J. Exp. Med. **162**:647.

16. Rubinstein, L.J., F.A. Bonilla, A.J. Manheimer, and C.A. Bona. 1985. Immunochemical anti-idiotype antibody carrying the internal image of bacterial levan in High Technology route to virus vaccines. Am. Soc. Microbiol. **1985**:167.

17. Moran, T., R. Mayer, and C. Bona. 1987. Effect of syngenic anti-idiotype antibodies on antihemagglutinin influence virus antibody response. *In* H. Kohler and J.F. Mohn, (eds.), Vaccines: New concepts and developments. In press.

18. Mazza, G., V. Guigou, D. Moinier, S. Corbet, P. Ollier, and M. Fougereau. 1987. Molecular interactions in the GAT idiotypic network: an approach using synthetic peptides. Ann. Immunol. (Inst. Pasteur), in press.

19. Bruck, C., S.M. Co, M. Slaoui, G.N. Gaulton, T. Smith, B.N. Fields, J.I. Mullins, and M.I. Green. 1986. Nucleic acid sequence of an internal image-bearing monoclonal anti-idiotype and its comparison to the sequence of the external antigen. Proc. Natl. Acad. Sci. USA **83**:6578.

20. Bona, C.A., S. Finley, S. Waters, and H.G. Kunkel. 1982. Anti-immunoglobulin antibodies III. Properties of sequential anti-idiotypic antibodies to heterologous anti-λ-globulins. Detection of reactivity of anti-idiotype antibodies with epitopes of Fc fragments (homobodies) and with epitopes and idiotopes (epibodies). J. Exp. Med. **156**:986.

21. Bona, C.A., C.-Y. Yang, H. Kohler, and M. Monestier. 1986. Epibody: the image of the network created by a single antibody. Immunol. Rev. **90**:115.

22. Chen, P.P., R.A. Houghten, S. Fong, and D. Carson. 1984. Characterization of an epibody-on anti-idiotype which reacts with both the idiotype of rheumatoid factors (RF) and the antigen recognized by RF. J. Exp. Med. **161**:323.

23. Dweyer, D., M. Vakil, and J.F. Kearney. 1986. Idiotypic network connectivity and possible cause of myasthenia. J. Exp. Med. **164**:1310.

3
Comparison of the Anti-PC Response to Nominal and Idiotope Antigens

JYH-HSIUNG HUANG, RONALD WARD, AND
HEINZ KOHLER

Introduction

Protective immunity against bacterial, parasitic, and viral infections depends in part on the quality and quantity of the antibodies induced by vaccination. An effective vaccine must be able to induce antibodies of high affinity and specificity as well as long-term memory of the immunity. These issues are becoming increasingly important in the current attempts to develop new approaches in vaccine production. One of these new developments is the so-called idiotype vaccine (1).

From our recent results in the PC-T15 murine response we have begun to understand some of the parameters which govern the idiotype, isotype, and epitope specificities. We have been using two anti-idiotypic monoclonal antibodies to induce an anti-PC (phosphorylcholine) response and have compared this response to the antibodies induced by conventional PC antigens. Ideally, one would like to have complete control over parameters, such as idiotype, isotype, and epitope specificity. This control would enable one to design idiotype vaccines and conventional vaccines which maximize protection. It is, for example, known that the idiotype and the antigen specificity changes in the secondary response to PC-KLH (2). We know also that the T15 idiotype is the most effective antibody for protection against streptococcal infection in mice. Thus, we can use the PC-T15 system (3) to study the parameters controlling the quality of the antibody response. Furthermore, our laboratory has previously demonstrated the effectiveness of an idiotype vaccine against experimental infection with *Streptococcus pneumoniae* (4).

Idiotype vaccines may be ideal biological reagents to direct the immune response into desired idiotypic and isotypic expressions because they can manipulate the immune network in an idiotype/epitope-specific manner. In addition, idiotype vaccines have distinct advantages over conventional vaccines because they do not contain an infectious organism or subunits of it. Thus they are very safe. Idiotype vaccines could also be used in concert with conventional, antigen-based vaccines to modulate the response for achieving optimum protection.

In the following we describe briefly our findings on the use of two monoclonal idiotype vaccines in the PC-T15 mouse model. We compare the idiotype-induced response to that induced by a conventional PC-carrier antigen. By doing this, we learn more about the complex nature of the involved network interactions. It was observed that two idiotype antigens stimulate different idiotype- and epitope-specific responses which in turn differ also from the PC antigen-induced response. These results support the notion of using idiotype and nominal antigens together as synergistic measures to ensure a most effective protection against infection.

Results and Discussion

We have previously generated and characterized two monoclonal anti-idiotypic antibodies for T15 (5) which differ in idiotype specificity. F6 is an anti-T15 which is not hapten inhibitable, while 4C11 is hapten inhibitable. Thus, F6 is $Ab_{2\alpha}$ and 4C11 is $Ab_{2\beta}$. The latter was confirmed by us to be an internal image PC antigen because 4C11-KLH induces specific anti-PC antibodies and protective immunity (4).

Here we used all three PC antigens, PC-OVA (ovalbumin), F6-Hy (hemocyanin), and 4C11-Hy, in mice and analyzed, in the major isotype IgG1 class, the T15 idiotype expression and the PC epitope specificity. BALB/c mice were immunized with 100 µg of either PC-OVA, F6-Hy, or 4C11-Hy in CFA (complete Freund adjuvant). The sera obtained on day 30 were analyzed for antibodies binding to PC-BSA (bovine serum albumin) plates, the percent of T15Id, and the degree of inhibition by free PC. The results of such experiments are shown in Fig. 3.1.

Next we analyzed the response to the above antigens in A/St mice. The data are shown in Fig. 3.2. The A/st mice differ from BALB/c mice in their response to PC by expressing the T15 idiotype less dominantly, i.e., 30–50% of total anti-PC antibodies. Thus, it is interesting to see how the A/st mouse would respond to anti-idiotype antigens. The results from the response in A/st mice indicate differences in idiotype and epitope expression upon immunization with PC-OVA and 4C11-Hy. The late primary response to PC-OVA lacks T15 and PC inhibitable antibodies. This is in concordance with a similar observation by Chang et al. (2). In contrast, the response to 4C11 is completely inhibitable by PC and about 60% T15 positive. This shows that an internal image antigen can prevent the idiotype shift and the loss of epitope specificity otherwise observed with PC-carrier antigens. The response to F6-Hy is very similar to the response in BALB/c, producing large amounts of T15-positive but PC-negative antibodies.

Finally, we analyzed the response in CBA/N mice which carry the xid defect. CBA/N mice are known not to respond to the conventional PC

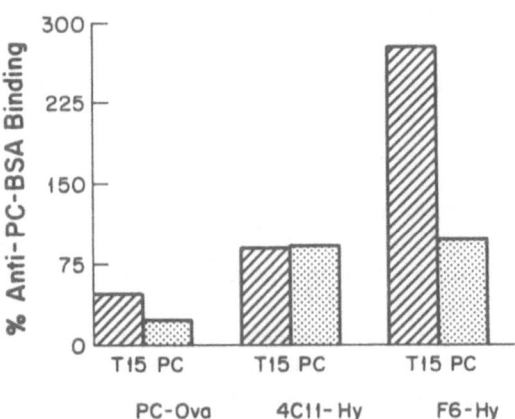

FIGURE 3.1. Response to PC antigens in BALB/c mice. Groups (4–6/group) of BALB/c mice were each immunized with 100 μg of PC-OVA, 4C11-Hy, or F6-Hy in CFA. On day 30 the sera were obtained and analyzed for antibodies binding to PC-BSA or F6(Fab′)$_2$-coated plates. The PC-specific antibodies were determined by inhibition of binding to PC-BSA with 10^{-4} M PC. The percent of T15 and PC-specific antibodies is expressed as percent of total PC-BSA binding antibodies. The assays were standardized using a polyclonal anti-PC, T15-positive antibody induced in BALB/c mice by PC-HGG.

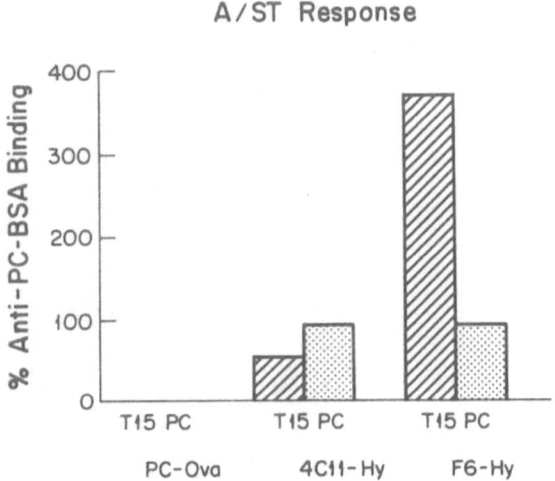

FIGURE 3.2. Response to PC antigens in A/St mice. See the legend to Fig. 3.1 for experimental details.

antigens of either T-dependent or T-independent forms (6,7). The results
are shown in Fig. 3.3.

As expected, the CBA/N mice do not produce PC-specific, T15-posi-
tive antibodies in response to PC-OVA and 4C11-Hy. Although F6-Hy
could stimulate a high titer of non-PC-inhibitable, F6-binding antibodies, it is not clear whether these antibodies are true T15 positive or Ab₃
(anti-F6 antibody). This question was addressed in subsequent experiments.

In the last set of experiments we analyzed the T15 expression in mice
immunized with Ab_2. The three sera from BALB/c, A/St, and CBA/N
mice immunized with F6-Hy were analyzed in a T15 inhibition assay. Plates
were coated with F6 F(ab')₂ and sera were added with serial dilutions of
purified T15. Binding antibodies were detected by enzyme-linked goat
anti-mouse IgG1. A canonical inhibition profile was generated using affinity-purified polyclonal T15 positive IgG1 anti-PC antibody. The results
are shown in Fig. 3.4.

Increasing concentrations of T15 decreases the binding of polyclonal
anti-PC antibody completely, establishing a reference inhibition profile
for a 100% T15-positive, anti-PC standard. When compared to the inhibition profile with BALB/c F6-Hy-induced antibodies, no significant inhibition is observed. This indicates that the majority of this antibody is a
non-T15 Ab_3-type antibody which binds to F6 ($Ab_{2\alpha}$). According to the
definition of Jerne et al. (8), this Ab_3 expresses the "alpha"-type Ab_3. In

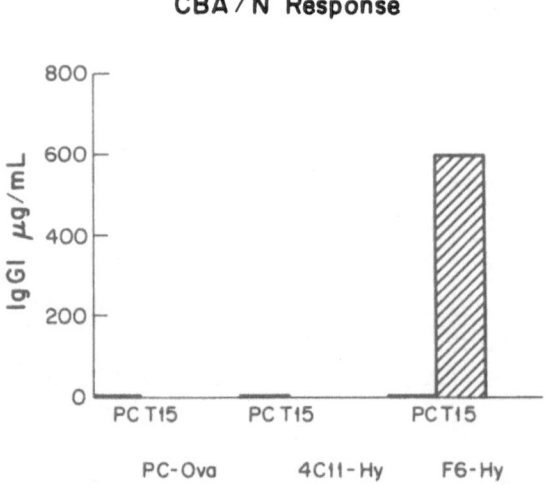

FIGURE 3.3. Response to PC antigens in CBA/N mice. See the legend to Fig. 3.2
for experimental details.

T15 Inhibition

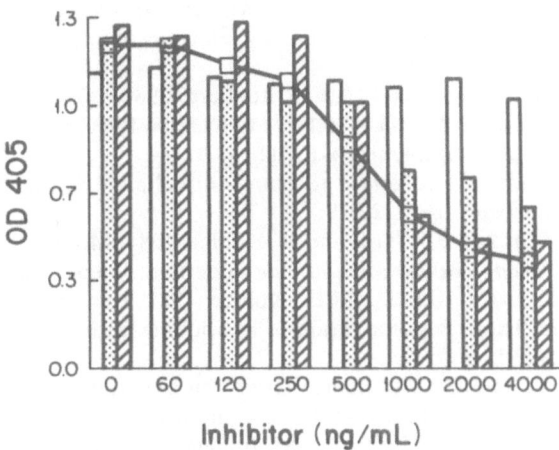

FIGURE 3.4. (Fab')₂ of F6 was insolubilized to plates (100 ng/well). As a standard, polyclonal anti-PC was added together with dilutions of T15. The ELISA binding is shown by the solid line. The T15 inhibition of sera from F6-Hy-immunized mice is shown in the column triplet for each inhibitor amount. The left bar in the triplet shows data from BALB/c mice, the middle bar from A/St mice, and the right bar from CBA/N mice.

contrast, the sera from F6-Hy-immunized CBA/N mice can be completely inhibited by T15. This finding suggests that the majority of the induced Ab_3 is T15 specific but not PC specific and therefore of the "beta" type. The sera from A/St mice produce an intermediate inhibition profile. Such sera contain a mixture of $Ab_{3\alpha}$ and $Ab_{3\beta}$ antibodies.

Summary

The presented data show that both F6 ($Ab_{2\alpha}$) and 4C11 ($Ab_{2\beta}$) can stimulate PC-specific, T15-positive IgG1 antibodies in BALB/c and A/St mice. While F6 can stimulate T15-dominant antibodies, 4C11 ($Ab_{2\beta}$) can induce epitope-specific antibodies. For the purpose of vaccine development the differential effects of anti-idiotype vaccines offer the opportunity to direct and manipulate a given immunity to obtain the expression of those antibodies which give optimal protection. We would like to propose the development of combination vaccines which include nominal and idiotope antigens. Another important aspect for idiotype vaccine production is the question of the induction of immunological memory in the B- and T-cell compartments. Such studies are under way in our laboratory.

References

1. Kennedy, R.C., G.R. Dreesman, and H. Kohler. 1985. Vaccines utilizing internal image anti-idiotypic antibodies that mimic antigens of infectious organisms. Bio Techniques 3:404.
2. Chang, S.P., M. Brown, and M.B. Rittenberg. 1982. Immunologic memory to phosphorylcholine. II. PC-KLH induces two antibody populations that dominate different isotypes. J. Immunol. 128:702.
3. Kohler, H. 1975. The response to phosphorylcholine: dissecting an immune response. Transplant. Rev. 27:29.
4. McNamara, M.K., R.E. Ward, and H. Kohler. 1984. Monoclonal idiotope vaccine against *Streptococcus pneumoniae* infection. Science 226:1325.
5. Wittner, M.K., M.A. Bach, and H. Kohler. 1982. Immune response to phosphorylcholine. IX. Characterization of hybridoma anti-TEPC15 antibodies. J. Immunol. 128:595.
6. Scher, I. 1982. The CBA/N mouse strain: an experimental model illustrating the influence of the X-chromosome on immunity. Adv. Immunol. 33:1.
7. Kohler, H., S. Smyk, and J. Fung. 1981. Immune response to phosphorylcholine. VIII. The response of CBA/N mice to PC-LPS. J. Immunol. 126:1790.
8. Jerne, N.K., J. Roland, and P.A. Cazenave. 1982. Recurrent idiotypes and internal images. EMBO J. 1:243.

4
Anti-Idiotypes as Probes of Immunoglobulin Structure

ROBERT L. RAISON AND ALLEN B. EDMUNDSON

Introduction

Idiotypes, which represent antigenic determinants located in immunoglobulin (Ig) variable regions, may be used as probes of antibody combining sites. These determinants range from those shared by other immunoglobulins exhibiting similar ligand binding properties to private idiotopes expressed only on the original immunogen. Although no clear generalizations on the structural correlates of these serologically defined determinants can as yet be made, detailed analyses of several families of murine antibodies (reviewed in Ref. 1) have allowed some progress in this area. Similar studies with human immunoglobulins have not generally been possible because of a lack of available proteins sharing identical or cross-reactive binding properties. However, in a number of cases, paraproteins arising from B-cell neoplasia exhibited significant binding characteristics. For example, monoclonal IgMs with rheumatoid factor (anti-IgG) activity have been described in Waldenstrom's macroglobulinemia (2) and Bence-Jones proteins exhibiting tissue binding specificity have been identified in amyloidosis (3). As these binding specificities are mediated by the variable regions, an examination of the structure and distribution of the idiotypes characterizing these paraproteins may extend our understanding of the phenomena. Recently, 10 of 12 monoclonal IgM rheumatoid factors (RFs) were shown to possess a cross-reactive idiotope by virtue of reactivity with an anti-idiotype raised against a synthetic peptide derived from the second hypervariable region of the light chain of the IgM-RF Sie (4). However, similar studies with an anti-idiotype produced against a peptide of the third hypervariable region of the heavy chain defined a private, dominant idiotope present only on the immunoglobulin from which the peptide sequence was derived (5). Direct structural correlation of these idiotopes with the known anti-Ig specificity of the paraproteins has not been possible although the data suggest that the light chain second hypervariable region may determine the binding specificity.

Dimers of the Bence-Jones λ chain from patient Mcg (6) bind a variety of ligands, including bis(DNP)lysine, fluorescein, lucigenin, and rhodamine derivatives (7–9). Three-dimensional characterization of the Mcg combining site has demonstrated a surprising degree of flexibility which, in vivo, may be reflected in the amyloidogenic properties (tissue affinity) of this paraprotein. Here, we describe an idiotope on the Mcg dimer and examine its relationship to the ligand binding properties of this model combining site. The data are discussed in the light of extensive crystallographic analyses of this protein.

Materials and Methods

Preparation of Bence-Jones Proteins and Hybrid Dimers

The Bence-Jones proteins were prepared from urine of patients with multiple myeloma and/or amyloidosis as previously described (8). Covalently linked dimers of Mcg and a second κ or λ light chain were produced by sulfonation of the interchain disulfide bond of the Mcg dimer followed by reaction with the reduced form of the second chain (10).

Monoclonal Antibodies

Monoclonal antibodies to the Mcg dimer were produced by cell fusion techniques as previously applied to the production of an anti-κ reagent (11).

Direct and Competitive Enzyme-Linked Immunoassay

Antibody-secreting hybridomas were screened and selected using an indirect enzyme immunoassay (EIA) in which the wells of plastic microtiter trays were coated with the Mcg dimer at 10 μg/ml. Binding of mouse immunoglobulin was assessed using alkaline phosphatase-conjugated sheep anti-mouse Ig (Sigma Chemical Co., St. Louis, Mo.) followed by substrate incubation and measurement of absorbance at 405 nm.

Reactivity of a number of polypeptides, including immunoglobulins and Bence-Jones proteins, with mAb M3.9 was determined by competitive EIA as previously described (12). Proteins were mixed with antibody, the concentration of which was approximately 70% of that required for saturation in the indirect EIA versus Mcg. After incubation for 1 h at 37°C and then 16 h at 4°C, duplicate aliquots were transferred to Mcg-coated wells for analysis by indirect EIA. The inhibition of binding of the monoclonal antibody to Mcg was determined with reference to a standard curve constructed for each experiment and consisting of a range of concentrations of Mcg dimer.

Hapten Binding Experiments

The effect of ligand occupancy of the Mcg combining site on the binding
of mAb M3.9 was determined for both rhodamine 123 and bis(DNP)lysine.
In each case Mcg dimers were first incubated with varying concentrations
of the ligand and then tested by inhibition EIA for their ability to react
with the antibody. The ligand-protein complexes were prepared under
conditions similar to those developed earlier for binding studies in crystals
and in solution (7–9,13).

Results and Discussion

The Binding Region of the Mcg Dimer

X-ray analyses of Mcg L chain crystals to 2.3-Å resolution allowed elucidation of the three-dimensional structure of the combining site along
the interface of the V domains (7,14,15). The main cavity is shaped like
a truncated cone 15 Å across at the cavity entrance and 17 Å deep. The
rim of the main cavity is formed by tyrosines 34, 51, and 93 and glutamic
acid 52 of both monomers and by aspartic acid 97 of only one of the chains
(designated monomer 2 in Fig. 4.1). Twelve of the 21 residues which line
the interior of the main cavity are aromatic (eight tyrosines and four phenylalanines), and it is therefore not surprising that the dimer binds a wide
variety of aromatic and hydrophobic aliphatic compounds. At the base
of the main cavity is a much smaller ellipsoidal pocket lined by glutamine
40, proline 46, and tyrosine 89 of both monomers.

Binding Specificity of M3.9 Monoclonal Anti-Idiotype

Three cloned cell lines exhibiting putative anti-Mcg idiotype specificity
were obtained from two fusions using splenocytes from mice immunized
with the Mcg protein. Of these, clone M3.9 was further characterized and
used exclusively in this study. The binding specificity of M3.9 as assessed
by direct EIA against human L chains and Igs is shown in Table 4.1. The
antibody reacted only with the immunizing Mcg L chain dimer and with
no other Ig or Ig polypeptide chain. Using a competitive EIA, M3.9 was
shown to bind to Mcg IgG in addition to the Mcg dimer (Fig. 4.2). However, while 0.3 μM Mcg IgG (45 μg/ml) resulted in 50% inhibition of
the binding of M3.9 to solid-phase-adsorbed Mcg dimer, 1 μM Mcg
dimer (50μg/ml) was required to achieve an equivalent level of inhibition.
Thus, the IgG molecules expressed at least twice the number of M3.9
defined idiotopes as the L chain dimer when compared on a molar basis. Furthermore, the identification of the idiotope on Mcg IgG indicated

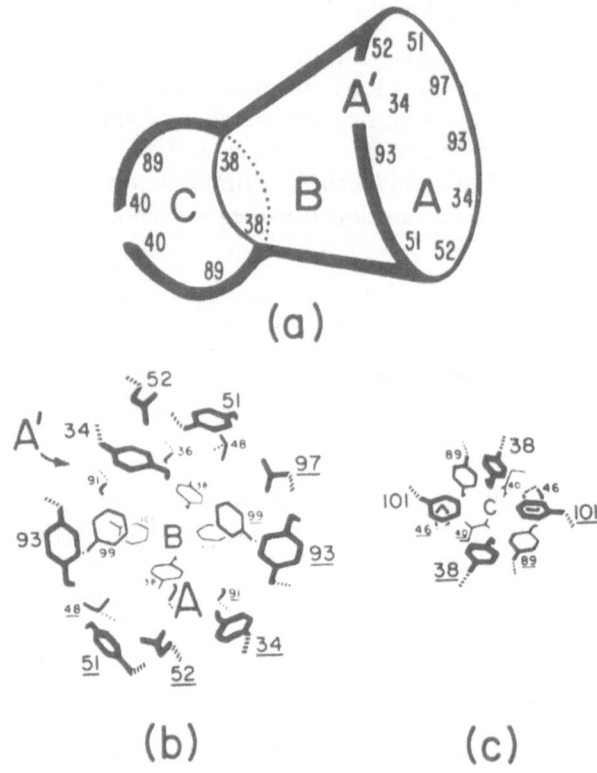

FIGURE 4.1. Schematic views of the binding regions for DNP compounds in the combining site of the Mcg Bence-Jones dimer. (a) Side view. Key side chains are numbered. Bis(DNP)lysine bridged sites A and B, with one DNP ring in each site. Under appropriate conditions, ligand was also detected in site C. Access to site A was blocked by packing interactions in the crystal but should be available for ligand binding in solution. (b) Perspective drawing of the side chains lining the main binding cavity containing sites A, A', and B. The residues contributed by monomer 2 are underlined. The α-β carbon bonds are represented as dotted lines. (c) Side chains lining the deep pocket, reached through the floor of the main cavity and containing site C. (Drawings by Kathryn R. Ely; taken from Ref. 10). Reprinted with permission from Biochemistry *19*:2827. Copyright 1980, American Chemical Society.

that only a single V_L region is required to express the determinant. In the Mcg dimer, one of the L chain partners (monomer 1) adopts a conformation characteristic of the heavy chain in an Fab (14,15). It therefore appears that the M3.9 idiotope is expressed on monomer 2, which adopts a conformation analogous to that of the L chain component of an Fab.

TABLE 4.1. Binding specificity of M3.9
antibody.

Target	Reactivity with M3.9[a]
Mcg dimer	+
Vor κ	−
Mos λ	−
Normal pool L chains	−
Normal pool IgG	−

[a]Determined by indirect EIA against target antigen.

Localization of the M3.9 Idiotope

Two human λ chains of known amino acid sequence (Vil, Ref. 16; Weir,
Ref. 17) were tested for expression of the M3.9 idiotope by inhibition EIA
(Table 4.2). Both proteins failed to react with mAb M3.9. Since Mcg and
Vil share identical sequences (18) in the first complementarity determining
region (CDR 1; see Fig. 4.3), it is clear that this region does not encode
the idiotope. In CDR2, Vil and Weir differ from Mcg at two and three
positions, respectively. Two of these residues (54 and 55) are located in

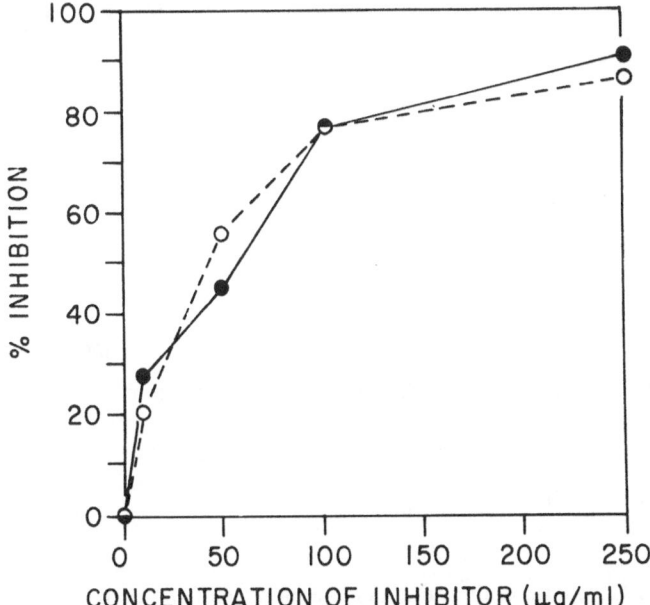

FIGURE 4.2. Expression of the M3.9 idiotope on Mcg IgG. Mcg light chain dimers
(•) and Mcg IgG (○) were tested for reactivity with mAb M3.9 by competitive EIA
as described in Materials and Methods.

TABLE 4.2. Hapten binding and idiotope expression of light chain hybrids.

Protein	Light chain isotypes	Binding constant[a] for bis(DNP)lys (M^{-1})	Relative binding of M3.9[b]
Mcg × Mcg	λ × λ	1.3×10^5	1.0
Vil × Vil	λ × λ	8.2×10^4	0
Mcg × Vil	λ × λ	4.1×10^4	0.97
Weir × Weir	λ × λ	Not detectable	0
Mcg × Weir	λ × λ	1.9×10^5	1.05
Tew × Tew	κ × κ	2.0×10^4	0.25
Mcg × Tew	λ × κ	2.1×10^5	0.99

[a]Data from Ref. 10. Value given is for single site occupancy.
[b]Calculated relative to a value of 1.0 assigned to inhibitory activity of Mcg dimer. All proteins were assayed at 250 µg/ml.

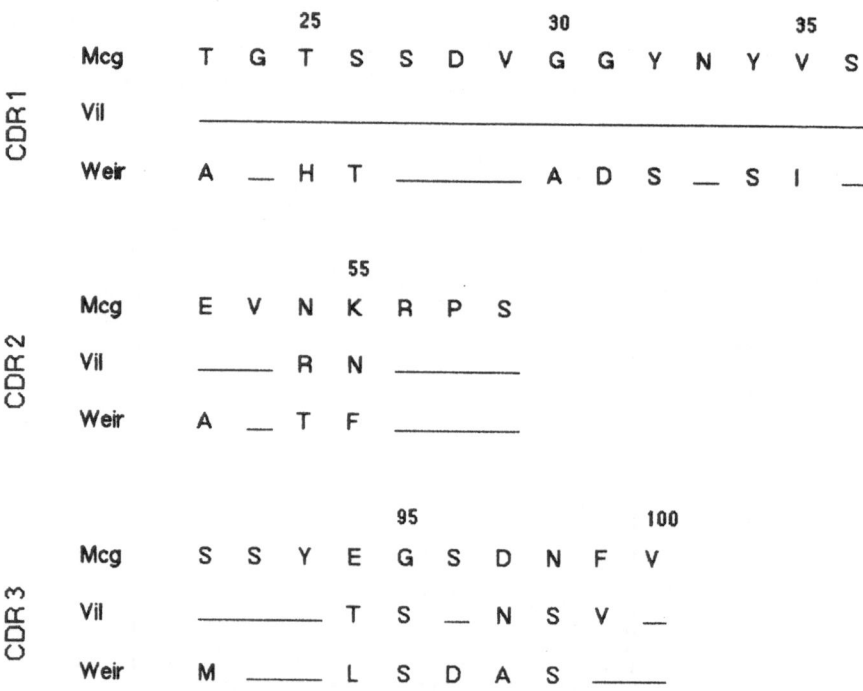

FIGURE 4.3. Comparison of amino acid sequences of the complementarity determining regions (CDRs 1, 2, and 3) of λ Bence-Jones proteins Mcg (Ref. 6), Vil(Ref. 16), and Weir (Ref. 17). Amino acid residues are designated by the one-letter code. Lines indicate homology with the Mcg sequence.

the interior of the domain in the Mcg structure and are therefore unlikely to be available to act as part of an epitope. While residue 52 is alanine in the Weir sequence, Vil and Mcg both have a glutamic acid side chain in this position. This residue also cannot be assigned to an exclusive Mcg epitope. Tyrosine 51 is an important contact residue in the Mcg dimer combining site, but was excluded as a dominant influence in the M3.9 idiotope because of the identity of the idiotope-negative protein Weir at this position. The greatest variability in sequence lies in CDR3, where Weir and Vil share only four and five identities with the corresponding 10 residues of Mcg. From sequence arguments alone we expected to find the M3.9 epitope in this region.

The M3.9 idiotope was further characterized through examination of hybrid dimers consisting of one Mcg chain and a heterologous light chain (10). The composition, binding specificity, and reactivity of these hybrids with Mab 3.9 are shown in Table 4.2. When tested at 250 μg/ml in the inhibition EIA, Mcg hybrids containing either Weir or Vil bound the M3.9 anti-idiotype as effectively as the homologous Mcg dimer. Of particular interest was the κ Bence-Jones protein Tew, which also binds DNP. The Tew protein, which shares the crucial DNP binding residues tyrosine 34 and 38 and phenylalanine 101 (19) in common with Mcg, displayed partial reactivity to the M3.9 antibody (25% binding compared to Mcg). Three other Mcg hybrids containing heterologous λ chains (Hud, Bla, and May) bound M3.9 with relative binding values near that of the Mcg dimer. Homologous dimers of these three proteins failed to react with mAb 3.9.

The expression of the M3.9 idiotope on all Mcg hybrid dimers at levels equivalent to the Mcg dimer provides further evidence that this idiotope is expressed on a single chain of Mcg. Structural analysis of the Mcg X Weir hybrid indicates that the Mcg chain in the hybrid probably assumes a conformation similar to that of monomer 2 in the Mcg dimer (20). Collectively, these results strongly suggest that the idiotope is localized on monomer 2 in the Mcg dimer.

Effect of Ligand Binding on the Expression of the M3.9 Idiotope

A wide variety of hydrophobic ligands can be bound by the combining site formed by the association of V_L domains in the Mcg dimer. The locations, occupancies, and structures of these ligands were determined by crystallographic analyses (7,8,21). In osmotically shocked crystals, rhodamine 123 was found to bind in the outer part of the main cavity of the Mcg dimer (see Fig. 4.4). The stabilizing interactions with this ligand involved only the side chains along or just inside the rim of the cavity. In solutions of 0.1 M phosphate, pH 6.2, rhodamine 123 was bound with an association constant of $2 \times 10^4 \ M^{-1}$(9). To determine the effects of ligand binding on expression of the M3.9 idiotope, ligand-protein complexes were produced by incubating rhodamine 123 in concentrations of 0.2, 0.5, and

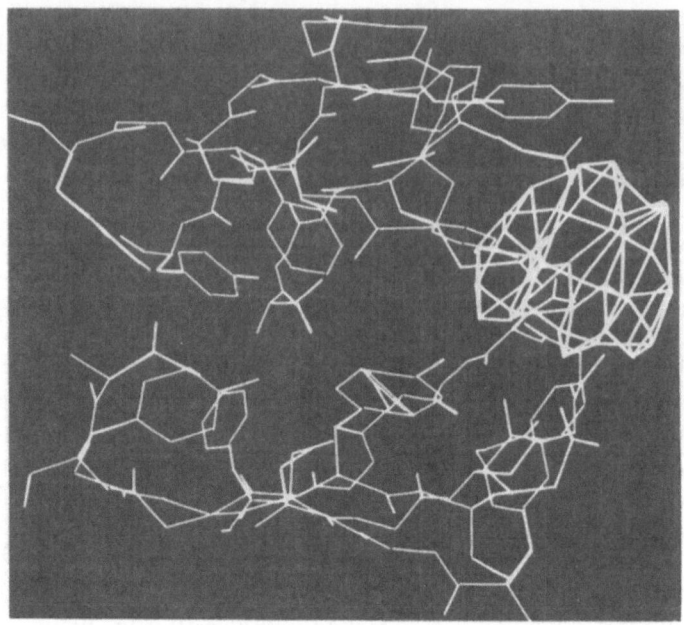

FIGURE 4.4. Binding of rhodamine 123 in a crystal of Mcg dimer at 6.5-Å resolution. The cage electron density for the ligand (heavier lines) is codisplayed with a side view of the skeletal model of the main binding cavity of the Mcg dimer. Taken from Ref. 8.

1.0 mM with Mcg covalent dimers at 100 and 250 μg/ml for 16 h at 37°C. Duplicate aliquots from each incubation mixture were then tested for the ability to block the binding of M3.9 to solid-phase-adsorbed Mcg (Fig. 4.5). At rhodamine 123 concentrations greater than 0.5 mM, effective blocking of the M3.9 idiotope was seen when Mcg was tested at both 100 and 250 μg/ml. Rhodamine at a concentration of 0.2 mM partially blocked the idiotope when Mcg was tested at 100 μg/ml and failed to block at the higher concentration of Mcg. These data indicate that occupancy of the outer rim area of the combining site by rhodamine 123 significantly alters or blocks access to the idiotypic determinant.

A second ligand, bis(DNP)lysine, can be made to bind either in the main cavity or in a deep pocket of the Mcg dimer, depending on the experimental conditions (7). In the present study, outer-site binding was achieved by incubating the covalent dimer with a 10-fold molar excess of bis(DNP)lysine. The ability of this preparation to bind M3.9 was tested in the presence of free DNP to ensure occupancy of the main cavity. In contrast to rhodamine 123, bis(DNP)lysine did not block the M3.9 idiotope (Fig. 4.6).

Irreversible binding of bis(DNP)lysine in the deep pocket was achieved

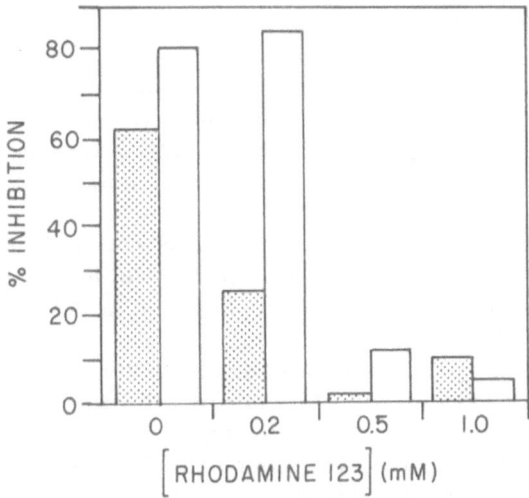

FIGURE 4.5. Effect of rhodamine 123 occupancy of the Mcg binding cavity on the expression of the M3.9 idiotope. Rhodamine 123 was incubated at varying concentrations with Mcg dimer at 100 (stippled bars) and 250 μg/ml (open bars). The ability of the ligand/Mcg complex to bind mAb M3.9 was then determined by competitive EIA.

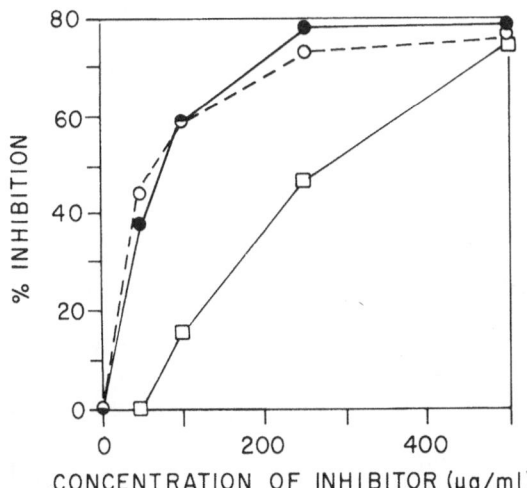

FIGURE 4.6. Effect of bis(DNP)lysine occupying distinct sites in the Mcg binding cavity on the expression of the M3.9 idiotope. Mcg dimer was preincubated with bis(DNP)lysine under conditions resulting in occupancy of either the deep pocket site (C in Fig. 4.1) (•) or the outer cavity sites (A and B in Fig. 4.1) (○). The ability of these ligand/Mcg complexes to bind mAb M3.9 was determined by competitive EIA and compared to the competitive binding of noncomplexed Mcg (□).

using the noncovalent form of the Mcg dimer (A.B. Edmundson et al., unpublished work). After prolonged incubation with hapten, the complex was dialyzed to remove bis(DNP)lysine from the main cavity. The ligand in the deep pocket remained firmly bound. The final complex contained 1.0 mole of bis(DNP)lysine per mole of Mcg dimer, as determined by absorbance at 365 nm. This preparation showed a markedly reduced ability to bind the M3.9 antibody, as indicated by low inhibition levels obtained in the assay (Fig. 4.6). The above effect was not due to loss of the idiotope on the noncovalent form of the Mcg dimer since the inhibition curve obtained with a nonliganded control was identical to that obtained with the covalent dimer (Fig. 4.7).

The finding that the M3.9 idiotope is blocked when the main cavity is occupied by rhodamine 123 but is not affected by the binding of bis(DNP)lysine suggests that different contact residues are involved in the binding of these two ligands. This conclusion is consistent with the crystallographic studies of the binding of these ligands. The observation that the binding of bis(DNP)lysine in the deep pocket significantly influences the idiotope was very surprising to us. However, the binding of bis(DNP)lysine to the Mcg dimer was previously shown to be accompanied by distal conformational changes (13,22). Although the mechanism is unclear, it now appears that the conformational changes also exert more immediate effects on the idiotypic determinant on the outer rim of the cavity.

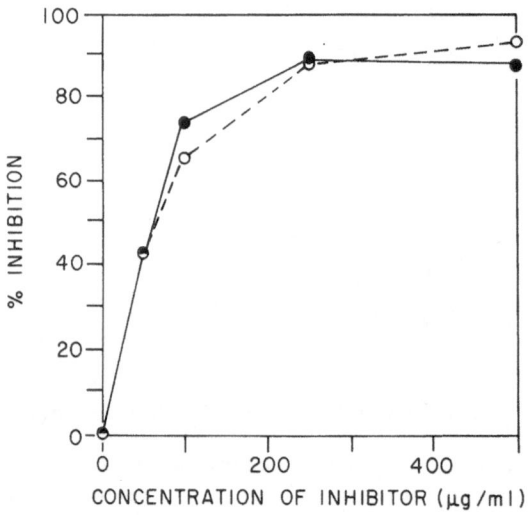

FIGURE 4.7. Expression of the M3.9 idiotope on noncovalent dimers of Mcg. Covalent (•) and noncovalent (○) dimers of Mcg were assayed for reactivity with mAb M3.9 by competitive EIA.

Conclusion

The importance of specific V_H-V_L pairing has been clearly demonstrated in correlations of antibody structures with the expression of idiotypic determinants (23). However, exceptions have been reported (24,25) particularly when synthetic peptides with CDR-like sequences have been used to elicit anti-idiotypic responses (4). It is clear from the data presented in this article that the M3.9 idiotope expressed on the Mcg dimer is localized within the variable region of only one of the L chains. The companion polypeptide chain influences the expression of this idiotope indirectly through interactions leading to proper positioning of the determinant.

As the L chain conformation in an Fab fragment of Mcg IgG is similar to that of monomer 2 in the Mcg L-L dimer, it is probable that the idiotope is expressed on this conformational isomer of the Mcg λ chain. In developing a simple model for the structure of the M3.9 idiotope, we have restricted our survey to monomer 2 residues which are (i) accessible for interactions with anti-idiotype; (ii) likely contact residues for the ligand rhodamine 123, the binding of which blocks the idiotope; and (iii) unique to the Mcg λ chain sequence. Of the nine residues which form the outer rim of the binding cavity of the dimer, aspartic acid 97 is the only residue contributed exclusively by monomer 2. This residue is not found in a comparable location in monomer 1 and is not present in the sequences of Vil and Weir, which fail to express the idiotope. The negatively charged side chain of aspartic acid is appropriate for interactions with the positively charged rhodamine ligand. Finally, the positioning of aspartic acid 97 in the third hypervariable loop places its side chain in a region which is very accessible for external interactions. While we currently believe that aspartic 97 is critical, further immunological and crystallographic studies will be required to obtain a comprehensive understanding of the structural features involved in the expression of the M3.9 idiotope.

Of particular interest is the finding that the idiotope is significantly altered by conformational effects accompanying the binding of ligand in a site remote from the idiotypic determinant. This observation should be considered in arguments concerning mimicry of antigen by anti-idiotype.

Acknowledgment. This work was supported by Grant CA19616 awarded by the National Cancer Institute, Dept. of Health and Human Services.

References

1. Davie, J.M., M.V. Seiden, N.S. Greenspan, C.T. Lutz, T.L. Bartholow, and B.L. Clevinger. 1986. Structural correlates of idiotopes. Annu. Rev. Immunol. 4:147.
2. Kunkel, H.G., V. Agnello, F.G. Joslin, R.J. Winchester, and J.D. Cooper.

1973. Cross-idiotypic specificity among monoclonal IgM proteins with anti-gamma-globulin activity. J. Exp. Med. **137**:331.

3. Bertram, J., R.J. Gualtieri, and E.F. Osserman. 1980. Amyloid related Bence-Jones proteins bind dinitrophenyl L-lysine (DNP), p. 351. *In* G.G. Glenner, (ed.), Amyloid and amyloidosis. Excerpta Medica, Amsterdam.

4. Chen, P.P., F. Goni, R.A. Houghten, S. Fong, R. Goldfien, J.H. Vaughan, B. Frangione, and D.A. Carson. 1985. Characterization of human rheumatoid factors with seven antiidiotypes induced by synthetic hypervariable region peptides. J. Exp. Med. **162**:487.

5. Goldfien, R.D., P.P. Chen, and D.A. Carson 1985. Synthetic peptides corresponding to third hypervariable region of human monoclonal IgM rheumatoid factor heavy chains define an immunodominant idiotype. J. Exp. Med. **162**:756.

6. Fett, J.W., and H.F. Deutsch. 1974. Primary structure of the Mcg.L chain. Biochemistry **13**:4102.

7. Edmundson, A.B., K.R. Ely, R.L. Girling, E.E. Abola, M. Schiffer, F.A. Westholm, M.D. Fausch, and H.F. Deutsch. 1974. Binding of 2,4-dinitrophenyl compounds and other small molecules to a crystalline lambda type Bence-Jones dimer. Biochemistry **13**:3816.

8. Edmundson, A.B., K.R. Ely, and J.N. Herron 1984. A search for site-filling ligands in the Mcg Bence-Jones dimer: crystal binding studies of fluorescent compounds. Mol. Immunol. **21**:561.

9. Herron, J.N., K.R. Ely, and A.B. Edmundson. 1985. Pressure-induced conformational changes in a human Bence-Jones protein (Mcg). Biochemistry **24**:3453.

10. Peabody, D.S., K.R. Ely, and A.B. Edmundson. 1980. Obligatory hybridization of heterologous immunoglobulin light chains into covalently linked dimers. Biochemistry **19**:2827.

11. Boux, H.A., R.L. Raison, K.Z. Walker, G.E. Hayden, and A. Basten. 1983. A tumour associated antigen specific for human kappa-myeloma cells. J. Exp. Med. **158**:1769.

12. Raison, R.L., and H.A. Boux. 1985. Conformation dependence of a monoclonal antibody defined epitope on free human kappa chains. Mol. Immunol. **22**:1393.

13. Firca, J.R., K.R. Ely, P. Kremser, F.A. Westholm, K.J. Dorrington, and A.B. Edmundson. 1978. Interconversion of conformational isomers of light chains in the Mcg immunoglobulins. Biochemistry **17**:148.

14. Schiffer, M., R.L. Girling, K.R. Ely, and A.B. Edmundson. 1973. Structure of a λ-type Bence-Jones protein at 3.5-Å resolution. Biochemistry **12**:4620.

15. Edmundson, A.B., K.R. Ely, E.E. Abola, M. Schiffer, and N. Panagiotopoulos. 1975. Rotational allomerism and divergent evolution of domains in immunoglobulin light chains. Biochemistry **14**:3953.

16. Ponstingl, H., and N. Hilschmann. 1971. Die vollstandige Primar-struktur einer monoklonalen immunoglobulin-1-Kette von 1-Typ, Subgruppe II (Bence-Jones Protein Vil). Hoppe-Seyler's Z. Physiol. Chem. **352**:859.

17. Jabusch, J.R., and H.F. Deutsch. 1982. Primary structure of a human λ-chain (Weir) of the Mcg type. Mol. Immunol. **19**:901.

18. Kabat, E.A., T.T. Wu, and H. Bilofsky. 1977. Unusual distributions of amino acids in complementarity-determining (hypervariable) segments of heavy and light chains of immunoglobulins and their possible roles in specificity of antibody-combining sites. J. Biol. Chem. **252**:6609.

19. Putnam, F.W., E.J. Whitley, Jr., C. Paul, and J.N. Davidson. 1973. Amino acid sequence of a κ Bence-Jones protein from a case of primary amyloidosis. Biochemistry 12:3763.
20. Ely, K.R., D.S. Peabody, T.R. Holm, B.D. Cheson, and A.B. Edmundson. 1985. Accessible intrachain disulfide bonds in hybrids of light chains. Mol. Immunol. 22:85.
21. Edmundson, A.B., and K.R. Ely. 1985. Binding of N-formylated chemotactic peptides in crystals of the Mcg light chain dimer: similarities with neutrophil receptors. Mol. Immunol. 22:463.
22. Ely, K.R., J.R. Firca, K.J. Williams, E.E. Abola, J.M. Fenton, M. Schiffer, N.C. Panagiotopoulos, and A.B. Edmundson. 1978. Crystal properties as indicators of conformational changes during ligand binding or interconversion of Mcg light chain isomers. Biochemistry 17:158.
23. Hopper, J.E., and A. Nisonoff. 1971. Individual antigenic specificity of immunoglobulins. Adv. Immunol. 13:57.
24. Cannon, L.E. and R.T. Woodland. 1983. Rapid and sensitive procedure for assigning idiotypic determinants to heavy or light chains: application to idiotopes associated with the major cross-reactive idiotype of A/J anti-phenylarsonate antibodies. Mol. Immunol. 20:1283.
25. Fulton, R.J., and J.M. Davie. 1984. Influence of the immunoglobulin heavy chain locus on expression of the $V\kappa_1{}^{GAC}$ light chain. J. Immunol. 133:465.

5
A Molecular Recognition Code: Its Use for the Purification of ACTH, Endorphin, and LHRH Receptors

KENNETH L. BOST, LAWRENCE R. SMITH, AND
J. EDWIN BLALOCK

Introduction

In recent years a great deal of emphasis has been placed on the isolation and characterization of receptors that are found on the surface of cells. An understanding of receptor-mediated biological responses would not be complete without an understanding of receptor structure and ligand-induced signal transmission. In an effort to dissect receptor structure and function, many investigators have used anti-receptor antibodies, the advantages of which are numerous (1). A problem with this experimental approach is that monospecific anti-receptor antibodies can be difficult to make by conventional methodologies. As demonstrated by other reports in this publication, antibodies directed against idiotopes on anti-ligand antibodies can be used as probes for receptor structure. Conceptually, it is possible to envision that the anti-idiotypic antibodies represent conformational homologues of the respective ligands and therefore have the ability to bind receptors in much the same way as does the ligand. Thus, the ability of the immune system to produce internal images has been exploited to generate anti-receptor antibodies. Recent investigations in our laboratory have been directed at defining a molecular recognition code that may help to explain the ability of the immune system to produce these internal images. These studies have also resulted in the production of antibodies to and subsequent characterization of receptors for corticotropin (ACTH), endorphin, and luteinizing hormone-releasing hormone (LHRH).

Binding of Peptides Encoded by Complementary RNAs

The ability to produce anti-receptor antibodies was founded on the observation that amino acids encoded by complementary RNAs were anti-complementary with respect to their hydropathic character. Specifically,

in the 5' to 3' direction, codons for hydrophilic amino acids were complemented by those for hydrophobic amino acids, and codons for hydrophobic amino acids were complemented by those for hydrophilic amino
acids (2). Furthermore, the majority of "uncharged" (slightly hydrophilic)
amino acids were found to be complemented by similar "uncharged" amino acids. Interestingly, the same pattern was found when complementary
codons were read in the 3' to 5' direction (3). Since only the second base
of the codon never changes regardless of the direction of reading (3' to
5' versus 5' to 3'), it is the middle base which determines the hydropathic
nature of the amino acid. This apparently results from the preponderance
of hydrophilic amino acids having A as their second base, whereas most
hydrophobic amino acids have U as their second base, and these bases
are, of course, complementary.

Since amino acids encoded by complementary codons were inverse with
respect to their hydropathicity, we speculated that peptides encoded by
complementary strands of nucleic acids may be inverse with respect to
the conformations that they would assume in aqueous solutions. If, in
fact, these complementary peptides were internal images of each other,
then it would be predicted that peptides encoded by complementary RNAs
should bind one another. This possibility was confirmed when we demonstrated that ACTH and endorphin could bind with specificity and high
affinity to peptides encoded by the respective complementary RNA sequences for ACTH-(1-24) (Fig. 5.1) and γ-endorphin (4). While the original

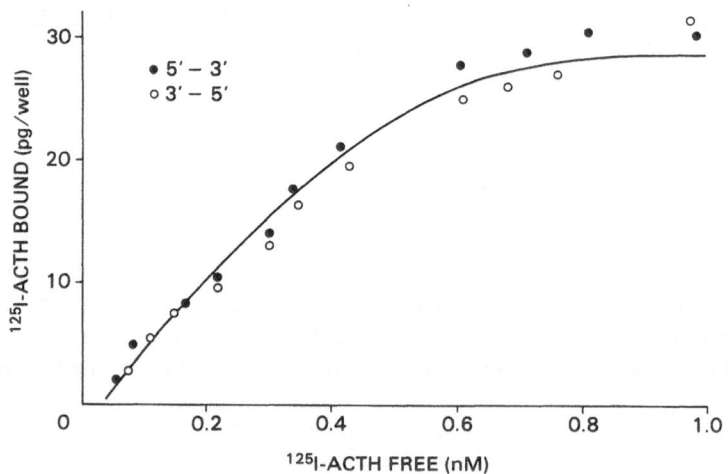

FIGURE 5.1. Binding of ^{125}I-ACTH to the 5' to 3' or 3' to 5' complementary mRNA
strand peptides. Dilutions of ^{125}I-ACTH were incubated in wells coated with 5' to
3' complementary mRNA strand peptide (•) (HTCA) or 3' to 5' complementary
mRNA strand peptide (○), and the amount of specifically bound radiolabel was
determined. Unlabeled ACTH blocked greater than 84% of the ^{125}I-ACTH binding.
Reproduced with permission from Ref. 3.

complementary peptide for ACTH was encoded in the 5′ to 3′ direction, it would be predicted that a complementary peptide encoded in the 3′ to 5′ direction would also bind ACTH since a similar interchange of hydrophilic and hydrophobic amino acids occurs in this direction. In support of this possibility, it was found that there was essentially no difference (K_d 0.3 nM) between the abilities of the 3′ to 5′ and the 5′ to 3′ complementary RNA strand peptides to bind ACTH or to block ^{125}I-ACTH binding (Fig. 5.1). Thus peptides encoded by complementary strands of nucleic acids have the ability to bind one another with specificity and high affinity.

Generation of Anti-ACTH Receptor Antibodies

Since the 5′ to 3′ complementary peptide to ACTH (designated HTCA) behaved like an ACTH receptor binding site in its ability to bind ACTH (Fig. 5.1), we hypothesized (4) that antibodies directed against HTCA would recognize an ACTH receptor binding site. To test this hypothesis, HTCA was coupled to a carrier protein, keyhole limpet hemocyanin (KLH) and used to immunize a rabbit. Total immunoglobulin was extracted from the immune sera, and the anti-KLH antibodies were removed by immunoaffinity chromatography. The monospecific anti-HTCA antibody was then shown to bind to the surface of Y-1 adrenal cells which express ACTH receptors. More importantly, addition of ACTH blocked the ability of the antibody to bind to Y-1 cells in a dose-dependent fashion, indicating that the antibody and ACTH were competing for the same binding site (Fig. 5.2). In addition, anti-HTCA was able to induce a steroidogenic response when applied to Y-1 adrenal cells, confirming that the antibody did in fact recognize an ACTH receptor binding site in an agonistic fashion. Immunoaffinity columns conjugated with anti-HTCA antibodies allowed the isolation of a molecule from the surface of Y-1 adrenal cells. This protein had the ability to bind ^{125}I-ACTH after being isolated and was subsequently used to immunize a rabbit to produce an anti-"whole" ACTH receptor antibody.

Immunoaffinity columns conjugated with anti-whole receptor antibodies were then used to isolate and characterize the Y-1 adrenal ACTH receptor (5). Using sodium dodecyl sulfate-polyacrylamide gel electrophoresis, a single protein with a molecular weight of approximately 225,000 daltons was detected under nonreducing conditions. Using reducing conditions of 2-mercaptoethanol and urea, four subunits with molecular weights of 83,000, 64,000, 52,000 and 22,000 were detected. Further studies demonstrated that the 83,000- and 52,000-dalton subunits were covalently linked, whereas the 64,000- and 22,000-dalton subunits were noncovalently associated. ^{125}I-ACTH was used to demonstrate that the 83,000-dalton subunit contained the binding site and that the purified receptor possessed binding affinities of 3.4×10^{10} M^{-1} and 1.0×10^{9} M^{-1}. These binding affinities were similar to those obtained by Payet and Escher (6) for ^{125}I-

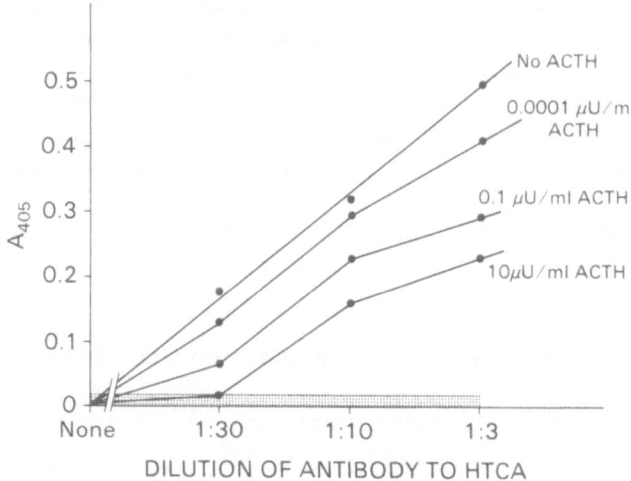

FIGURE 5.2. Blocking of anti-HTCA binding to mouse adrenal (Y-1) cells by synthetic ACTH-(1-24). Anti-HTCA binding to glutaraldehyde-treated cells was determined in the absence of ACTH or in the presence of synthetic ACTH-(1-24) at various concentrations. Antibody to KLH (shaded region) did not bind to Y-1 cells. U, ACTH unit. Reproduced with permission from Ref. 4.

ACTH binding to rat adrenal glomerulosa cells and similar to those reported by McIlhinney and Schulster (7) for ^{125}I-ACTH binding to whole rat adrenal cells. Thus, by using antibodies directed against a peptide encoded by the complementary RNA for ACTH, we were able to isolate and characterize the ACTH receptor. These studies supported our initial hypothesis that complementary peptides resemble receptor binding sites.

Generation of Anti-Endorphin Receptor Antibodies

As noted previously, a peptide encoded by the complementary RNA for γ-endorphin was shown to specifically bind γ-endorphin (Fig. 5.3). Using methodology similar to that described above, an antibody against this complementary peptide for γ-endorphin was made (8). This antibody, as well as the complementary peptide itself, blocked binding of ^{125}I-β-endorphin to a cell line, NG108-15, which expresses delta opiate receptors (9). Furthermore, antibody binding to NG108-15 cells could be blocked with naloxone or with β-endorphin. Taken together, these experiments strongly suggest that the antibody generated against the complementary peptide for γ-endorphin recognized an opiate receptor binding site.

Anti-receptor antibodies were then used to immunoaffinity purify the opiate receptor from the surface of NG108-15 cells (8). Under reducing

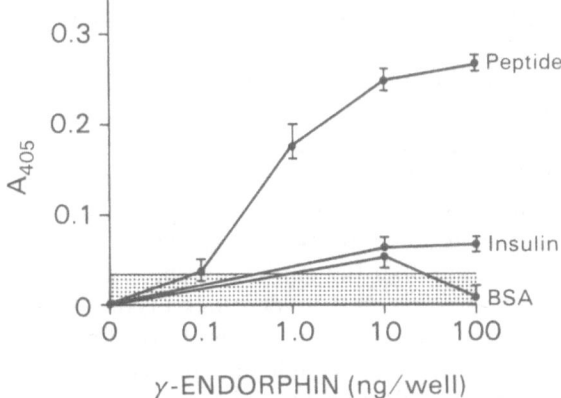

FIGURE 5.3. γ-Endorphin binding to a peptide encoded by the complementary RNA for bovine γ-endorphin. γ-Endorphin at various concentrations was incubated on microtiter plates coated with the complementary peptide (40 μg per well), insulin (20 units per well), or bovine serum albumin (BSA; 200 μg per well). Plates were incubated for 1 h, then washed three times with phosphate-buffered saline/Tween. The amount of γ-endorphin bound was detected by an ELISA using a monospecific antiserum to γ-endorphin. The shaded region demonstrates the amount of γ-endorphin binding to the solid-phase complementary peptide in the presence of excess soluble peptide (200 μg/ml). Reproduced with permission from Ref. 4.

conditions, polyacrylamide gel electrophoresis demonstrated four subunits of molecular weights 68,000, 58,000, 45,000, and 30,000 daltons. When molecules on the surface of NG108-15 cells were cross-linked prior to immunoaffinity purification, a single protein of approximately 210,000 daltons was detected on polyacrylamide gels. Since, until our studies, very little was known about the molecular structure of the ACTH receptor, it was not possible to compare our findings with others. However, several studies dealing with the molecular structure of the opiate receptor on NG108-15 cells had been done using diverse methodologies. This allowed us to compare our results with those of other investigators using different techniques. Interestingly, McLawhon et al. (10) reported a molecular weight of approximately 200,000 daltons for the opiate receptor on NG108-15 cells as estimated by radiation target size analysis. Furthermore, Howard et al. (11) reported that four subunits with approximate molecular weights of 65,000, 53,000, 38,000, and 25,000 daltons could be covalently cross-linked to ^{125}I-β-endorphin. Another group (12) demonstrated that a 58,000-dalton subunit on NG108-15 cells was the binding site for a specific delta opiate affinity reagent. Thus the results obtained by us and others, using vastly different techniques, are quite similar and serve to support the use of antibodies directed against complementary peptides as a useful methodology for receptor purification.

Generation of Anti-LHRH Receptor Antibodies

Another example of the use of a complementary peptide to isolate the respective receptor is that of LHRH (13). As before, antibodies to a peptide encoded by the complementary RNA for LHRH were generated and purified. This antibody specifically stained only the gonadotropes from dispersed rat anterior pituitary, and this staining could be blocked by the addition of the complementary peptide or an LHRH agonist. These results, together with the finding that the antibody suppressed LHRH-stimulated luteinizing hormone release in a reverse hemolytic plaque assay, leaves little doubt that the antibody recognizes the LHRH receptor binding site. Furthermore, when solubilized rat pituitary proteins were electrophoresed on polyacrylamide gels, blotted onto nitrocellulose, and then immuno-peroxidase stained, proteins at 60,000 and 51,000 daltons were detected. These protein molecular weights are similar to those previously described for the LHRH receptor using photoaffinity labeling techniques (14). Thus, the peptide specified by the complementary RNA for LHRH mRNA is similar, if not identical, to the LHRH binding site since the peptide bound LHRH and the anti-peptide antibody bound the LHRH receptor.

Binding Sites Encoded by Complementary Segments of Nucleic Acids

From the three examples given above, it is quite clear that peptides encoded by complementary RNAs represent binding sites for the respective ligands. These findings, however, raise a question as to the relationship between ligands, their binding sites, and the genomic DNA from which they are ultimately translated. From a present-day perspective, there are no examples to suggest that a ligand is coded from one DNA strand and the appropriate receptor is coded from the complementary DNA strand. In fact, it might be deleterious for a single cell to translate peptides or proteins from complementary RNAs since, in addition to nucleic acid hybridization, the proteinaceous molecules would have the potential to bind to one another. While the segment of DNA complementary to a particular ligand may not be translated, the information necessary to code for a binding site is nonetheless present. Stated simply, a potential binding site is encoded by the complementary DNA strand of that ligand. If this theory is generally applicable, then every known ligand has encoded in its complementary DNA the potential for a ligand binding protein, and every receptor binding site has encoded in its complementary DNA the potential for an endogenous ligand. Investigating whether or not this potential is expressed in the form of a translated product may lead to the discovery of heretofore unidentified ligands and receptors.

To support the idea that portions of proteins capable of binding one another could be encoded by complementary segments of DNA, we questioned whether sequences from known ligand-receptor pairs possessed regions of complementarity (4). At the time the analysis was done, only the nucleotide sequences for three ligand-receptor pairs (i.e., epidermal growth factor, interleukin-2, and transferrin) were published. Interestingly, in each pair highly significant regions of complementarity were found in portions of the proteins that would be compatible with ligand binding regions. For example, two regions of six amino acids each within epidermal growth factor and its receptor were found to have a high degree of complementarity. These 12 amino acids constituted approximately 23% of the total sequence of epidermal growth factor. Furthermore, the complementary regions in the receptor were located in the extracellular, ligand binding domain and not in the intracellular domain. These complementary regions were unique to epidermal growth factor and its receptor since out of 3,060 proteins, representing 616,748 six-amino-acid test segments, only epidermal growth factor contained either sequence. If our hypothesis is correct, these complementary sequences should turn out to be the contact points for the ligand-receptor interaction. In retrospect, if the receptor sequences for epidermal growth factor, interleukin-2, or transferrin had not been known, antibodies directed against complementary peptides for each ligand would have recognized the respective receptor. This outcome is certain since each of these ligand-receptor pairs possessed significant regions of complementarity. Thus, by making anti-complementary peptide antibodies, these receptors could also have been purified much like the ACTH, endorphin, and LHRH receptors were purified.

It is tempting to speculate from an evolutionary point of view how these complementary segments first came to be expressed in ligand-receptor pairs. It is certainly possible that the two interacting molecules evolved independently of one another, requiring a great deal of time to develop sufficiently high affinity for the interaction to be biologically significant. However, a more attractive hypothesis is that ligand-receptor pairs co-evolved over a much shorter time frame. If two cells were to translate segments of complementary DNA, they would immediately be making proteins which had the potential to bind one another with high affinity.

Antibodies Directed against Complementary Peptides Have an Idiotypic-Anti-Idiotypic Relationship

We have recently observed (15) that immunization with peptides specified by complementary strands of nucleic acids results in the generation of idiotypic and anti-idiotypic antibodies, respectively. Antibodies generated against ACTH bound specifically to antibodies generated against the complementary peptide, HTCA, and vice versa. Furthermore, the anti-HTCA antibody effectively competed with ACTH for binding the anti-ACTH antibody. These experiments demonstrated that not only could the two antibodies bind one another but that the binding took place at or near the binding site region. Thus, by definition the two antibodies have an idiotypic-anti-idiotypic relationship. Similar results were obtained using antibodies generated against γ-endorphin and its complementary peptide. Thus, these findings represent the first possible explanation at a molecular level for the network theory of immune regulation (16). It may be that idiotopes and anti-idiotopes represent complementary sequences in the hypervariable regions of such immunoglobulin pairs. Therefore, it may be possible to generate anti-idiotypic antibodies in a predetermined rather than random fashion.

Potential Uses and Theoretical Implications of the Molecular Recognition Code

From the studies presented above, it is obvious that peptides complementary to known ligand sequences can be used to generate anti-receptor antibodies. As we have shown, these antibodies can be quite useful in the isolation and characterization of the particular receptor. However, there are other potential uses for complementary peptides which have not yet been discussed.

From a diagnostic point of view, peptides which are relatively small with high binding affinity for a specific ligand would be advantageous in clinical assay procedures. Presently, radioimmunoassays and ELISA procedures depend upon the production of a specific, high-affinity antibody. In many cases, these antibodies are costly to make, there can be significant differences between different lots of antibodies, and the antibodies sometimes have limiting shelf lives. On the other hand, a complementary peptide specific for a particular ligand can be synthesized, maximizing cost efficiency and reproducibility. As an example, the complementary peptide for ACTH (i.e., HTCA) has been used effectively in a solid-phase radioligand binding assay to detect serum ACTH levels (unpublished observation).

From a clinical point of view, complementary peptides could be used to block a variety of biological responses by, for example, blocking ligand binding to its receptor. In theory, at least, any protein or peptide of biological importance could be inactivated by complexing the molecule with its complementary peptide. While there are many considerations that need to be taken into account in order to maximize clinical treatments, we have preliminary evidence that in vivo biological responses can be blocked by the appropriate complementary peptide. After a simple intramuscular injection of HTCA (i.e., the complementary peptide for ACTH), mice were suppressed in their ability to mount a steroidogenic response to a mild stress.

We have demonstrated previously that the complementary peptides for ACTH, γ-endorphin, and LHRH have the ability to block binding of the ligand to its respective receptor. The ability to block ligand-receptor interactions gives the complementary peptide technology a tremendous potential for antiviral therapeutics. Many viruses bind to specific receptors prior to entry into the cell. If these receptor sequences or the sequences on the viral attachment protein were known, then it may be possible to construct a complementary peptide which would block viral attachment.

Theoretically, the implications evoked by the finding that peptides from complementary strands of nucleic acid bind one another are potentially far reaching. As noted previously, this observation provides a simple and foolproof method for the evolution of interacting peptide or protein pairs. All that might be required for the generation of a ligand and its primordial receptor would be the translocation of the ligand's complementary DNA sequence to a coding strand. Furthermore, complementary sequences may provide a basis for internal imaging in the immune system and circuit formation within the central nervous system. Our findings may also be important in understanding cellular differentiation, since cells translating complementary sequences could recognize and communicate with one another via the peptide products.

Acknowledgments. We would like to thank Diane Weigent for secretarial assistance in preparing the manuscript. This research was sponsored in part by a grant from Triton Biosciences, Inc.

References

1. Yavin, E. 1982. Perspectives for monoclonal antibodies in receptor research, p. 67. *In* L.D. Kohn (ed.), Hormone receptors. John Wiley & Sons, Inc., New York.
2. Blalock, J.E., and E.M. Smith. 1984. Hydropathic anti-complementarity of amino acids based on the genetic code. Biochem. Biophys. Res. Commun. **121**:203.
3. Blalock, J.E., and K.L. Bost. 1986. Binding of peptides that are specified by complementary RNAs. Biochem. J. **234**:679.
4. Bost, K.L., E.M. Smith, and J.E. Blalock. 1985. Similarity between the corticotropin (ACTH) receptor and a peptide encoded by an RNA that is complementary to ACTH mRNA. Proc. Natl. Acad. Sci. USA **82**:1372.
5. Bost, K. and J.E. Blalock. 1985. Molecular characterization of a corticotropin (ACTH) receptor. Mol. Cell. Endocrinol. **44**:1.
6. Payet, N.G., and E. Escher. 1985. ACTH receptors in rat adrenal glomerulosa cells. Endocrinology **117**:38.

44   Kenneth L. Bost, Lawrence R. Smith, and J. Edwin Blalock

7. McIlhinney, R.A.J., and D. Schulster. 1975. Studies on the binding of ^{125}I-labeled corticotropin to isolated rat adrenocortical cells. J. Endocrinol. **64**:175.
8. Carr, D.J.J., K.L. Bost, and J.E. Blalock. 1986. An antibody to a peptide specified by a RNA that is complementary to γ-endorphin mRNA recognizes an opiate receptor. J. Neuroimmunol. **12**:329.
9. Simonds, W.F., T.R. Burke, A.E. Jacobson, and W.A. Klee. 1985. Purification of the opiate receptor of NG108-15 neuroblastoma-glioma hybrid cells. Proc. Natl. Acad. Sci. USA **82**:4974.
10. McLawhon, R.W., J.C. Ellory, and G. Dawson. 1983. Molecular size of opiate (enkephalin) receptors in neuroblastoma-glioma hybrid cells as determined by radiation inactivation analysis. J. Biol. Chem. **258**:2101.
11. Howard, A.D., S. de La Baume, T.L. Gioannini, J.M. Hiller, and E.J. Simon. 1985. Covalent labeling of opioid receptors with radioiodinated human b-endorphin. J. Biol. Chem. **260**:10833.
12. Klee, W.A., W.F. Simonds, F.W. Sweat, T.R. Burke, A.E. Jacobson, and K.C. Rice. 1982. Identification of a M_r 58000 glycoprotein subunit of the opiate receptor. FEBS Lett. **150**:125.
13. Mulchahey, J.J., J.D. Neill, L.D. Dion, K.L. Bost, and J.E. Blalock. 1986. Antibodies to the binding site of the LHRH receptor: generation with a synthetic decapeptide encoded by a RNA complementary to LHRH mRNA. Proc. Natl. Acad. Sci. USA, **83**:9714.
14. Iwashita, M., and K.J. Catt. 1985. Photoaffinity labeling of pituitary and gonadal receptors for gonadotropin-releasing hormone. Endocrinology **117**:738.
15. Smith, L.R., K.L. Bost and J.E. Blalock. 1987. Generation of idiotypic and anti-idiotypic antibodies by immunization with peptides encoded by complementary RNAs: A possible molecular basis for the network theory. J. Immunol. **138**:7.
16. Jerne, N. 1974. Towards a network theory of the immune system. Ann. Immunol. (Inst. Pasteur) **125C**:373.

6
Anti-Idiotypic Antibodies and Substance P Receptors

JEAN-YVES COURAUD

Introduction

Some fifty-five years ago Von Euler and Gaddum (1) described the presence in equine brain and intestine of a factor displaying hypotensive and spasmogenic activities. The peptidic nature of this factor, called substance P (SP), was established forty years later in 1971 (2). Since that date, SP has been one of the most extensively studied neuropeptides because of its involvement in biological functions, like pain transmission, modulation of central monoaminergic functions, control of blood pressure, and inflammatory reactions. However, the definite characterization of SP receptors has suffered from the relative inadequacy of the tools available for these studies, i.e., nonspecific and unstable peptidic agonists or antagonists, and remains until now rather tentative.

After a brief review of our present knowledge on substance P and other similar compounds present in mammals, as well as their receptors, we will present preliminary results concerning the production of anti-idiotypic antibodies which recognize the SP receptors displaying pharmacological actions, and consequently could serve to further characterize these receptors.

Generalities on Substance P, the Tachykinin Family, and Their Receptors

Substance P and the Tachykinin Family

Substance P is a peptide with 11 amino acids (Table 6.1) whose C-terminal extremity carries an amide group. SP belongs to the family of tachykinins, which all share a common C-terminal pentapeptide: Phe-X-Gly-Leu-Met-NH_2. This common pentapeptide has been demonstrated extensively to be involved in the interaction with the physiological receptor and hence to be responsible for most of the pharmacological actions of the tachy-

TABLE 6.1. The three mammalian neurokinins and their receptors.[a]

Neurokinin	Amino acid sequence	Receptor
	1 2 3 4 5 6 7 8 9 10 11	
Neurokinin P (NKP) = substance P (SP)	Arg-Pro-Lys-Pro-Gln-Gln-Phe-Phe-Gly-Leu-Met-NH$_2$	NK-P or NK$_1$ [SP-P]
Neurokinin A (NKA) [substance K (SK)]	His-Lys-Thr-Asp-Ser-Phe-Val-Gly-Leu-Met-NH$_2$	NK-A or NK$_2$ [SP-K]
Neurokinin B (NKB) [neuromedin K (NK)]	Asp-Met-His-Asp-Phe-Phe-Val-Gly-Leu-Met-NH$_2$	NK-B or NK$_3$ [SP-E]

[a]The new nomenclature which has been proposed for unifying the terminology concerning mammalian neurokinins and their receptors at the Symposium on Substance P and Neurokinins (Montreal, Quebec, Canada, July 21–24, 1986) is given here. The former terminology, still used, is indicated in brackets.

kinins. While SP occurs in tissues of all vertebrate species, other tachykinins seem to be restricted to lower vertebrate (physalaemin and kassinin in amphibians) or invertebrate tissues (eledoisin, in molluscans).

The main physiological properties of SP (and of tachykinins in general) are the following (see Ref. 3 for a review). SP has been shown to be an important neurotransmitter in the brain and the spinal cord. For example, SP is the transmitter (presumably not the only one) for sensory fibers of the C type which carry pain information from the periphery to the spinal cord. In the brain, SP acts as a neuromodulator, controlling, for example, the activity of the nigro-striatal pathways. SP has also multiple peripheral actions. It evokes the contraction of smooth-muscle fibers of the gastrointestinal (ileum), respiratory (trachea), and urogenital (bladder) tracts and, by contrast, is a relaxant for large arteries (carotid), which explains its potent hypotensive effect. SP controls the secretion of a variety of both endocrine (i.e., pancreas) and exocrine (i.e., parotid) glands. In addition, SP has been found recently to bind various immunocompetent cells, like macrophages, T lymphocytes, and mastocytes.

However, in addition to SP, two other tachykinins have been recently discovered in mammals (4,5) and may well be responsible, under normal conditions, for at least some of the physiological actions previously attributed to SP.

These new tachykinins were originally called substance K (SK) and neuromedin K (NK) but the names neurokinin A (NKA) and neurokinin B (NKB), respectively, are now recommended. In an attempt to unify this terminology, it has been proposed that substance P be renamed neurokinin P (NKP) and the term "neurokinins" be reserved for the tachykinins naturally occurring in mammals (6).

Two SP precursor molecules for neurokinins have been also identified (7)—an α-preprotachykinin (α-PTT) containing the sequence of SP only and a β-PTT containing a copy of SP and of NKA, both precursors probably deriving from a single gene. Very recent data indicate that a δ-PTT deriving from a distinct gene could be the precursor for NKB.

SP Receptors: A Historical Survey

METHODOLOGY

SP receptors have been studied until now through several kinds of experimental approaches but never through immunological methods. Most information available on SP receptors derives from classical pharmacological studies. Binding studies and cytochemical observations developed only recently with the availability of different radiolabeled tachykinins with a high specific radioactivity. Finally, a few studies have dealt with the second messenger systems associated to the interaction of SP with the receptor, namely, the hydrolysis of membranous phospholipids.

MULTIPLICITY OF SP RECEPTORS

The idea that multiple receptors for SP may exist is relatively recent. Erspamer et al. (1980–81) were the first to note that the various tachykinins known at that time were not equipotent in all bioassays. In 1982, Iversen and his group (8) selected out two rank orders of potency for the tachy-kinins on smooth-muscle preparations and used these as evidence for the existence of two receptor subtypes. In one group of peripheral organs, typified by the guinea pig ileum, the four peptides tested (SP, physalaemin, eledoisin, kassinin) were all approximately equipotent. This class was termed SP-P since physalaemin was generally the most potent agonist. In a second group, typified by the rat vas deferens, kassinin and eledoisin were much more potent than SP or physalaemin. This class was termed SP-E (E for eledoisin).

Confirming this heterogeneity in SP receptors, first binding studies using both SP and eledoisin as radiolabeled ligands clearly demonstrated, on brain membranes for example (9), the existence of two distinct high-affinity specific binding sites for the two tachykinins. Likewise, autoradiographic observations also confirmed the occurrence of two different binding sites for SP, on the one hand, and eledoisin (10) or kassinin (11), on the other hand.

Thus, data available in 1982–83 all converged to suggest the existence of two types of SP receptors, one more specific for SP and the other for eledoisin or kassinin. So the following question was raised: Since eledoisin and kassinin are not found in mammals, what is (or what are) the endog-enous ligand(s) for this second receptor subtype? Just as happened for opioid peptides, the discovery of receptors stimulated the search for en-dogenous ligands. In fact, at that time, both NKA and NKB were isolated from mammalian nervous tissues (see above).

As expected, pharmacological studies (12) as well as binding experiments (13) have rapidly indicated that while SP could be the endogenous ligand for SP-P receptors, either NKA or NKB could be that for the SP-E sub-type. By 1983–84, people were thus facing three endogenous ligands for only two receptor types. The distinction of a third receptor subtype was reported for the first time by Buck et al. in 1984 (14), on the basis of binding experiments. Using three radiolabeled tachykinins (SP, NKA, and eledoisin), they observed three different ligand selectivity patterns. They showed that the previously defined SP-E receptor subtype was hetero-geneous and distinguished, in addition to the SP-P site specific for SP, and SP-E site *sensu stricto* specific for NKB and a new defined SP-K site specific for NKA (or substance K).

Subsequently, autoradiographic experiments as well as binding studies have revealed that the binding sites of SP, NKA, and NKB have indeed different tissue localizations (see Ref. 15 as a recent review). For example, substantia nigra (a brain area which contains very high levels of SP-like immunoreactivity) is practically devoid of any binding site for SP (SP-P) but is enriched in sites for NKA (SP-K).

Finally, most recent pharmacological studies in peripheral tissues, especially from Regoli's laboratory (6,16,17), now indicate that these three binding sites do represent three distinct physiological receptors for the three neurokinins. These authors have clearly demonstrated that SP is a very weak agonist on preparations containing exclusively SP-K or SP-E receptors, so that this terminology has to be modified. They propose the terms NK-P or NK-1, NK-A or NK-2, and NK-B or NK-3 (NK for neurokinin) for the receptors of NKP (neurokinin P, or substance P), NKA, and NKB, respectively. In other words, and whatever the terminology may be, the idea prevailing is that mammalian tissues do not contain three subtypes of receptors for SP but rather three separate receptors for different neurotransmitters (Table 6.1).

CONCLUDING REMARKS

While the distinction of three receptors is acceptable, it is still highly possible that other neurokinins very recently described in mammals (elongated forms of SP and NKA, for example) each possess their own specific receptor. Furthermore, we know now that SP exerts some of its physiological functions (namely, in triggering immunocompetent cells) or compartmental effects through its N-terminal region, demonstrating the existence of a novel type of receptor, which we will not consider here.

The definite demonstration of three or more neurokinin receptors is still hindered by several difficulties. Some of these are related to biological preparations (presence of receptors for different neurokinins on the same tissue, indirect effects of a given neurokinin, etc.) The others are related to the drugs (agonists or antagonists) used for characterizing the receptors. All of these are small peptides subject to a rapid degradation in biological media, and highly selective antagonists are not available. This justified our trying to produce antibodies to the SP receptor (i.e., tools expected to be specific and stable), and the only possible route was the anti-idiotypic approach.

The Anti-Idiotypic Approach for the SP Receptor

Characterization of Anti-SP Antibodies with a Specificity Close to That of SP Receptors

POLYCLONALS

Since our final aim was to get anti-idiotypes behaving as "internal images" of the C-terminal part of SP (the epitope recognized by receptors), we tried to raise in a first step anti-SP antibodies (Ab$_1$) recognizing this peculiar epitope, i.e., with a specificity close to that of the SP receptor. We used SP conjugated to BSA through its N-terminal basic residues as an immunogen which was injected into ten rabbits (18). Antibodies (Abs) were detected in the sera using either a conventional radioimmunoassay (RIA, described in Ref. 18) or a highly sensitive enzyme immunoassay (EIA) newly developed in our laboratory (23). Briefly, the enzymatic tracer is obtained by coupling SP to the enzyme acetylcholinesterase (AChE), whose activity is revealed by the spectrophotocolorimetric method of Ellman. The presence of specific anti-SP rabbit Abs is detected on microtiter plates coated with swine antibodies to rabbit Ig (SAR), prepared in the laboratory. The sensitivity of this technique is as good as, and often better (by a factor of 4 to 5) than, that of classical RIA and allows detection of SP amounts smaller than fmol/well with best polyclonals.

Two polyclonal antisera with titers close to 1/60,000 were selected (rabbits no. 144 and 418) and their fine specificities have been tested (by RIA and EIA) by measuring the immuno-cross-reactivities for SP (100%) and various SP-related compounds—the series of (L-Ala)SP derivatives (in which each amino acid is successively replaced by an Ala residue), SP fragments, and other naturally occurring tachykinins. For each compound, the cross-reactivity was compared with the biological activities (the hypotensive effect in the rat in vivo and the spasmogenic action on the guinea pig ileum in vitro) to get an idea about the expected resemblance between anti-SP Abs and the SP receptor. As seen in Table 6.2, there is a good correlation between the three parameters (cross-reactivity and the two selected biological activities). These results indicate that both anti-SP antibodies and SP receptors (at least those involved in spasmogenic and hypotensive actions) recognize very similar epitopes on the SP molecule (i.e., the C-terminal pentapeptide, especially Phe^7, Phe^8, Leu^{10}, and Met^{11}-NH$_2$) and hence strongly suggest that the paratopes of anti-SP Abs and SP receptor binding sites carry similar structural features.

MONOCLONALS

Ten mice of the Biozzi strain were immunized with the same immunogen used for rabbits (see above). Three of them responded, and spleen fusions were performed by classical cell hybridization techniques. Screening tests were carried out on the culture supernatants using the EIA procedure described above ([with the modification that microtiter plates were coated with swine antibodies to mouse Ig (SAM)], which proved very timesaving. Five monoclonal antibodies (mAbs) were selected (24) with high affinities for SP (K_d close to 0.1 nM). As for polyclonals, their specificities were tested using various SP-related compounds. The results indicate that the specificities of the five monoclonals were very close. All of them are directed towards the C-terminal part of SP, essentially Phe^7 (but not Phe^8)

TABLE 6.2. Comparison of biological activities and immunological cross-reactivities with polyclonal and monoclonal anti-SP antibodies for substance P (100%) and the series of (L-Ala) SP derivatives.[a]

| Peptide | Biological activities[b] | | Immunological cross-reactivities[c] | | |
| | Rat blood pressure | Guinea pig ileum | Polyclonals | | Monoclonal[d] |
			No.144	No.418	
SP	100	100	100	100	100
(Ala-1)SP	100	100	73	32	91
(Ala-2)SP	100	100	102	43	123
(Ala-3)SP	100	100	75	23	82
(Ala-4)SP	100	100	80	49	126
(Ala-5)SP	100	100	92	50	86
(Ala-6)SP	100	100	69	45	61
(Ala-7)SP	1.5	0.3	1.7	1.4	0.5
(Ala-8)SP	3.0	22	2.2	1.9	62
(Ala-9)SP	100	100	49	20	55
(Ala-10)SP	7.4	55	0.2	0.08	22
(Ala-11)SP	8.0	20	0.5	<0.05	35

[a]Various other SP-related compounds have been tested (SP fragments, nonmammalian and mammalian neurokinins); for each of them, a close correlation between biological activities and immunological cross-reactivity with anti-SP antibodies was observed.
[b]Biological activities were taken from Refs. 20 and 21. Data are expressed as percent of biological activity of substance P, taken as 100%.
[c]Immunological cross-reactivities were measured by EIA. Data are expressed as the ratio x 100% of the respective concentrations (in M) of SP and analogue which displace 50% of the enzymatic tracer bound to anti-SP antibodies.
[d]Monoclonal antibody is SP 14. Four other mAbs (SP 2, SP 19, SP 24, and SP 31) have also been used for testing cross-reactivities. Results were very similar to those obtained with SP 14 (24).

and to a lower extent Leu[10] and Met[11]-NH$_2$. We observed a rather good correlation between cross-reactivities and biological activities (Table 6.2) except for the fact that the paratopes of the Abs do not recognize the cyclic amino acid in position 8 while the receptor recognizes it well. So, the receptor binding sites appear somewhat to be best mimicked by the paratopes of polyclonals rather than by those of monoclonals.

Characterization of Anti-SP Anti-Idiotypic Antibodies

Rabbit Polyclonal Anti-Idiotypic Antibodies Which Recognize SP Receptors

IMMUNOLOGICAL CHARACTERIZATION

Purified total IgG fraction from anti-SP antiserum (no. 144) was injected into ten allotype-matched rabbits. Two of them responded by producing in their sera large amounts of anti-idiotypic IgG antibodies (18), which were able to inhibit the binding of the radioactive or enzymatic tracer to anti-SP IgG from rabbit 144 in a dose-dependent fashion. In accord with the concept of "internal images," these anti-idiotypes (anti-Ids) were also shown to compete with the enzymatic tracer for binding to not only a large variety of rabbit polyclonal anti-SP (Ab₁) antibodies (Fig. 6.1) but also all the anti-SP mAbs described above (Fig. 6.2). Therefore, these anti-Ids may be classified as Ab_2 of the β type.

Receptor Recognition by Anti-Idiotypic Antibodies

We investigated three receptors specific for SP (NK-P) occurring in distinct mammalian tissues.

Rat Parotid Cell Glandular Receptor

In collaboration with B. Rossignol from the University of Paris Sud (Orsay, France), we could visualize, by indirect immunofluorescence, specific membranous binding sites for purified anti-Id IgG on a preparation of dis-

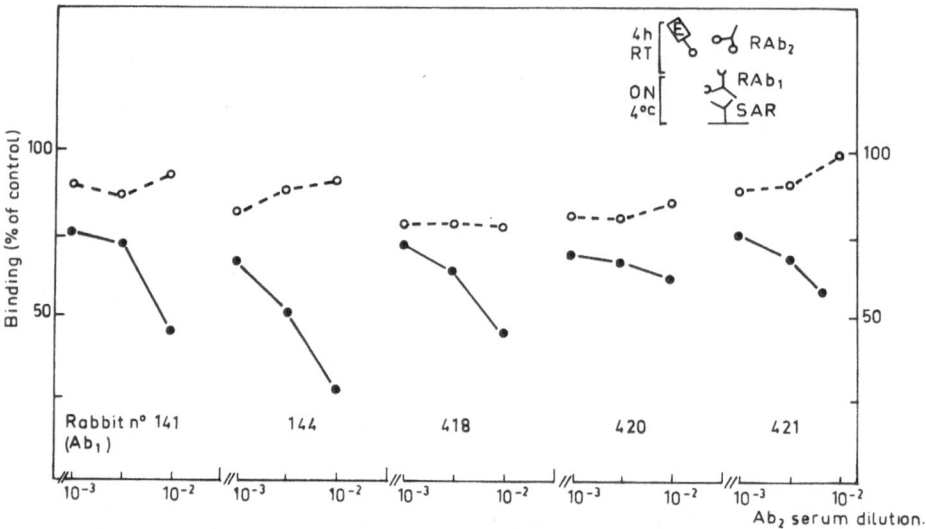

Figure 6.1. Inhibition of binding of the enzymatic tracer (SP-AChE) to various rabbit anti-SP antisera (RAb₁) by rabbit anti-idiotypic antibodies (RAb₂). Experiments were performed on microtiter plates coated with swine antibodies to rabbit Ig (SAR). RAb₁ were incubated overnight at 4°C. After rinsing, the enzymatic tracer and anti-idiotypic serum (RAb₂) at different dilutions (•) or rabbit normal serum (○) or control buffer were added for 4 h at room temperature. After rinsing, the Ellman medium, which reveals AChE activity, was added and after an incubation of a few hours, optical density was recorded at 412 nm. Results are expressed as percent of the maximal binding of the tracer ($B_0 = 100\%$) obtained in the presence of control buffer. RAb₂ inhibit the binding of the tracer to all anti-SP antibodies and especially to those from rabbit 144, which served as immunogens for their production.

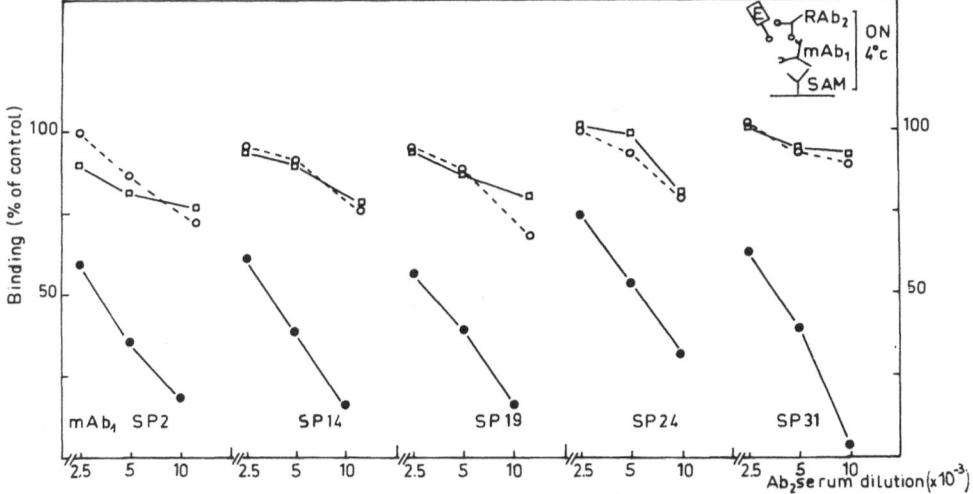

FIGURE 6.2. Inhibition of binding of the enzymatic tracer (SP-AChE) to five monoclonal anti-SP antibodies (mAb₁) by rabbit anti-idiotypic antibodies (RAb₂). Experiments were performed on plates coated with swine antibodies to mouse Ig (SAM). The tracer, mAb₁, and RAb₂ at different dilutions (or control buffer for measuring maximal binding, B_0) were incubated together overnight at 4°C. After rinsing, the Ellman medium was added for a few hours. The end of the experiment was as described in the legend to Fig. 6.1. Nonspecific inhibition of binding was measured using normal rabbit serum (○) or the preimmune bleeding of the rabbit which produced Ab₂ (□). Anti-idiotypic antibodies (•) strongly inhibited the binding of SP-AChE to all anti-SP monoclonals.

persed parotid gland cells (Fig. 6.3). The staining was not observed with control nonimmune IgG and was considerably lowered when cells were preincubated with SP or with an SP antagonist, suggesting that labeled membranous sites were actually SP receptors. In this preparation, as in a few other ones, SP is known to enhance the turnover of membranous phosphatidylinositol. Likewise, anti-Id IgG at a concentration of 3.5 nM SP equivalent (as measured by RIA) was demonstrated to specifically stimulate, like SP itself, the incorporation of [^{32}P]phosphate in phospholipids (18). The stimulating actions induced both by SP and anti-Ids were abolished by an SP antagonist. Preincubation of anti-Ids with anti-SP Abs prior to the incubation with the cells also prevented the stimulating effect of anti-Ids. In this test, anti-Ids thus behaved as agonists of SP.

Guinea Pig Ileum Muscular Receptor

SP is known to be a potent spasmogenic agent on guinea pig ileum (GPI) smooth muscle. At a concentration as low as 2×10^{-10} M SP equivalent, anti-Id IgG was found to inhibit by 50% the contraction of GPI induced

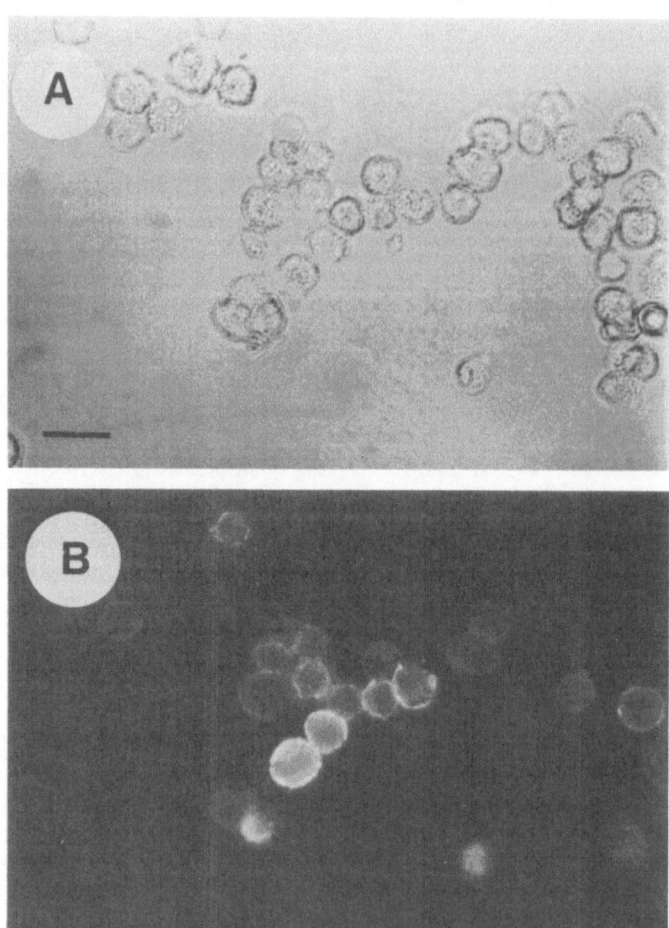

FIGURE 6.3. Binding of rabbit anti-idiotypic antibodies to the rat parotid gland cells, as revealed by indirect immunofluorescence. Isolated cells of the rat parotid gland were incubated for 1 hour at O°C in the presence of anti-idiotypic IgG (7.5 mg/ml). After washing, the cells were resuspended and labeled with goat anti-rabbit-IgG FITC conjugate under similar conditions. The same cells are shown under incandescent light (top panel A) and fluorescent light (bottom panel B). Scale bar = 20 μm.

by 2×10^{-9} M SP (Table 6.3). Control nonimmune IgG had no effect on the spasmogenic action of SP. When tested in the absence of SP, neither anti-Id nor control IgG had any effect on muscle contraction. So, contrary to what was observed on parotid cells, anti-Ids behaved as potent antagonists of SP in this muscular preparation. Experiments were done in collaboration with D. Regoli in Sherbrooke, Quebec, Canada.

TABLE 6.3. Antagonist effect of anti-idiotypic antibodies for the spasmogenic action of substance P on the isolated guinea pig ileum.[a]

Incubation medium	Ileum contraction (mm)[b]
SP alone (1.6×10^{-9}M)	29.0 ± 3.2
SP + control buffer	26.1 ± 2.0
SP + nonimmune IgG	25.8 ± 2.0
SP + anti-Id IgG (2×10^{-10} M SP equivalent)	13.2 ± 1.6

[a]Modified from Ref. 18.
[b]The contraction of ileum was reduced by 40–50% and ($p < 0.01$) by anti-Id IgG (0.75 mg/ml final concentration, i.e., 2×10^{-10} M equivalent concentration of SP, as measured by RIA) but not by control nonimmune IgG (0.75 mg/ml). Tests were carried out on the same ileum. The inhibition of contraction by anti-Id was rapidly and completely reversible upon rinsing. Values are means $\pm$ SD of at least 4 determinations. In the absence of SP, neither anti-Id nor control IgG had any effect on ileum contraction. Experiments were performed in the laboratory of Dr. D. Regoli, Sherbrooke, Quebec, Canada.

Rat Spinal Cord Neuronal Receptor

In spinal cord, SP is one of the neurotransmitters for small-diameter sensory fibers, whose cell bodies are located in dorsal root ganglia and which are involved in the transmission of pain. The fine localization of SP receptors in spinal cord has been recently reported by Shults et al. (19), using conventional autoradiographic techniques with radiolabeled SP. By incubating rat spinal cord sections, taken at the cervical level, with the anti-Id IgG (or control nonimmune IgG) and then with peroxidase-labeled goat antibodies to rabbit Ig, we have been able to observe specific binding sites for anti-Ids (M. Conrath et al., manuscript in preparation). No label was observed with control IgG. The distribution of peroxidase labeling with anti-Ids is superimposable on that observed in autoradiography experiments using ^{125}I-SP. Anti-Id binding sites are located essentially in dorsal horn, but also in the medial border of lamina IV and around the central canal, so that anti-Ids are likely to reveal actual SP receptors.

Mouse Polyclonal Anti-Idiotypic Antibodies

One anti-SP mAb (SP 14) has been chosen as an immunogen on the basis of the structural resemblance between its paratope and the SP receptor (Table 6.2) and was injected into 15 Biozzi mice. Four of them responded and anti-Ids were detected in their successive bleedings (unpublished results).

These anti-idiotypes (Ab_2) strongly inhibited the binding of the enzymatic tracer to SP 14, the immunogen mAb (Fig. 6.4A), or SP 19, another anti-SP mAb (not shown).

In addition, mouse anti-idiotypic antisera (MAb_2) from successive bleedings were able to recognize and bind anti-SP antibodies from rabbits (RAb_1) (Fig. 6.4b). From a comparison of Figs. 6.4A and 6.4B, it seems

56   Jean-Yves Couraud

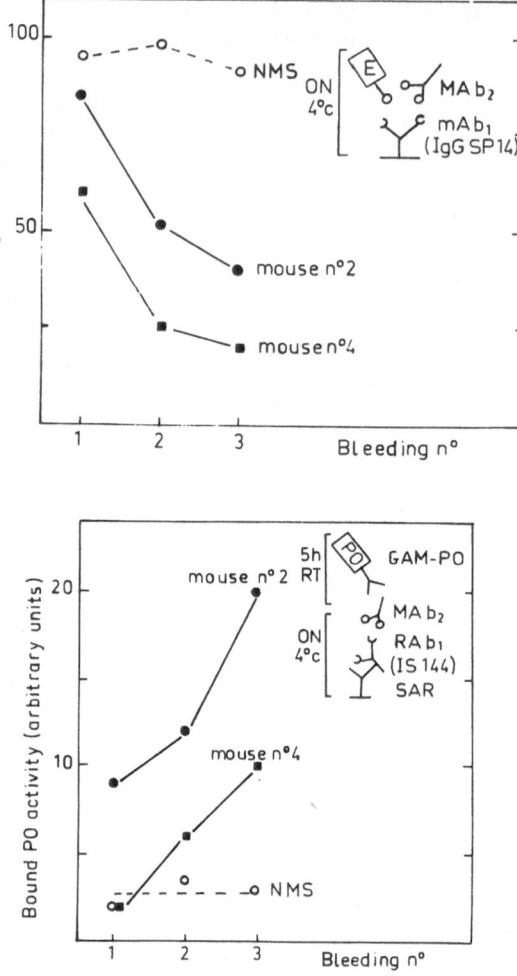

FIGURE 6.4. Immunological characteristics of polyclonal mouse anti-idiotypic antibodies (MAb$_2$) of two representative animals. (A) Inhibition of binding of the enzymatic tracer (SP-AChE) to monoclonal anti-SP IgG (mAb$_1$) by successive mouse anti-idiotypic antisera (MAb$_2$). Microtiter plates were coated with monoclonal anti-SP antibody mAb$_1$ (IgG SP 14). The enzymatic tracer and successive anti-idiotypic antisera MAb$_2$ (diluted 1/40) from mice nos. 2 and 4 or normal mouse serum (NMS) at the same dilution or buffer (for measuring maximal binding, B_0) were incubated together overnight at 4°C. After rinsing and addition of the Ellman medium during 2–3 h, optical density was measured at 412 nm. Results are expressed as percent of maximal binding (B_0). (B) Direct binding of successive mouse anti-idiotypic antisera (MAb$_2$) to rabbit anti-SP antibodies (RAb$_1$). Microtiter plates were coated with swine antibodies to rabbit Ig (SAR). Rabbit anti-SP antibodies (RAb$_1$) from rabbit no. 144 and successive anti-idiotypic antisera MAb$_2$ (diluted 1/40) from mice nos. 2 and 4 or normal mouse serum (NMS) were incubated overnight at 4°C. After rinsing, peroxidase-labeled goat antibodies to mouse Ig (GAM-

that MAb_2 which are more effective in binding homologous mouse Ab_1 (MAb_1) and those more effective in binding heterologous rabbit Ab_1 (RAb_1) are distinct. In addition, it is worth noting that the interspecies binding of MAb_2 to RAb_1 did not prevent RAb_1 from binding the enzymatic tracer (data not shown). These preliminary experiments suggest once more that the paratopes of polyclonal (RAb_1) and monoclonal (MAb_1) anti-SP may be slightly different. Due to the low amounts of anti-idiotypic mouse sera available, no attempt was made to investigate whether mouse anti-Ids did recognize SP receptors.

Up to now, four fusions have been performed from spleens of responding animals, but we were unable to detect any positive hybridoma supernatant out of about 5,000 using three different screening tests (competition with the enzymatic tracer for binding homologous SP 14 Ab_1, direct binding to homologous SP 14 Ab_1 and to heterologous rabbit Ab_1). Anti-Ids could have escaped detection either because they are not released free in the culture medium or because they display a too low affinity for Ab_1. Alternatively, too few responding spleen lymphocytes could have been involved in fusions with myeloma cells.

Conclusions

Anti-idiotypic antibodies can be regarded as promising tools for fundamental research to characterize further the neurokinin receptors. In this regard, the physiological activity of polyclonal anti-Ids have been tested only on receptors specific for SP (NK-P receptors). It would be now of great interest to ascertain whether anti-Ids can discriminate the receptors for the three mammalian neurokinins SP, NKA, and NKB. Anti-Ids could also be used as specific and stable ligands in binding experiments and cytochemical studies to map the tissular localization of SP receptors. Such experiments are in progress. Of course, anti-Ids, preferentially monoclonals, might be ideal tools for purifying the SP receptor by immunoaffinity chromatography.

Finally, in view of the involvement of SP and, more generally, neurokinins in such fundamental physiological functions as pain transmission, control of blood pressure, and inflammatory processes, it would be worth testing whether anti-Ids could exert any in vivo physiological effect. If they would appear to modulate in vivo the normal functioning of the neu-

---

PO) were added for 5 h at room temperature. After another rinsing, peroxidase activity was measured using the classical spectrophotocolorimetric method. Results obtained in A and B reveal the presence in these successive mouse bleedings of increasing concentrations of anti-Id antibodies recognizing both mouse and rabbit anti-SP (Ab_1) antibodies.

rokininergic system, then our hope is that anti-Ids could be used in clinics as therapeutic reagents.

Acknowledgments. I am greatly indebted to all members of the laboratory who were involved in this research program: Dr. Y. Frobert, M. Plaisance and P. Lamourette for raising monoclonal antibodies, M.C. Nevers for expert technical assistance, Dr. J. Grassi for developing the EIA procedures, and Dr. P. Pradelles for constant interest and helpful discussions. I acknowledge also very much Dr. D. Renzi for invaluable help in numerous experiments. Finally, F. Wierniezky must be thanked for typing the manuscript.

References

1. Euler, U.S.V., and J.H. Gaddum. 1931. An unidentified depressor substance in certain tissue extracts. J. Physiol. **72**:74.
2. Chang, M.M., S.E. Leeman, and H.D. Niall. 1971. Amino-acid sequence of substance P. Nature (London) New Biol. **232**:86.
3. Pernow, B. 1983. Substance P. Pharmacol. Rev. **35**:85.
4. Kangawa, K., N. Minamino, A. Fukuda, and H. Matsuo. 1983. Neuromedin K: a novel mammalian tachykinin identified in bovine spinal cord. Biochem. Biophys. Res. Commun. **114**:533.
5. Maggio, J.E., B.E.B. Sandberg, C.V. Bradley, L.L. Iversen, S. Santikarn, D.H. Williams, J.C. Hunter, and M.R. Hanley. 1983. Substance K: a novel tachykinin in mammalian spinal cord, p. 20. *In* P. Skrabanek and D. Powell (eds.), Substance P, Dublin 83. Boole Press, Dublin.
6. Regoli, D., G. Drapeau, S. Dion and P. Juste-D'Orleans. 1987. Miniview: pharmacological receptors for substance P and neurokinins. Life Sci. **40**:109.
7. Nawa, H., T. Hirose, H. Takashima, S. Inayama, and S. Nakanishi. 1983. Nucleotide sequences of cloned cDNAs for two types of bovine brain substance P precursor. Nature (London) **306**:32.
8. Lee, C.M., L.L. Iversen, M.R. Hanley and B.E. Sandberg. 1982. The possible existence of multiple receptors for substance P. Naunyn-Schmiedeberg's Arch. Pharmakol. **318**:281.
9. Cascieri, M.A., and T. Liang. 1984. Binding of [^{125}I] Bolton Hunter conjugated eledoisin to rat brain cortex membranes—evidence for two classes of tachykinin receptors in the mammalian central nervous system. Life Sci. **35**:179.
10. Ninkovic, M., J.C. Beaujouan, Y., Torrens, M. Saffroy, M.D. Hall, and J. Glowinski. 1985. Differential localization of tachykinin receptors in rat spinal cord. Eur. J. Pharmacol. **106**:463.
11. Mantyh, P.W., J.E. Maggio, and S.P. Hunt. 1984. The autoradiographic distribution of kassinin and substance K binding sites is different from the distribution of substance P binding sites in rat brain. Eur. J. Pharmacol. **102**:361.
12. Hunter, J.C., and J.E. Maggio. 1984. Pharmacological characterization of a novel tachykinin isolated from mammalian spinal cord. Eur. J. Pharmacol. **97**:159.

13. Torrens, Y., S. Lavielle, G. Chassaing, A. Marquet, J. Glowinski, and J.C. Beaujouan. 1984. Neuromedin K, a tool to further distinguish two central tachykinin binding sites. Eur. J. Pharmacol. **102**:381.
14. Buck, S.H., E. Burcher, C.W. Shults, W. Lovenberg, and T.L. O'Donohue. 1984. Novel pharmacology of substance K-binding sites: a third type of tachykinin receptor. Science **226**:987.
15. Quirion, R. 1985. Multiple tachykinin receptors. Trends Neurosci. **8**:183.
16. Regoli, D., E. Escher, G. Drapeau, J.P. D'Orleans, and J. Mizrahi. 1984. Receptors for substance P. III. Classification by competitive antagonists. Eur. J. Pharmacol. **97**:179.
17. Mizrahi, J., S. Dion, J.P. D'Orleans, E. Escher, G. Drapeau, and D. Regoli. 1985. Tachykinin receptors in smooth muscles: a study with agonists (substance P, neurokinin A) and antagonists. Eur. J. Pharmacol. **118**:25.
18. Couraud, J.Y., E. Escher, D. Regoli, V. Imhoff, B. Rossignol, and P. Pradelles. 1985. Anti-substance P anti-idiotypic antibodies. Characterization and biological activities. J. Biol. Chem. **260**:9461.
19. Shults, C.W., R. Quirion, B. Chronwall, T.N. Chase, and T.L. O'Donohue. 1984. A comparison of the anatomical distribution of substance P and substance P receptors in the rat central nervous system. Peptides. **5**:1097.
20. Couture, R., A. Fournier, J. Magnan, S. Saint-Pierre, and D. Regoli. 1979. Structure-activity studies on substance P. Can. J. Physiol. Pharmacol. **57**:1427.
21. Fournier, A., R. Couture, J. Magnan, M. Gendreau, D. Regoli, and S. Saint-Pierre. 1980. Synthesis of peptides by the solid-phase method. V. Substance P and analogs. Can. J. Biochem. **58**:272.
22. Erspamer, V. 1981. The Tachykinin family. Trends Neurosci. **4**:267.
23. Renzi, D., J.-Y. Couraud, Y. Frobert, M.C. Nevers, P. Geppetti, P. Pradelles, and J. Grassi. 1987. Enzyme immunoassay for substance P using acetylcholinesterase as label. *In* Trends in Cluster Headache, F. Sicuteri, et al., editors. Elsevier New York. pp. 125–134.
24. Couraud, J.Y., Y. Frobert, M. Conrath, D. Renzi, J. Grassi, G. Drapeau, D. Regoli and P. Pradelles. Monoclonal antibodies to substance P: production, characterization of their fine specificities and use in immunocystochemistry. J. Neurochem, in press.

7
Anti-Idiotypic Antibodies Approach to the Study of the TSH Receptor

NADIR R. FARID

Introduction

Following the binding of thyroid-stimulating hormone (TSH) to its receptor, the receptor undergoes conformational changes which trigger a number of signals including the activation of adenylate cyclase and phospholipase C, hydrolysis of phospholipids, and movement of calcium ions. The ultimate result would be the synthesis and secretion of thyroid hormones. TSH effect is modulated by other hormone and growth factors, the latter contributing to thyroid cell growth in most species. TSH action is terminated by uncoupling of receptor from the cyclase system and disappearance of the high-affinity binding site on thyroid cell as well as action of protein kinase C (1).

TSH Receptor

On the basis of results obtained using a variety of techniques, we have suggested that the porcine TSH receptor is made up of a basic unit of M_r 95,000 transmembrane acidic glycoprotein. Membrane-bound proteases digest this moiety into M_r 70,000 and M_r 45,000 TSH-binding fragments. The appearance of an M_r 45,000 fragment is seemingly triggered by TSH interaction with the receptor. It is presently unclear whether this phenomenon is a route for mediating TSH action or is a terminal event. Strong noncovalent interactions between TSH receptor units lead to the formation of conjugates. The complex most frequently encountered in the experiments to be described is a dimer of M_r 200,000, although higher (M_r 280,000) forms are noted (1). Conjugate formation appeared to enhance both the capacity and affinity for TSH binding.

Anti-TSH Anti-Idiotype

At the inception of studies looking at raising anti-hormone anti-idiotypes capable of interacting with the receptor, it was of interest to investigate the possibilitỳ that such antibodies may, in addition, trigger postbinding events (2,3). Investigation of biological activity of anti-TSH anti-idiotypes was attractive from a number of viewpoints. TSH mediates many of its actions through a well-defined second messenger, unlike insulin (2), and up until that time, ligands of as high a molecular weight as TSH (28,000) or of its complex glycoprotein nature had not been tested. An anti-idiotypic approach allowed us to test whether an antibody which theoretically should only recognize the TSH binding site of the receptor could simulate attributes of Graves' IgG, a naturally occurring anti-receptor antibody (4).

Anti-human TSH anti-idiotype was found to inhibit the binding of ^{125}I-bTSH to thyroid plasma membrane in a dose-dependent manner, and the high-affinity component of the antibody preparation had a dissociation constant only one order of magnitude less than that of TSH itself. Binding of idiotype was specifically inhibited by 160 mU/ml bTSH to an extent of 65%. The anti-idiotypes were stimulated by thyroid plasma membrane adenylate cyclase in a cyclic-GMP-dependent manner as well as to mediate two processes which are at least in part dependent on cyclic-AMP generation: uptake of radioiodine by isolated thyrocytes and their ability to organize into follicular structures (4). Like TSH, the idiotype showed a bimodal effect on iodide uptake by thyroid epithelial cells (Fig. 7.1). The anti-idiotypes bound to a M_r 200,000 peptide on protein blots of thyroid plasma resolved on SDS-PAGE under nonreducing conditions but not in the presence of reductant. This interaction is inhibited by bTSH but not by insulin or human chorionic gonadotropin. The anti-idiotype also inhibited the cross-linking of ^{125}I-bTSH to a high-molecular-weight component of M_r 195,000 and, to a much lesser degree, to molecular-weight components resolved under nonreducing conditions and a single M_r 58,000 peptide resolved under reducing conditions (5) (Fig. 7.2). This finding suggested that the idiotype recognized a high-affinity site for TSH binding generated by glycoconjugate.

The sum total of these results suggested that the anti-TSH anti-idiotype was an agonist which bound to a high-affinity membrane component. As the anti-idiotype simulated all the attributes of Graves' IgG, we have taken this as evidence that an anti-TSH receptor antibody can account for all the biological effects of Graves' IgG (4,6). We suggested that Graves' IgG may, at least in some instances, arise as an anti-idiotype to natural TSH-binding IgG (6,7). Possible mechanisms whereby anti-hormone anti-idiotypes may initiate and sustain autoimmunity directed against the TSH receptor have been proposed (7,8).

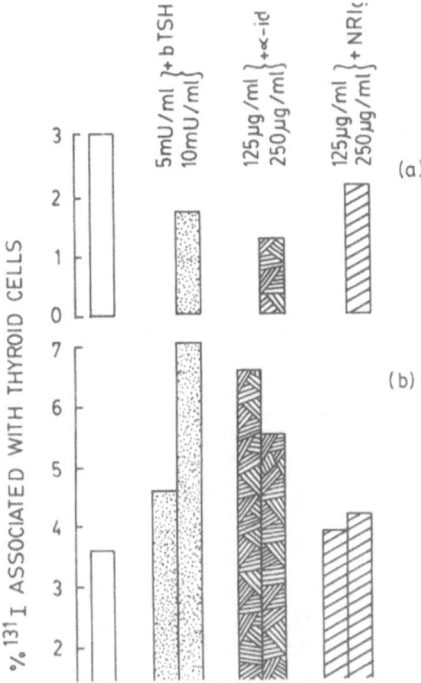

FIGURE 7.1. The influence of anti-idiotypic antibody on uptake of radioiodine by porcine thyrocytes. (a) Radioiodine uptake was studied after incubation of thyroid cells with bTSH, NR IgG, or anti-Id for 10 min. Na^{131}I (10^5 cpm) was then added and incubation was continued for a further 10 min. Counts associated with pelleted cells are expressed as percent of total counts added. Only the higher dosage of the experimental agents was used in these experiments. Both bTSH and anti-Id caused a decrease in the ^{131}I uptake by thyrocytes. (b) The percent of radioiodine associated with thyrocytes 4 h after incubation with bTSH, NR IgG, or anti-Id. Anti-Id (125 μg/ml) caused a significant stimulation of ^{131}I uptake ($p < 0.05$). The higher concentration (250 μg/ml) caused less stimulation, raising the possibility that the anti-Id preparation contains in lesser relative concentration an antibody species which inhibits iodine concentration. Reproduced with permission from Farid et al., J. Cell. Biochem. **19**:305(1982).

Anti-TSH Subunit Anti-Idiotype

Like the glycoprotein hormones FSH, LH, and human chorionic gonadotropin, TSH consists of a common subunit and a specific subunit. Although isolated subunits have minimal or no biological activity, several lines of evidence suggest that the subunit is associated with hormone specificity. It was, however, uncertain whether the hormone-specific signal

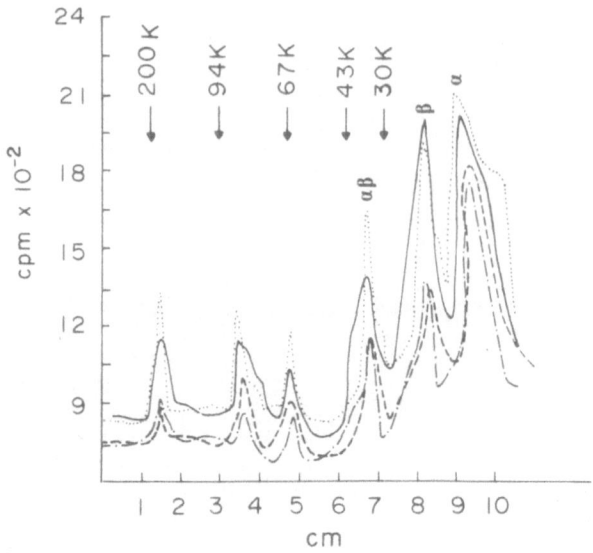

FIGURE 7.2. The effect of Graves' IgG and anti-TSH anti-idiotype on the photo-cross-linking of ¹²⁵I-labeled bTSH to its receptor. Thyroid plasma membranes were photocross-linked with ¹²⁵I-labeled bTSH in the presence of N-hydroxysuccinim-idyl-N-azidobenzoate, in the presence or absence of 500 μg/ml of IgG. After washing, the samples were resolved on a 7.5–15% SDS-PAGE in the absence of reductant. The gels were cut at 1-mm intervals and counted. Electrophoretic pro-files of plasma membrane components cross-linked with ¹²⁵I-labeled bTSH in the absence of native bTSH (.....) or the presence of normal IgG (_____), Graves' IgG (----), or anti-TSH anti-idiotype (α - Id IgG) (-•-•-•) are shown. The radioactivity peaks correspond to the dissociated α and β subunits and the αβ dimer, as well as cross-linked receptor components. These include M_r 193,000, 87,000, and 59,000 peaks. The arrows indicate migration of unlabeled M_r ($\times$ 10⁻³) markers. Normal human IgG and to a much greater extent Graves' IgG resulted in a decrease of overall radioactivity associated with plasma membranes. Whereas cross-linked radioactivity peaks remained substantital after treatment with normal IgG, their labeling was greatly decreased with Graves' IgG. α-Id IgG caused a proportionately greater inhibition of the labeling of the M_r - 193,000 peak compared with M_r 87,000 or 59,000 peaks.

was imparted by α or β subunit while the alternate subunits bound specifi-cally to the receptor or whether two signals were necessary for the mediation of hormone action. We approached this question by raising anti-idiotypes to rat monoclonals specific for the α and β subunits of TSH (9). Table 7.1 outlines the properties of these monoclonals. The binding of the anti-idiotype raised against the β subunit-specific monoclonal to its monoclonal idio-type was ligand inhibitable whereas the binding of the idiotype to the α sub-unit monoclonal was not, suggesting that the latter idiotype was probably

TABLE 7.1. Anti-TSH subunit specific monoclonals.

Property	Anti-α	Anti-β
Species	Rat	Rat
Ig isotype	IgM	IgG
Binding to ^{125}I-hTSH	+	+
Binding affinity (K_D, M^{-10})	3.4	3.7
Displacement of bound ^{125}I-hTSH by bTSH	< 5%	< 3%
Glycoprotein hormone causing displacement of bound ^{125}I-hTSH	hTSH, hFSH, hLH, hCG	hTSH
Inhibition of membrane-bound ^{125}I-bTSH	−	+
Inhibition of adenylate cyclase activation	+	+

directed against the framework determinant. Neither α nor β subunit anti-idiotypes nor their combination was able to inhibit ^{125}I-bTSH binding to thyroid plasma membranes. Each of the anti-idiotypes studied separately had minimal stimulating effect on thyroid plasma membrane basal cyclase, whereas an equimolar mixture, however, activated cyclase to the extent observed with 200 mU/ml bTSH (Fig. 7.3). The two anti-idiotypes clearly interacted at the receptor in a cooperative manner. β-Subunit anti-idiotype produced a measure of ^{131}I concentration, α-subunit anti-idiotype was without effect, and their combination produced near-maximal ^{131}I concentrating capacity. Similar results were obtained when the ability of anti-idiotypic antibodies to induce follicular structures was investigated (9).

^{125}I-anti-idiotype bound to thyroid plasma membranes to an extent significantly greater than that found with normal rabbit IgG, and the ability of the two anti-idiotypes to bind to thyroid plasma membrane was additive. Binding of anti-α and anti-β anti-idiotypes and their combination was inhibited in a dose-dependent manner by bovine TSH. Residual binding of ^{125}I-anti-idiotype was observed at the highest concentration of bovine TSH used (500 mU/ml), with the α subunit anti-idiotype and particularly in the case of the combination of anti-α and anti-β anti-idiotypes. This residual binding may represent irreversible association with the receptor. By contrast, only an equimolar combination of the two anti-idiotypes was found to interact with the M_r 200,000 receptor peptide on protein blots (Fig. 7.4). This seeming paradox may be explained by the fact that protein blots are treated with 200 mM sodium chloride to reduce nonspecific background binding. Only antibodies binding with high affinity to protein blots would withstand the rigor of this treatment, again providing evidence of a cooperative effect of the anti-α and anti-β anti-idiotypes on the TSH receptor. This finding, of course, presumes a degree of modulation of receptor covalently associated with nitrocellulose paper. By contrast, in the case of the anti-TSH anti-idiotype described in the previous section, we consistently found faint bands at the level of M_r 100,000 and M_r 65,000–70,000 with the combination of anti-idiotypes (9).

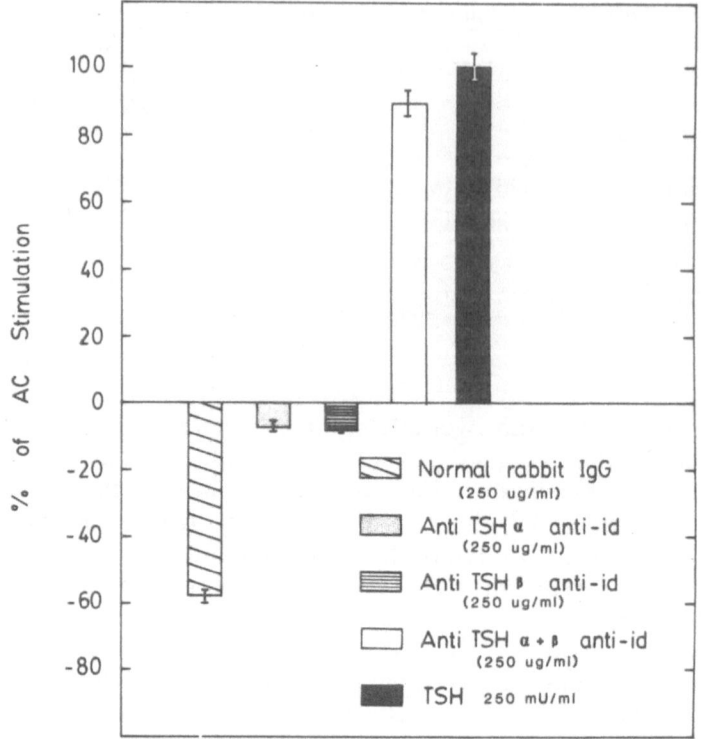

FIGURE 7.3. Guanyl nucleotide-dependent stimulation of adenylate cyclase activity by anti-TSH-α and anti-TSH-β anti-idiotypic antibodies. The effect of combining anti-α and anti-β anti-idiotypic antibodies is more than additive and comparable with the effect of 200 mU/ml of TSH.

To my knowledge, this study remains the only example of the interaction of two anti-idiotypic antibodies in binding to a receptor. Cooperative interaction between these anti-idiotypic antibodies is all the more surprising because only the β subunit anti-idiotype was ligand inhibitable. The particular α subunit preparation used was specific in that its cooperative effect with the anti α anti-idiotype could not be substituted by normal rabbit IgG nor by the IgG obtained from other rabbits immunized with the subunit-specific monoclonal.

Our results indicate that hormone specificity was associated with the TSH β subunit and that the two TSH subunits deliver two cooperative signals. However, given the discrepancy in molecular mass between IgG and TSH, one must invoke aggregation of the TSH receptor through the engagement of antibodies with receptor epitopes which interact, respectively, with the α and β subunits of the hormone. This might explain the lack of ^{125}I-bTSH inhibition by the anti-idiotypes but not their ability to

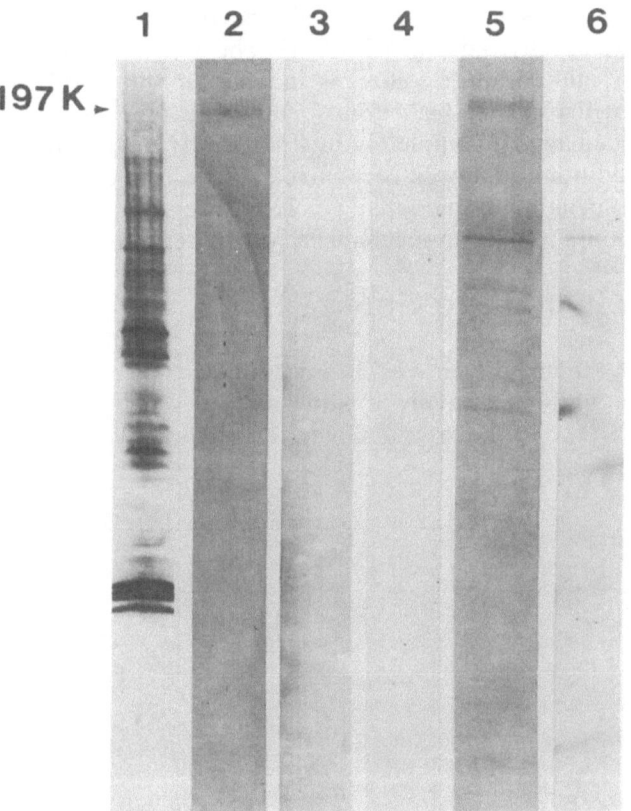

FIGURE 7.4. The interaction of TSH α and β anti-idiotype mixture with the TSH receptor. Lane 1: Protein bands resolved by SDS-PAGE under nonreduced band transferred onto nitrocellulose paper and stained with Coomassie blue. Lane 2: Interaction of bTSH (1 U/ml) with the blot identified by high-affinity anti-TSH antibody and peroxidase-conjugated anti-rabbit IgG. Lane 3: Anti-β subunit anti-idiotype (1 mgm/ml); no band seen. Lane 4: Anti-α subunit anti-idiotype; same as lane 3. Lane 5: The binding of a mixture (0.5 mg/ml each) of anti-idiotype with M_r -200,000 peptide which also binds bTSH. Lane 6: Normal rabbit IgG.

stimulate cyclase, and it has been suggested that adenylate cyclase activation is affected by microaggregation of receptor (10).

We have used the combination of anti-idiotypes to identify receptor-specific bands in [³H]leucine-labeled lysates of thyroid cells. We used this approach to examine the effect of TSH on the rate of receptor synthesis and, in pulse chase experiments, on TSH receptor half-life (11). To summarize our findings, TSH appeared to enhance the rate of synthesis of its receptor until an optimum rate was reached, following which a gradual decrease of receptor rate synthesis was found for up to 96 hours of thyroid cell culture in presence of bTSH. TSH receptor half-life was accelerated by incubation of thyroid cells with TSH. The most marked effect occurred between 12 and 16 hours following incubation with TSH. Interestingly, over that period of time high-affinity, low-capacity ^{125}I-bTSH binding sites were replaced almost completely by low-affinity, high-capacity sites (11). Thus, in addition to the well-recognized mechanism of desensitization involving dissociation or uncoupling of TSH receptor from cyclase, we now provide a more long-term mechanism for the termination of TSH action.

An Anti-Idiotype to Graves' IgG

Apparently, stimulatory Graves' IgG is a homogeneous antibody with some 90% of the antibodies involved being restricted to IgG1 carrying a light chain (see Ref. 6 for a review). Given this finding, we wondered whether

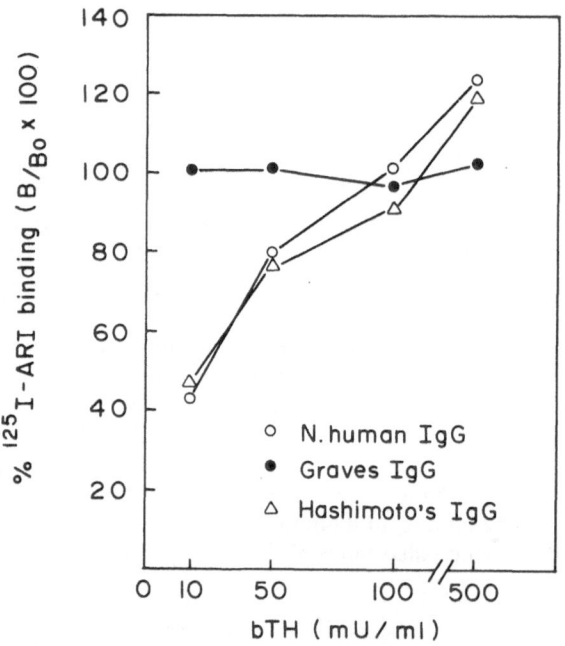

FIGURE 7.5. The lack of inhibition of Graves' IgG binding of ARI by bTSH. ARI (10 µg) was iodinated by the lactoperoxidase method. Binding of IgG to ^{125}I-ARI was tested by incubating 10µg/ml IgG with 10,000-cpm ^{125}I-ARI overnight. Bound from free counts were separated by 7.2% polyethylene glycol. Binding was done in the absence and presence of bTSH.

Graves' IgG binding to ^{125}I-ARI is not inhibited by bTSH. The other two IgGs were inhibited to -50% of basal by 10 mU/ml TSH. Higher doses are associated with increased binding to TSH, consistent with the ability of these antibodies to bind ^{125}I-bTSH.

the common idiotypic determinant (framework) was shared by different patients with Graves' disease (12). Rabbit antibodies were raised to IgG from a patient with active Graves' disease, which had been specifically absorbed to a TSH receptor affinity column. The absorbed rabbit antibodies (ARI) were found to bind ^{125}I-bTSH. This binding was inhibited by the interaction of the ARI with Graves' IgG. ^{125}I-ARI counts precipitated by 10 μg IgG/ml gave good separation between normal human, Hashimoto's, and Graves' IgG. The binding of the ARI to Graves' IgG was not inhibited by concentrations of TSH of up to 500 mU/ml (Fig. 7.5). At a concentration of 10 mU/ml, bTSH inhibited the binding of normal and Hashimoto's IgG to ^{125}I-ARI precipitated, consistent with the ability of ARI to bind TSH. It thus appears that Graves' IgG, but not normal and Hashimoto's IgG, has an idiotypic determinant which, when engaged by ARI, interferes with its binding to ^{125}I-TSH. This bimodal effect of TSH binding was found to be specific for TSH in that neither insulin nor HCG appeared to influence ^{125}I-ARI binding to normal IgG (Fig. 7.6).

Using bTSH (at 10 mU/ml) noninhibitability of the interaction of ^{125}I-ARI with human IgG as a criterion for the presence of Graves' disease-related idiotype, we found a positive response in 10/11 patients with Graves' disease, 0/5 patients with Hashimoto's disease, and 0/7 controls.

We further obtained proof that the binding of Graves' IgG to ARI was saturable and that the ARI preparation contained a complementary antibody, in that increasing concentration of unlabeled ARI was associated with dose-dependent increase in ^{125}I-ARI precipitated (Fig. 7.7).

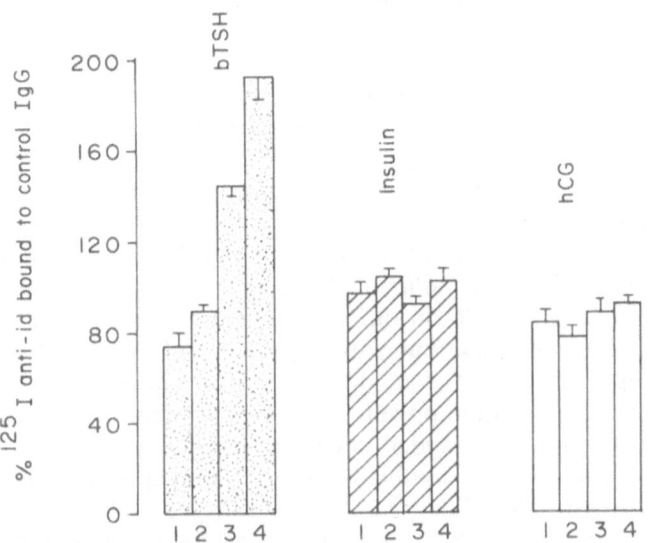

FIGURE 7.6. Specificity of binding of normal IgG to bTSH. 1 = 10 mU/ml; 2 = 50 mU/ml; 3 = 100 mU/ml; 4 = 500 mU/ml of bTSH ▨, insulin ▨ thCG ☐.

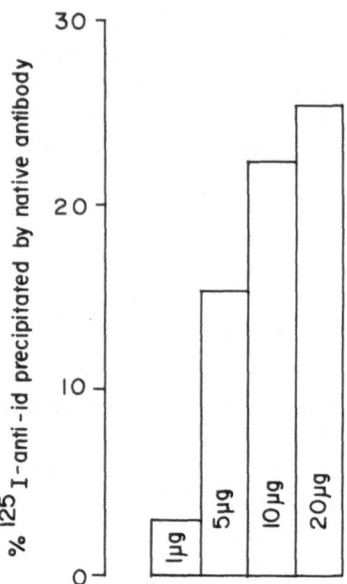

FIGURE 7.7. Presence of auto-anti-diotypes in ARI. In this experiment the ability of the indicated concentrations of unlabeled ARI indicated to precipitate 10,000-cpm ^{125}I-ARI was examined.

Assuming that the complementary antibody was raised to the determinant of ARI which bound ^{125}I-TSH, we tested the possibility that the auto-ARI anti-idiotype may have TSH receptor activity. ARI preparation was passed through a TSH affinity column and the breakthrough fraction tested for its ability to stimulate cyclase in the FRTL-5 (Table 7.2) and human thyroid cell systems. It is clear that the FRTL-5 system was a much more sensitive measure of the anti-receptor activity of the ARI breakthrough fraction than was the human system. Interestingly, in the human thyroid system the breakthrough fraction did not interfere with Graves' IgG effect but augmented the effect of TSH, similar to observations previously made on different Graves' IgG preparations (12). In essence, thus, immunizations with Graves' IgG have resulted in an auto-anti-idiotype with activities qualitatively, if not quantitatively, similar to those of the original immunogen. These experiments are strong proof that the network is not open-ended but rather cyclic and that this cyclicity of the idiotypic network is involved in the TSH receptor autoantibody. The fact that the auto-anti-idiotype (Ab$_3$) was raised to ^{125}I-TSH binding moieties lends some credence to the notion that TSH-binding immunoglobulin may be the origin of TSH receptor autoantibodies (6).

TABLE 7.2. Adenylate cyclase
activation by cloned FRTL-5 cell.

Substance tested	cAMP (fmol/ml)[a]
bTSH	
0 μU	75 ± 10
10 μU/ml	547 ± 28
50 μU/ml	6,010 ± 641
ARI[b]	
1.3 μg/100 ml	208 ± 13
13 μg	228 ± 20
130 μg	401 ± 11
Normal IgG	
100 μg/ml	136 ± 36

[a]Cyclic-AMP was measured in the medium
after 60-min incubation with bTSH or antibody.
[b]The lack of a dose-response curve with ARI
likely reflects the presence of antibodies other
than Ab' and possibly undissociated immune
complexes. TSH immunoreactivity was undetectable with a high-sensitivity TSH radioassay.

Concluding Remarks

The anti-idiotype route for raising anti-TSH receptor antibodies has allowed us to investigate the structure, biosynthesis, and turnover of the
receptor as well as the interaction of TSH subunits with that receptor.
We have learned that Graves' IgGs from different patients share common
idiotypic determinants and that a polyclonal response to these antibodies
includes auto-anti-idiotypes (Ab$_3$) with anti-receptor activity, thus affirming
that the idiotypic network is not open-ended.

Acknowledgments. This work is supported by grants from the Medical
Research Council of Canada and the Faculty of Medicine, Research and
Development Fund.

References

1. Farid, N.R., and G. Fahraeus-van Ree. 1987. About the porcine TSH receptor.
   Acta Endocrinol. (in press).
2. Sege, K., and P.A. Peterson. 1978. Use of anti-idiotypic antibodies as cell-
   surface receptor probes. Proc. Natl. Acad. Sci. USA **75**:2443.
3. Farid, N.R., and T.C.Y. Lo. 1985. Antiidiotypic antibodies as probes for receptor structures and function. Endocrine Rev. **6**:1.

4. Islam, M N., B.M. Pepper, R. Briones-Urbina, and N. R. Farid. 1983. Biological activity of anti-thyrotropin anti-idiotypic antibody. Eur. J. Immunol. **13**:57.
5. Bakó, G., M.N. Islam, and N.R. Farid. 1985. Photoaffinity labelling of the porcine thyrotropin receptor—effect of Graves' immunoglobulin G and anti-TSH anti-idiotypic antibodies. Clin. Invest. Med. **8**:126.
6. Farid, N.R., R. Briones-Urbina, and J.C. Bear. 1983. Graves' disease—the thyroid stimulating antibody and immunological networks. Mol. Asp. Med. **6**:355.
7. Farid, N.R. 1987. Graves' disease. In N.R. Farid (ed.), Immunogenetic aspects of endocrine disorders. Alan R. Liss, Inc., New York. In press.
8. Gaulton, G.N., and M.I. Green. 1986. Idiotypic mimicry of biologic receptors. Annu. Rev. Immunol. **4**:253.
9. Briones-Urbina, R., M.N. Islam, J. Ivanyi, and N.R. Farid. 1987. Use of anti-idiotypic antibodies as probes for the interactions of TSH subunits with its receptor. J. Cell. Biochem. (in press).
10. Avivi, A., D. Tramontano, F.S. Ambesi-Impiombato, and J. Schlessinger. 1981. Adenosine 3',5'-monophosphate modulates thyrotropin receptor clustering and thyrotropin activity in culture. Science **214**:1237.
11. Farid, N.R., and G. Fahraeus-van Ree. 1987. Non-isotopic study of the TSH receptor. In S. Pal (ed.), Immunoassay technology. In press.
12. Hawe, B.S., T F. Davies, and N.R. Farid. 1987. An anti-idiotypic antibody to Graves' IgG. Autoimmunity. (in press).

8
Mice Immunized to Insulin Develop Anti-Idiotypic Antibody to the Insulin Receptor

YORAM SHECHTER, DANA ELIAS, RAFAEL BRUCK,
RUTH MARON, AND IRUN R. COHEN

Introduction

Several diseases of man can be related to the formation of autoantibodies that interact with surface receptors. Examples are the antibodies that trigger TSH receptor in Graves' disease (1), antibodies that bind to and block the receptor for acetylcholine at the neuromuscular junction in myasthenia gravis (2), and antibodies that trigger insulin like responses which appear in certain (very rare) types of severe insulin-resistant diabetes (3).

Sege and Peterson (4) and Jerne (5) proposed possible explanations for receptor autoimmunity. According to their hypothesis, the immune system may be regulated by a network in which an antigen (such as a hormone) may induce production of idiotypes which, in turn, induce anti-idiotypes that can feed back to shut off or modify the original idiotypic response. Some of the anti-idiotypes against the hormone-binding region of the idiotypes may mimic the structure of the hormone itself. Such anti-idiotypic antibodies might bind to the hormone receptor and thus function as antibodies to the receptor. Accordingly, Sege and Peterson (4) immunized rabbits against affinity-purified insulin antibodies and obtained anti-idiotypic antibodies that bound to the insulin receptor.

The present article reviews published papers in which we demonstrate that immunization of mice to ungulate insulins can lead to the development not only of antibodies to insulin but of insulin like antibodies (ILA) that appear to recognize the insulin receptor and to trigger the bioresponses of insulin at low concentrations (6–9).

We further demonstrate that ILA are anti-idiotypic antibodies and that the specific idiotypic antibodies are confined to the early phase of the primary response, while anti-idiotypic receptor antibodies were detected only after the idiotypic antibodies had disappeared. Several other aspects of this system are demonstrated and discussed.

Materials and Methods

Animals

Female mice of the strains (C3H/ebxC57BL/6)F1 hybrid, C3H/eb, C3H.SW, SJL and BALB/c and male Wistar rats (70–100 g) were obtained from the Animal Breeding Center of the Weizmann Institute.

Materials

D[U-^{14}C]glucose (4–7 mCi/mol) was purchased from New England Nuclear, collagenase type I (134 U/mg) from Worthington, and protein A-Sepharose column from Pharmacia. Insulin-agarose was prepared as described in Ref. 10.

Methods and Procedures

The following procedures described in the literature were used with no modification: immunization of the mice to insulin (11), a solid-phase radioimmunoassay of insulin antibodies (12), sodium dodecyl sulfate-gel electrophoresis (13), affinity purification of insulin antibodies on an insulin-agarose column (1), separation of IgG from immune serum by protein A-Sepharose column (1), Ouchterlony assay of precipitation in agar for determining antibody class (14), preparation of isolated fat cells from epidydimal fat pads (15), binding of ^{125}I-insulin to fat cells (16), lipogenesis (17), and inhibition of lipolysis (1).

Results and Discussion

Kinetics of the Appearance of Insulin Antibodies and Anti-Idiotypic Insulin Receptor Antibodies

Mice immunized to bovine insulin produced insulin antibodies detectable 7–10 days after the primary injection. As shown in Fig. 8.1A, a booster injection was followed by a second rise in the titer of insulin antibodies to a high peak, and traces of these antibodies could be detected for at least another 25 days. Receptor antibodies were first detected 5 days after the booster injection (26 days after primary immunization), as determined by the lipogenic activity of the serum. The activity remained high for 9–12 days in all mice as shown in Fig. 8.1B. Receptor antibody activity then declined to undetectable levels over a period of 12–16 days, followed by a second peak of receptor antibody activity detectable about 40 days after

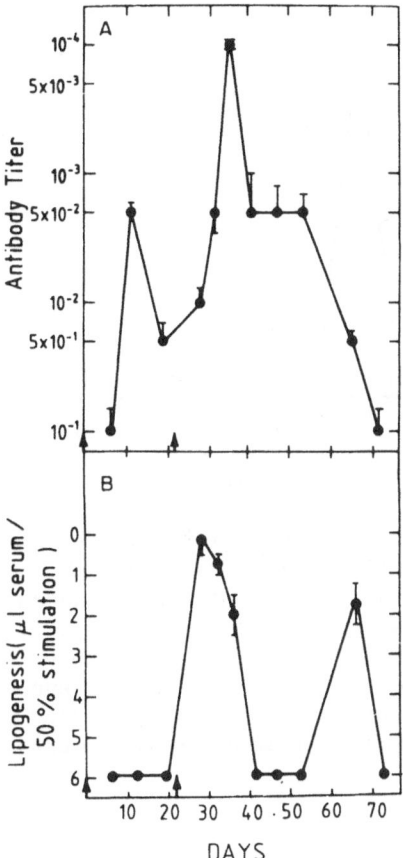

FIGURE 8.1. Development of (A) insulin antibodies and (B) anti-idiotypic receptor antibodies following immunization to insulin. Mice were imunized to insulin on day 0 and boosted on day 21. The titer of insulin antibodies on various days after the primary immunization was determined using a solid-phase radioimmunoassay. The magnitude of the insulinlike activity produced by the receptor antibodies was measured as the volume of serum (μl) required to induce 50% of the maximal lipogenesis produced by incubating adipocytes with an optimal concentration of insulin (100 ng/ml). Control sera had negligible titers of insulin antibody and insulinlike lipogenic activity.

the booster injection. This second peak of activity was shorter in duration and of lesser magnitude than the first peak. Primary immunization alone was sufficient to elicit the first peak of anti-idiotypic receptor antibodies; however, the second peak of receptor antibodies appeared to be more prominent in mice that had received a booster injection. Insulin antibodies bearing the specific idiotype were observed only in the early primary re-

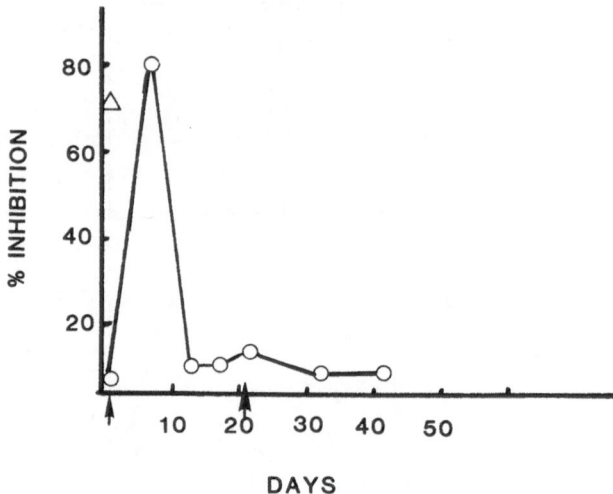

DAYS

FIGURE 8.2. Appearance of idiotype-positive insulin antibodies following immunization to insulin. Insulin antibodies were isolated by affinity chromatography from sera obtained on various days after immunization of mice to insulin as described in Fig. 8.1. The presence of specific idiotypes was detected by the ability of the insulin antibodies to inhibit the lipogenic activity of anti-idiotypic receptor antibodies in the sera of mice 30 days after primary immunization (○). The triangle indicates the degree of inhibition produced by idiotype-positive guinea pig insulin antibodies. Control mouse sera had no idiotypic insulin antibodies.

TABLE 8.1. Time schedule for the appearance of the various antibodies after immunizing mice with insulin.[a]

Days after immunization	Designation of antibody	Specification of antibody	Class of antibody
7–9	"Specific" idiotype	Anti-insulin "binding" site	IgM
9–70	Anti-insulins	Anti-insulin "external determinants"	IgG1
28–40	Anti-idiotype-ILA	Anti-anti-insulin "binding" site (receptor antibody)	IgG2
60–70[b]	Anti-idiotype-ILA	Anti-anti-insulin "binding" site (receptor antibody)	IgG2

[a]BALB/c mice were immunized with 25-μg bovine insulin in complete Freund adjuvant and boosted at day 21.
[b]Does not appear if mice are not boosted.

sponse to insulin, and as Fig. 8.2 shows, at later dates they were undetectable. Thus, the specific idiotype and the anti-idiotypic receptor antibody did not coexist freely in the same serum. The time schedule for the
appearance of the various antibodies after immunizing mice to insulin is
also summarized in Table 8.1.

Identification of the Serum Fraction with Insulin like Activity

Figure 8.3 demonstrates fractionation of an immune serum on a protein
A-Sepharose affinity column. We loaded an immune serum obtained 12–
14 days after booster injection (see Fig. 8.1B). This serum contained both
insulin antibodies in high titer (10^4 dilution^{-1}) and insulin like activity detectable at $0.4 \times 10^4 - 1 \times 10^4$ serum dilutions^{-1}.

We found that the fraction containing the insulin like activity was quantitatively adsorbed to the column at pH 7.0 and could be eluted at a lower
pH (pH 3.0). The fraction that did not bind at pH 7.0 contained the insulin
antibodies (Fig. 8.3). Since protein A binds specifically to the Fc portion
of antibodies, primarily of the IgG2 class at pH 7.0 (18), it seemed that
the insulinlike activity was a property of IgG2 molecules. This was further

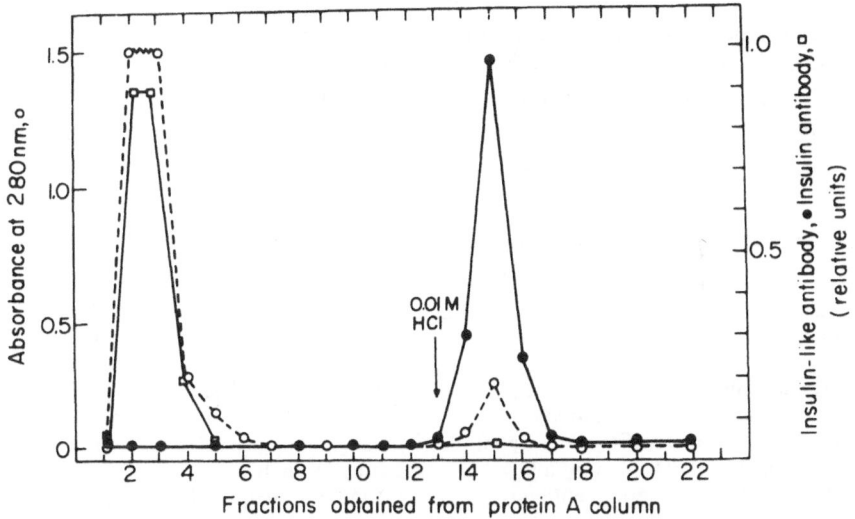

FIGURE 8.3. Purification of IgG2 from immune serum on protein A-Sepharose
column (Pharmacia, 2.0×0.5 cm) that was preequilibrated and washed with 0.05
M sodium phosphate (pH 7.0). Fractions of 0.4 ml were collected. The eluted
fractions were collected into tubes containing a sufficient amount of NaHOO₃ to
neutralize the acidity. The fractions were examined for their absorbance at 280
m (o), ILA (•), and insulin antibodies (□).

confirmed by means of polyacrylamide gel electrophoresis, which demonstrated complete absence of detectable protein(s) other than IgG (not shown). Ouchterlony precipitation of affinity-purified insulin antibody indicated that these antibodies were primarily IgG1. As mouse IgG1 is not adsorbed to protein A-Sepharose at pH 7.0 (19), this one step of purification was sufficient to separate insulin antibodies (IgG1) from the insulin like activity (IgG2).

Overall Properties of Protein A-Sepharose-Purified ILA

Table 8.2 and Figs. 8.4–8.6 summarize the characterization and properties of the purified ILA fraction. Briefly, ILA is an immunoglobulin of class IgG2 that does not contain free insulin, antibodies to insulin, or insulin-anti-insulin immune complexes. This IgG2 fraction produces insulin like bioactivities at concentrations in the microgram range (Table 8.2, Figs. 8.4 and 8.5). In vitro, bioactivities are blocked efficiently by affinity-purified guinea pig insulin antibodies or by mouse affinity-purified insulin antibodies (idiotypes), isolated at days 7–10 after the primary injection but not at later periods (Fig. 8.2). ILA also displaces labeled insulin from young fat cells (Table 8.2, Fig. 8.6). Thus, these antibodies seem to be directed to the region of the insulin receptor that binds insulin. The insulin like activity of an immune serum emerged quantitatively in the void volume of a Sephadex G-100 column under conditions that dissociate immune complexes (summarized in Table 8.2). Therefore, ILA activity could not be explained by insulin contaminating or complexed with the IgG2 antibodies.

TABLE 8.2. Characterization and properties of ILA, isolated from immune mouse serum by protein A-Sepharose column.[a]

Property	As judged by
Pure IgG	Sodium dodecyl sulfate-polyacrylamide gel electrophoresis
IgG2 antibodies	Ouchterlony analysis
Does not contain insulin antibody	Solid-phase RIA[b]
Stimulates lipogenesis (ED$_{50}$ = 4 μg ml^{-1})	Assay of lipogenesis in fat cells
Inhibits lipolysis (ED$_{50}$ = 3 μg ml^{-1})	Assay of lipolysis
Activity inhibited by insulin antibody	Affinity-purified insulin antibodies added to lipogenesis assay
Displaces insulin from receptor	Competes with the binding of labeled insulin to fat cells (50% displacement at 20 μg ml^{-1})

[a]Details on the isolation of IgG2 from immune serum are given in the experimental section.
[b]Insulin antibodies from the mouse were found to be of the IgG1 class (by Ouchterlony analysis). This type is not adsorbed to the agarose-protein A column under the conditions applied here. The void volume contained all ILA activity but not insulin.

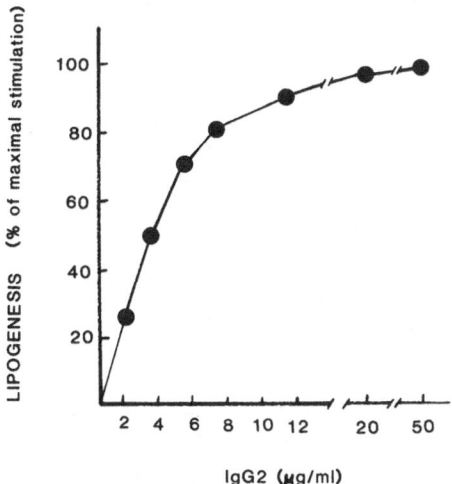

FIGURE 8.4. Stimulation of lipogenesis by IgG2 from immune serum. Lipogenesis was produced by the indicated concentrations of IgG2 that was purified from secondary serum by elution from a column of protein A-Sepharose.

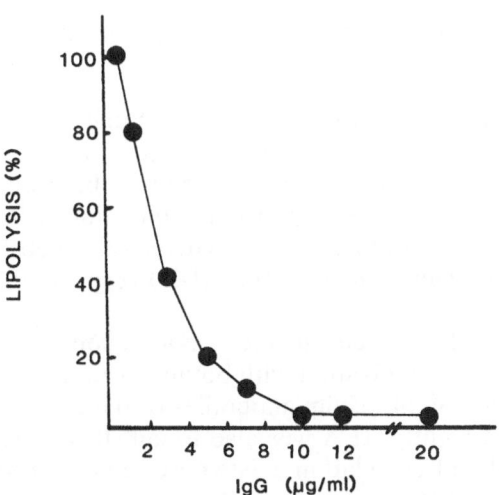

FIGURE 8.5. Inhibition of lipolysis by IgG2 from secondary serum. Assay of lipolysis was carried out in the presence of increasing concentrations from secondary serum.

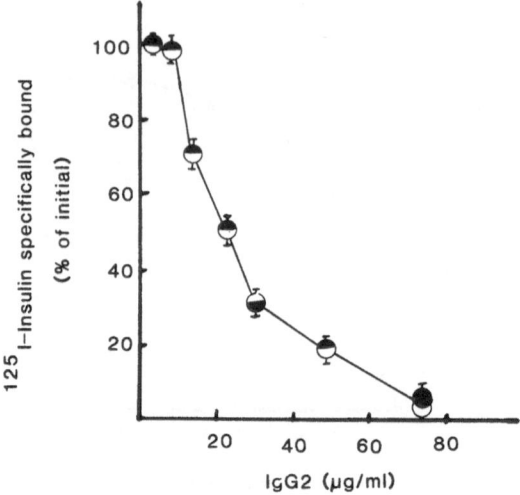

FIGURE 8.6. Displacement of ^{125}I-insulin from fat cells by IgG2 from secondary serum. Suspensions of fat cells were incubated with increasing concentrations of IgG2 from secondary serum (○) or IgG2 from secondary serum that was depleted of insulin antibodies, using a column of insulin-agarose (•). The percentage of initial binding is corrected for nonspecific binding that was not displaceable by a surplus of unlabeled insulin (15 μg/ml).

Production of Insulin Antibodies and ILA by Different Mouse Strains

Mice genetically incapable of mounting an immune response to ungulate insulins also did not produce insulinlike activity. For example, Table 8.3 shows that C3H.SW strain (H-2^b) mice produced both insulin antibodies and insulin like activity in response to immunization with bovine insulin but not with sheep or porcine insulin, whereas C3H/eb (H-2^k) mice responded to sheep insulin and BALB/c (H-2^d) responded to all three insulins.

The strain SJL (H-2^s) is considered a nonresponder strain to insulin (20). In response to immunization with bovine insulin, however, we have observed the formation of insulin antibodies of the IgM class exclusively (Table 8.3). Consequently, ILA was also found. It therefore can be concluded that (i) a direct correlation exists between the production of anti-insulin and receptor antibodies among the various strains of mice and (ii) the primary response bearing the specific idiotype of the IgM class is sufficient for the subsequent production of ILA in vivo.

TABLE 8.3. Production of insulin antibodies and ILA by different mouse strains.

Mouse strain	H-2	Type of insulin used for immunization	Insulin antibody titer[a] (dilution^{-1})	ILA[b] (fold dilution for obtaining 50% of maximal lipogenesis)
C3H/eb	k	Bovine	<10	10
		Sheep	10^3	10^3
		Porcine	<10	10
C3H.SW	b	Bovine	10^3	10^3
		Sheep	<10	10
		Porcine	<10	10
BALB/c	d	Bovine	10^3–10^4	10^3
		Sheep	10^3–10^4	10^3
		Porcine	10^3–10^4	10^3
SJL	s	Bovine	10^{2c}	10^2
		Porcine	<10	10^2

[a]Insulin antibody titer was determined on day 8 after the booster injection (see text).
[b]The ability of the serum to stimulate lipogenesis was tested 12 days after booster injection (see text).
[c]Insulin antibodies of the IgM class only were observed.

Critical Insulin Epitope Triggers Anti-Idiotypic Receptor Antibodies

To gain information about the antigenic determinant, or epitope, of insulin responsible for triggering the specific idiotypic-anti-idiotypic network, we chemically modified the insulin molecule at its three free amino groups (A1, B1, and B29) and compared the biological and immunological functions of the native and modified molecules. The three positions were cho-

TABLE 8.4. Development of receptor antibodies inhibited by alteration of insulin molecule.[a]

Immunizing insulin	Biological activity (%)	Antibodies to native insulin (titer)		Day 35 receptor antibodies (% lipogenic activity)
		Day 14	Day 35	
Native	100	10^3	10^4	150
Acet$_3$	10	10^3	10^4	10
Suc$_3$	2	10^3	10^4	7
TNB$_3$	<1	10^3	10^4	5

[a]Mice were immunized with native or modified insulins and their sera studied on days 14 and 35 for the titer of antibodies to native insulin and on day 35 for receptor antibody activity. The biological activity of the insulins was computed relative to that of native insulin (100%) from the dose-response curves.

sen because two of them (A1 and B1) have been shown to be essential for receptor activity (21). Table 8.4 shows the relative lipogenic activities of three modified insulins and their immunological potency in inducing insulin antibodies and receptor antibodies following primary and secondary immunizatiŏn. Acetyl$_3$ (Acet$_3$), succinyl$_3$ (Suc$_3$), and trinitrobenzene (TNB$_3$) insulins demonstrated 10%, 2%, and less than 1% of the lipogenic activity of insulin, respectively. However, all three modified insulins were immunogenic and stimulated comparable titers of antibodies binding to native insulin. Nevertheless, little or no receptor antibody was generated in response to the modified insulins. These results support the conclusion that induction of receptor antibody is not a function of the gross immune response to insulin but depends on the fine response to the particular epitope whose conformation involves the amino acid residues responsible for biological activity of insulin.

Lack of Production of ILA in Guinea Pigs That were Immunized to Insulin

One of the more important questions is why anti-idiotypic antibodies are produced in the immunized mice. The insulin molecule of mice has been conserved in the mainstream of insulin evolution in mammals and differs from pork and beef insulins by three and five amino acids respectively, that are structurally confined to the variable domain of the molecule that does not interact with the insulin receptor (23,24). In response to immunization of the mouse with ungulate insulins, the mouse produces antibodies that recognize the hormonal domain of the molecule, which is common to most mammalian insulins (22) and to mouse insulin itself. Hence, in response to ungulate insulin, the mouse probably makes autoantibodies. It is tempting to speculate that the mouse makes anti-idiotypic antibodies to regulate its autoantibodies to the hormonal domain of insulin. The price of these anti-idiotypes is their ILA activity at the insulin receptor.

Guinea pig insulin, in contrast, is an evolutionary deviation in the mammalian kingdom and differs from pork insulin in 17 of its 51 positions (22) and probably has a unique structure in its hormonal domain (21,25). Hence, in response to ungulate insulin, the guinea pig can make antibodies to the same antigenic determinants on the pork insulin as does the mouse. However, as guinea pig insulin differs radically from ungulate insulin, these antibodies may not be autoantibodies to its own insulin (25). The guinea pig is not constrained to regulate the idiotype which for it is not an autoantibody. Indeed, the results summarized in Table 8.5 demonstrate that guinea pigs immunized to either porcine or bovine insulin developed insulin antibodies of high titer. These antibodies also include the specific idiotypic antibody (Table 8.5). Anti-idiotypes having insulin like activity, however, could not be detected at any time point between 7 and 77 days after primary

TABLE 8.5. Antibody production in guinea pigs that were immunized to bovine or porcine insulin.[a]

Antigen	Insulin antibody titer	Production of insulinlike anti-idiotype	Production of specific idiotype
Porcine insulin	1/50,000	None	Positive[b]
Bovine insulin	1/100,000	None	Positive

[a]Guinea pigs were immunized with either porcine (40 μg/animal) or bovine (20 μg/animal) insulin in complete Freund adjuvant and boosted on days 21 and 50. Insulin antibody titer was determined by solid-phase RIA. Anti-idiotypes were assayed by lipogenesis.
[b]Defined as idiotype positive if stimulation of lipogenesis in rat adipocytes by ILA (of mouse origin) was inhibited.

immunization. Thus, "self antigen" rather than "foreign antigen" seems to be the relevant stimulus to elicit an anti-idiotypic response.

Effects of ILA Production on Glucose Metabolism and the State of Insulin Receptors

Another aspect undertaken here was to determine how large quantities of circulating ILA affect the physiology of glucose homeostasis and the state of insulin receptors. As mentioned earlier, the mice were able to adapt on a day-to-day basis when provided with food ad libitum. Subsequent to a fasting period, however, glucose homeostasis was markedly disturbed (Fig. 8.7). A period of hypoglycemia was observed in all mice

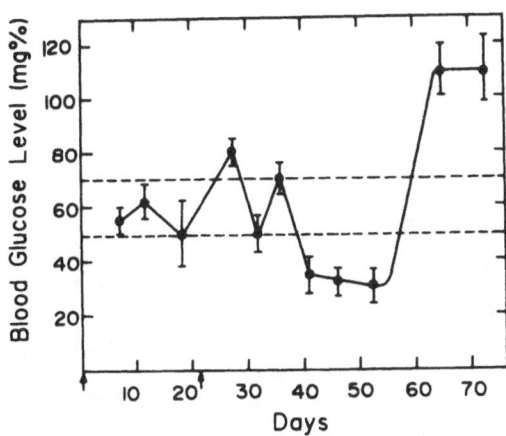

FIGURE 8.7. Concentrations of blood glucose in fasting mice with anti-idiotypic receptor antibodies. Mice at various times after immunization to insulin were deprived of food for 8 h and the concentration of blood glucose was measured. The dashed lines delineate the range of blood glucose concentration in control mice.

following the first peak of receptor antibodies. Fasting hypoglycemia was replaced by fasting hyperglycemia beginning at about the time of appearance of the second peak of receptor antibodies (Fig. 8.7). The ability of the mice to clear a glucose load was unimpaired one week after primary immunization (Fig. 8.8). However, glucose tolerance tests were abnormal at days 30 and 36. Note that on day 36, the insulin like receptor antibody activity was high and the insulin antibody titer was relatively low. On day 49, a time between two peaks of receptor antibodies, the glucose tolerance curve returned to normal. However, on day 58, the second peak of receptor antibodies was again accompanied by abnormal glucose tolerance, despite a very low titer of insulin antibodies. In view of the technical difficulties involved in measuring glucose tolerance in individual mice, we felt that it would not be wise to draw conclusions from the shape of the abnormal tolerance curves. Nevertheless, it was clear that abnormality of the response to a load of glucose appeared to be associated more with the measure of receptor antibodies than with the titer of insulin antibodies.

To determine whether exposure to receptor antibodies affected insulin receptors, we investigated the binding of insulin to adipocytes from im-

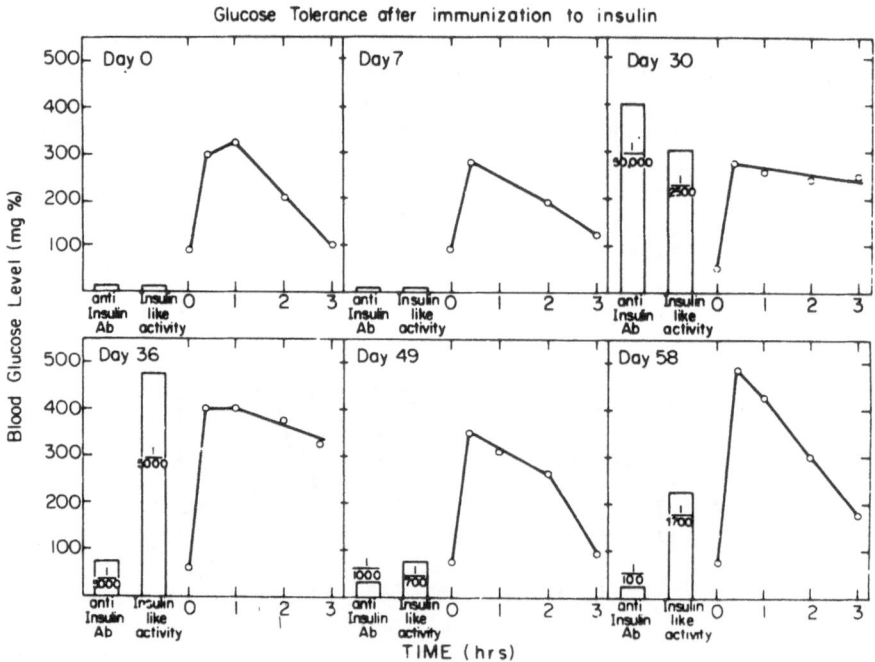

FIGURE 8.8. Glucose tolerance tests. Mice before or at various times after immunization to insulin were assayed for their tolerance to a load of glucose and for the relative amounts of insulin antibodies and insulinlike anti-idiotypic receptor antibodies in their sera.

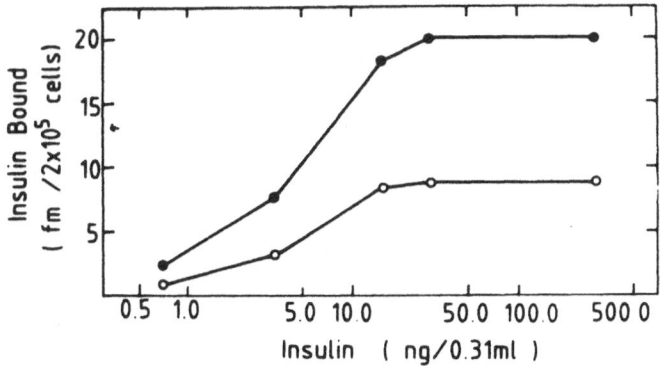

FIGURE 8.9. Down-regulation of insulin receptors. Adipocytes obtained from control mice (•) or from mice that had been immunized to insulin 30 days earlier and had anti-idiotypic receptor antibodies (○) were studied for their concentrations of insulin receptors, determined by the specific binding of radiolabeled ^{125}I-insulin that could be displaced by cold insulin.

munized and control mice. Figure 8.9 shows that adipocytes taken from mice with receptor antibodies had only about 40% of the insulin-binding capacity of control adipocytes. Adipocytes prepared from mice at the stage between the two peaks of receptor antibody had only a slightly reduced insulin-binding capacity (not shown). Therefore, insulin binding by adipocytes was associated inversely with the activity of receptor antibodies in the blood of the mice from which the adipocytes were taken. The reduced binding of insulin did not appear to result from a change in the binding affinity of insulin to its receptor, as half-maximal binding occurred at the same concentrations of insulin in test and control mice (Fig. 8.9). These findings are compatible with a typical down-regulation, usually associated with hyperinsulinemia in mammals, and occurs both in vivo and in vitro (reviewed in Ref. 26).

A decrease of 60% in available insulin receptors on the target tissues of immunized mice cannot by itself account for the diabetes like state of the mice because occupancy of only about 2% of the insulin receptors is sufficient to trigger a full biological response (27). To detect insulin resistance, we measured the lipogenic response of adipocytes as a function of the concentration of insulin. Figure 8.10 shows the response of adipocytes obtained from control mice and from test mice during the first peak of receptor antibodies. Resistance to insulin was evidenced by an increase in the insulin concentration necessary for producing a half-maximal response ($ED_{50} = 0.15$ ng/ml for control and 0.45 ng/ml for the test mice). Moreover, the test adipocytes responded to a large excess of insulin (500 ng/ml) by producing only 70% of the control maximal effects (Fig. 8.10). This implies that in addition to the altered sensitivity to insulin, the efficacy of the available receptors was reduced. In contrast to their re-

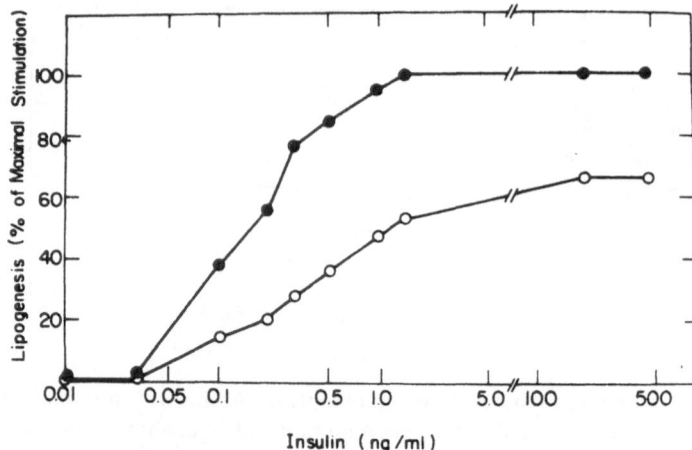

FIGURE 8.10. Insulin resistance. The lipogenic response to insulin was measured using adipocytes obtained from control mice (•) or from mice that had been immunized to insulin and had anti-idotypic receptor antibodies (o).

sistance to insulin, test adipocytes manifested no significant decrease in the maximal lipogenic response to the lectin concanavalin A (Table 8.6). Therefore, the reduced lipogenesis of the test adipocytes was specific for insulin and could not be attributed to a decrease in the enzyme systems responsible for lipogenesis.

Production of Monoclonal ILA-Receptor Antibodies by Mice That were Immunized to Insulin

Antibodies to plasma membrane hormone receptors are powerful tools in research. These, however, are difficult to prepare by conventional methods, as the purification of receptor protein is problematic due to its low

TABLE 8.6. Decrease in maximal lipogenic responses to insulin or concanavalin A of adipocytes obtained from mice with anti-idiotypic receptor antibodies.[a]

Days after immunization	Anti-idiotypic receptor antibodies	Decrease (%) in maximal lipogenic response to [b]:	
		Insulin, 0.5 µg/ml	Con A, 50µg/ml
0	No		
28	Yes	1	0
33	Yes	26	4
35	Yes	31	0
41	No	11	9

[a]Groups of five mice were immunized to insulin, and at various times their sera were tested for anti-idiotypic receptor antibodies (see Fig. 8.1) and the maximal lipogenic response of their adipocytes to insulin or concanavalin A was assayed.
[b]Computed relative to that obtained before immunization (day 0).

TABLE 8.7. Monoclonal anti-idiotypes.[a]

Clone	Lipogenesis	Binding to idiotype (cpm)
2D1/10-E12	0	10,400
3F1/10-H2	75	5,500
3F1/10-H5	70	6,000
3F1/10-G10	80	7,400
2C7/10-F5/2	74	6,200

[a]Monoclonal antibody supernatants were assayed in lipogenesis
(100 μl) and in solid-phase RIA with insulin antibody idiotype
bound to the plate.

concentration in normal target tissues. Also, the binding affinity of receptors towards the hormone is often lost after receptor extraction from plasma membranes, making further purification impossible. In view of this and as a tool in assisting us in our future studies, we have examined whether mice that were immunized to insulin could be utilized to form monoclonal anti-receptor antibodies. Hybridomas were prepared 2 days prior to the peak production of ILA and were screened for their (i) binding to insulin antibodies and (ii) having insulinlike activity in "in vitro" systems. Five clones (out of 700) were found to bind to anti-insulin (Table 8.7) and to displace labeled insulin from fat cells (not shown), and four clones produced insulin like effects in an in vitro system (Table 8.7). One clone (2D1/10-E12, Table 8.7) lacked the latter activity. Since this clone displaced labeled insulin from rat adipocytes, 2D1/10-E12 seems to be an insulin antagonist. This was further confirmed by the ability of this clone to inhibit efficiently the effects of insulin in stimulating bioactivities in fat cells (manuscript in preparation).

Possible Occurrence of Anti-Idiotypic Receptor Antibodies in Humans

Intensive studies by Kahn, Flier, and colleagues (28) have demonstrated that insulin receptor antibodies of the IgG class are rare in diabetic patients with *Acantosis-nigricance* type B (28). There is no evidence at present that these antibodies are anti-idiotypes. Their stimulating activity in in vitro systems could not be blocked by insulin antibodies raised in guinea pigs which also contain the idiotype positive antibody (Y. Shechter, unpublished observations).

In a recent study (29) we found in PF, a male child suffering from severe hypoglycemia, IgM antibodies that mimic the action of insulin in in vitro systems. The IgG of PF as well as control human IgM had no activity in stimulating lipogenesis in fat cells. The IgM fraction of PF also competed with insulin for binding to insulin receptors (Fig. 8.11). As the patient had low or normal secretion of insulin, we assumed that the severe hypoglycemia was associated with those insulin like IgM antibodies. Following treatment with glucocorticoids, both the receptor antibodies and the hy-

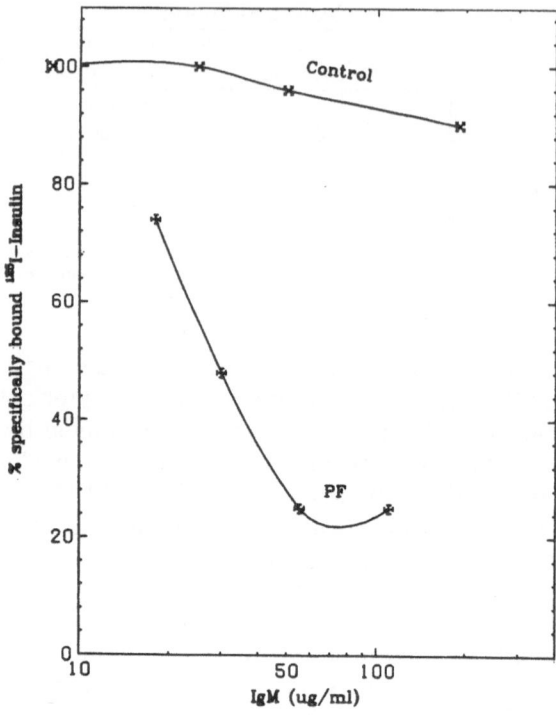

FIGURE 8.11. Displacement of labeled insulin from rat adipocytes by IgM of patient PF.

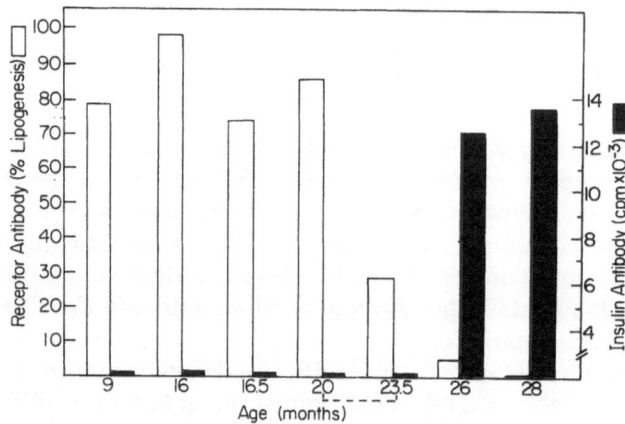

FIGURE 8.12. Anti-receptor antibodies are replaced by anti-insulin antibodies in patient's serum following steroid treatment. The presence of anti-receptor antibodies in the serum was determined by lipogenesis (black columns). Anti-insulin antibodies were measured by solid phase RIA (open columns). The dotted lines signify the 3 months prednisone treatment.

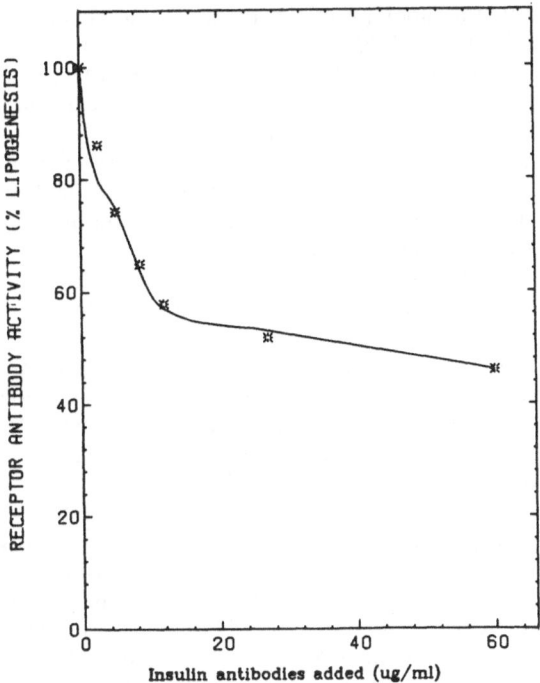

FIGURE 8.13. Recognition between anti-receptor antibodies and anti-insulin antibodies, the lipogenic activity of anti-receptor antibody was inhibited by adding increasing amounts of affinity purified anti-insulin antibodies from the patient's serum.

poglycemia disappeared, and antibodies to insulin appeared in the patient's serum (Fig. 8.12). The anti-insulin antibodies were isolated by affinity chromatography and were found to inhibit the receptor antibodies that were present earlier (Fig. 8.13).

Thus, in this patient the receptor antibodies may be idiotypes to anti-insulin (anti-idiotypes) which appeared subsequent to the treatment with glucocorticoids.

In conclusion, the possibility that the anti-idiotypic network also operates in human patients is attractive and should be further investigated, as it may have important implications for the pathophysiology of the disease. In mice, however, anti-idiotypic receptor antibodies are a relatively short-lived species (Fig. 8.1). Therefore, careful screening during the course of the human disease may be desirable.

References

1. Hearn, M.T.W. 1980. Graves' disease and the thyrotrophin receptor. Trends Biochem. Sci. 5:75.

2. Fuchs, S. 1979. Immunology of the nicotinic acetylcholine receptor. Curr. Top. Microbiol. Immunol. **85**:1.

3. Flier, J.S., C.R. Kahn, D.B. Jarrett, and J. Roth. 1976. Characterization of antibodies to the insulin receptor: a cause of insulin-resistant diabetes in man. J. Clin. Invest. **58**:1442.

4. Sege, K., and P.A. Peterson. 1978. Use of anti-idiotypic antibodies as cell-surface receptor probes. Proc. Natl. Acad. Sci. USA **75**:2443.

5. N.K. Jerne. 1974. Towards a network theory of the immune system. Ann. Immunol. (Inst. Pasteur) **125C**:373.

6. Schechter, Y., R. Maron, D. Elias, and I.R. Cohen. 1982. Autoantibodies to insulin receptor spontaneously develop as anti-idiotypes in mice immunized with insulin. Science **216**:542.

7. Schecter, Y., D. Elias, R. Maron, and I.R. Cohen. 1983. Mice immunized to insulin develop antibody to the insulin receptor. J. Cell. Biochem. **21**:179.

8. Schecter, Y., D. Elias, R. Maron, and I.R. Cohen. 1984. Mouse antibodies to the insulin receptor developing spontaneously as anti-idiotypes. I. Characterization of the antibodies. J. Biol. Chem. **259**:6411.

9. Elias, D., R. Maron, I.R. Cohen, and Y. Schecter. 1984. Mouse antibodies to the insulin receptor developing spontaneously as anti-idiotypes. II. Effects on glucose homeostasis and the insulin receptor. J. Biol. Chem. **259**:6416.

10. Cuatrecasas, P. 1972. Affinity chromatography and purification of the insulin receptor of liver cell membranes. Proc. Natl. Acad. Sci. USA **69**:1277.

11. Cohen, I.R., and J. Talmon. 1980. H-2 genetic control of the response of T lymphocytes to insulins. Priming of nonresponder mice by forbidden variants of specific antigenic determinants. Eur. J. Immunol. **10**:284.

12. Eshhar, Z., G. Strassmann, T. Waks, and E. Mozes. 1979. In vitro and in vivo induction of effector T cells mediating DTH responses to a protein and a synthetic polypeptide antigen. Cell. Immunol. **47**:378.

13. Laemmli, U.K. 1970. Cleavage of structural proteins during the assembly of the head of bacteriophage T4. Nature (London) **227**:680.

14. Ouchterlony, O. 1958. Diffusion-in-gel methods for immunological analysis. Prog. Allergy **5**:1.

15. Rodbell, M. 1964. Metabolism of isolated fat cells I. Effects of hormones on glucose metabolism and lipolysis. J. Biol. Chem. **239**:375.

16. Cuatrecasas, P. 1971. Insulin-receptor interactions in adipose tissue cells: direct measurement and properties. Proc. Natl. Acad. Sci. USA **68**:1264.

17. Moody, A.J., M.A. Stan, M. Stan, and J. Gliemann. 1974. A simple free fat cell bioassay for insulin. Horm. Metab. Res. **6**:12.

18. Kessler, S.W. 1975. Rapid isolation of antigens from cells with a staphylococcal protein A-antibody adsorbent: parameters of the interaction of antibody-antigen complexes with protein A. J. Immunol. **115**:1617.

19. Ey, P.L., S.J. Prowse, and C.R. Jenkin. 1978. Isolation of pure IgG1, IgG2a, and IgG2b immunoglobulins from mouse serum using protein A-sepharose. Immunochemistry **15**:429.

20. Keck, K. 1975. Ir-gene control of immunogenicity of insulin and A-chain loop as a carrier determinant. Nature (London) **254**:78.

21. Blundell, T., G. Dodson, D. Hodgkin, and D. Mercola. 1972. Insulin: the structure in the crystal and its reflective chemistry and biology. Adv. Protein Chem. **26**:279.

22. Dayhoff, M.O. (ed.). Atlas of protein sequence and structure. D186. National Biomedical Research Foundation, Washington, D.C.
23. De Meyts, P., E. Van Obberghen, J. Roth, A. Wollmer, and D. Brandenburg. 1978. Mapping of the residues responsible for the negative cooperativity of the receptor-binding region of insulin. Nature (London) **273**:504.
24. Ranghino, G., J. Talmon, A. Yonath, and I.R. Cohen. 1981. Structural aspects of recognition and assembly in biological macromolecules, p. 263. *In* M. Balaban (ed.), The conformation of antigenic determinants of insulin and H-2 gene control of the immune response. International Science Services, Philadelphia.
25. Neville, R.W., B.J. Weir, and N.R. Lazarus. 1973. Insulins of hystricomorph rodents. Diabetes **22**:851.
26. Roth, J., C.R. Kahn, M.A. Lesniak, P. Gorden, P. de Meyts, K. Megyesi, D.M. Neville, Jr., J.R. Gavin III, A.H. Soll, P. Freychet, I.D. Goldfine, R.S. Bar, and J.A. Archer. 1975. Receptors for insulin, NSILA-s, and growth hormone: applicative to disease states in man. Recent Prog. Horm. Res. **31**:95.
27. Kono, T., and F.W. Barham. 1971. The relationship between the insulin-binding capacity of fat cells and the cellular response to insulin. Studies with intact and trypsin-treated fat cells. J. Biol. Chem. **246**:6210.
28. Kahn, C.R., K. Baird, J S. Flier, and D.B. Jarrett. 1977. Effects of autoantibodies to the insulin receptor on isolated adipocytes. Studies of insulin binding and insulin action. J. Clin. Invest. **60**:1094.
29. Elias, D., I.R. Cohen, Y. Shechter, Z. Spirer and A. Golander. 1987. Antibodies to insulin receptor followed by anti-idiotype: antibodies to insulin in child with hypoglycemia. **36**:348.

9
Molecular Characterization of Anti-Neuroleptic Idiotypes and Anti-Idiotypes

MICHAEL B. BOLGER, MARK A. SHERMAN, AND
D. SCOTT LINTHICUM

Introduction

Recent advances in biotechnology have created tremendous opportunities for research at the interface between chemistry and modern molecular biology. A fundamental question in immunology is that of the specific molecular interactions which occur in the ligand-antibody complex. One would like to be able to predict the effects of specific changes in immunoglobulin amino acid sequence on affinity, specificity, and kinetics for interactions with both small ligands and peptides.

The molecular recognition of small drugs and neurotransmitters by antibodies is based on the summation of many weak noncovalent interactions between the chemical substituents of the drug and amino acid side chains found in the immunoglobulin binding site. Examples of these interactions are hydrogen bonding, hydrophobic bonding, electrostatic interactions, and van der Waals forces. It is well known that minor changes in amino acid substitutions of the hypervariable region in antibodies can produce marked differences in ligand affinity and binding kinetics. The same principles of noncovalent bonding control the interaction of idiotype and anti-idiotype antibodies and, consequently, may be very important in regulation of the B-cell immune network as well as in T-cell regulation of B-cell clonal expansion.

If one can develop rules by which to predict the effect of specific amino acid substitutions on the ability of the hypervariable region of monoclonal antibodies to bind with high affinity to synthetic haptens, to neuroactive peptides, and to the complementary determining regions of other antibodies, then it might become possible to engineer specific macromolecular receptors with unique binding properties that can recognize foreign substances or be used as synthetic vaccines (1). This approach would be valuable for producing protective antibodies to substances such as neuroactive agents which in and of themselves are too toxic to be used in a classical way as immunogens.

The recent development of anti-idiotypic antibodies to neurotransmitter receptor antagonists has shown that amino acid loops found in hypervariable regions can "mimic" the molecular reactions of neurotransmitters and their antagonists. Examples of this type of approach will be examined in detail in other sections of this book. This chapter will focus on the molecular characterization of idiotypic specificity for a number of antibodies which are able to bind to the dopaminergic D-2 antagonist haloperidol. The tools developed and techniques discussed have significance in relationship to their application to anti-idiotypes in order to further understand the concept of "molecular mimicry."

Considerations in Synthesis of Immunogen

Agonists vs. Antagonists

It has been hypothesized that in order for an antibody (Ab_1) to stimulate the production of an internal image anti-idiotype ($Ab_2\beta$) which mimics some property of the original antigen, the Ab_1 must have molecular properties of interaction with the antigen which are similar to the properties of interaction of the antigen with its biological receptor. For example, a monoclonal antibody (Ab_1) directed against alprenolol was chosen for induction of anti-idiotypes (Ab_2) because it showed stereospecificity and binding characteristics similar to those of the β-adrenergic receptor (2). However, when comparing the relative binding affinity of agonists and antagonists, the degree of quantitative similarity may not need to be absolute. In the case of Ab_1 to alprenolol, it was found that agonists had three orders of magnitude lower binding affinity than would be expected for binding to the β-adrenergic receptor but were still able to stimulate production of $Ab_2\beta$ which cross-reacted with their biological receptors (3).

Given the specific requirements of Ab_1, one can ask if the original hapten should be an agonist or an antagonist which is coupled to a carrier protein. When a binding site possesses the required structure for high-affinity agonist binding, some studies suggest that this will also confer on it the necessary properties to bind antagonists with high affinity (4). In addition, it will likely recognize the entire spectrum of synthetic ligands stereoselectively and with a potency order similar to that of the biological receptor. Thus, one would be tempted to utilize agonists for immunogen production as a first step in generation of $Ab_2\beta$. This approach has been extremely successful for the high-affinity, chemically stable agonists of the opiate receptor such as morphine (5,6).

However, other considerations are important in deciding on the initial hapten. If the agonist has relatively low binding affinity ($K_d > 1\mu M$) for its biological receptor, then the Ab_1 combining site will also likely have low binding affinity. Thus, the molecular nature of this binding site may not possess sufficient numbers of specific molecular contacts to stimulate production of $Ab_2\beta$ which is capable of specific cross-reactivity with the biological receptor. Also, agonists, such as the catecholamines, which are susceptible to chemical degradation or metabolism cannot be used as hapten because the chemical nature of the agonist conjugate which is presented to the immune system in vivo will most likely have no relationship to the original agonist molecule. In addition to choosing a chemically stable hapten, it is important to develop chemical coupling procedures that will maintain the important physicochemical features of neurotransmitter binding. For example, our procedures for conjugation of catechol and indole-ethyl-amine ligands to carrier protein produced good incorporation of hapten, maintained the integrity of the aromatic hydroxyl groups, and preserved a positively charged amino group. Unfortunately, immunization with dopamine and serotonin resulted in antibodies which primarily recognized the linker arm used to couple hapten to carrier protein and which were not able to bind to the free agonist ligand. This was probably the result of oxidation of the catechol hydroxyls to a quinone structure during immunization (7). Other groups have had similar difficulties in trying to immunize with catecholamines and indolamines (8–11).

In contrast to the difficulty of obtaining antibodies to catecholamine neurotransmitters, a number of antibodies to several synthetic neurotransmitter antagonists have been successfully produced (3,4,12,13). Generally, the chemical stability of synthetic antagonists is much better than for naturally occurring agonists. However, the method of attachment to carrier protein is more critical than for agonists. This is because it is desirable to maintain the pharmacophore, the portion of the antagonist which mimics the natural neurotransmitter, in an unsubstituted form. Theoretically, this will greatly increase the chances of obtaining Ab_1 antibodies which resemble the neurotransmitter receptor protein.

Haloperidol, a butyrophenone derivative, is a potent dopaminergic antagonist which binds with high affinity to the D-2 dopamine receptor in the central nervous system (14). In part because previous investigators had been successful in producing polyclonal antisera to haloperidol (15,16), we decided to try to make monoclonal idiotypic and anti-idiotypic antibodies to this antagonist. In order to make a rational choice of immunogen coupling chemistry, it became important to consider the pharmacophore of dopaminergic agents. Dopamine, dopamine agonists, and neuroleptic drugs all possess a basic nitrogen atom separated by a 5 to 7-Å chain or framework from an aromatic ring which constitutes the basic pharmacophore (17). Unfortunately, the pharmacophoric region of haloperidol has never been definitively identified. Opinion is equally divided among those who believe that the fluorophenyl ring and the tertiary nitrogen overlap the structure of dopamine and those who see the dopaminergic ring overlap as being with the chlorophenyl ring, as illustrated in Fig. 9.1

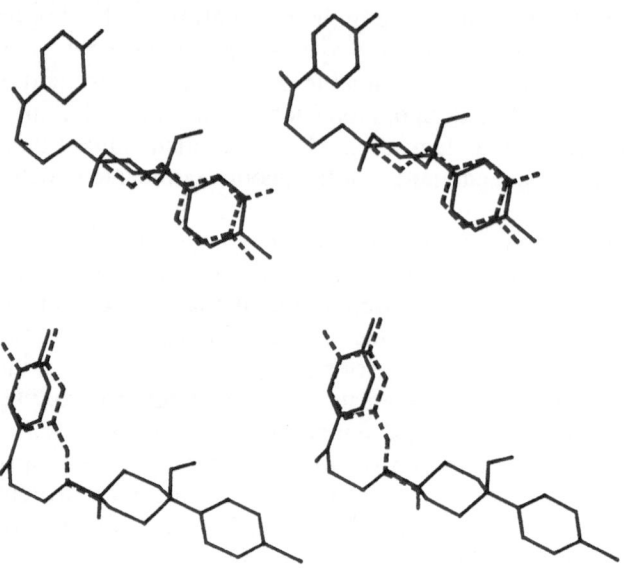

FIGURE. 9.1. Structure of haloperidol in stereo with dopamine superimposed over fluorophenyl ring and chlorophenyl ring.

(18,19). We therefore chose to produce antibodies against two different haloperidol-protein conjugates, each presenting one of the tentative pharmacophoric regions to the immune system as an antigenic determinant. The structures of these two conjugates are displayed in Fig. 9.2. The first conjugate uses a succinic acid linker arm to attach haloperidol to BSA via the tertiary alcohol function and is therefore referred to as the succinic acid conjugate (SUCC-CONJ). This conjugate was able to elicit antibodies which specifically recognize the butryrophenone ring and the piperidyl nitrogen. The second conjugate arises from attachment of haloperidol to BSA through the aryl keto group using (aminooxy)acetic acid to form the linker arm. This conjugate is therefore called the oxime conjugate (OX-CONJ) and primarily gives antibodies which recognize the chlorophenyl-piperidinyl structure. Design and synthesis of the SUCC-CONJ had already been accomplished by Dr. Pierre Laduron of Janssen Pharmaceutica, and this conjugate was kindly made available to us through his generosity. We developed a new synthetic route to the OX-CONJ, which has been described elsewhere (13).

By utilizing the two conformationally distinct immunogens described above, we have produced a library of 37 monoclonal antibodies to haloperidol. Each haloperidol-BSA conjugate was used to immunize BALB/c

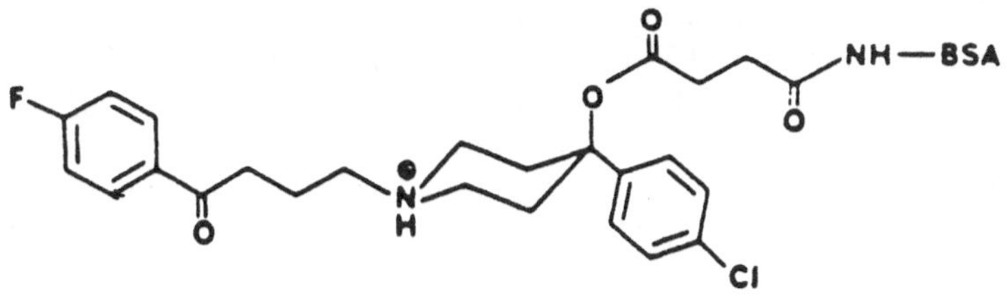

SUCCINIC ACID CONJUGATE

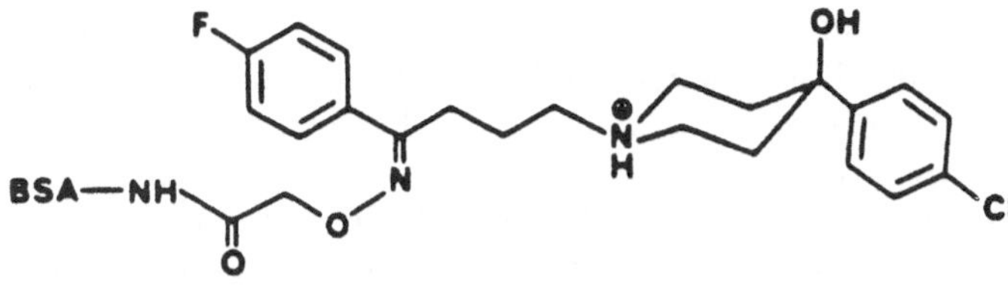

OXIME CONJUGATE

FIGURE 9.2. Structure of two haloperidol conjugates. M. Sherman, D. Scott Linthicum, M. Bolger, Mol. Pharm. *29:*589–598, 1987, © by the American Society for Pharmacology and Experimental Therapeutics.

mice for subsequent production of monoclonal antibodies which bind haloperidol. Immunization and hybridoma fusions were performed according to protocols which have been fully described elsewhere and are therefore not included in this discussion (7). The SP2/0 myeloma cell line, a variant of the P3NS-1/1Ag4-1 cell line which fails to synthesize immunoglobulin, was used as the fusion partner.

Pharmacodynamic Characterization

A total of 12 different monoclonal antibodies (mAbs) were produced against the SUCC-CONJ. Hybridoma cell lines secreting the mAb are assigned a number which describes conditions of hybridoma fusion. Thus, hybridoma 185(2)-1 represents fusion number 185, cloned twice, clone number 1. The initial characterization of each antibody involves determination of titer and equilibrium binding affinity on unpurified ascites fluid. For all of the pharmacodynamic measurements except for fluorescence quenching, a simple charcoal adsoption procedure was used to separate bound and free radioligands. This technique can be problematic for systems in which the binding affinity of radioligand is low or for lower molecular weight proteins which adsorb to BSA-coated charcoal. However, comparisons of [^{3}H]haloperidol binding to mAb gave identical results using either the charcoal assay, gel filtration, or direct quenching of tryptophanyl fluorescence. The 12 mAbs raised against the SUCC-CONJ fall into three classes: those with high, intermediate, and low affinity for free haloperidol. Interestingly, it was found that those with low affinity for free haloperidol often possess high affinity for the haloperidol-BSA conjugate, illustrating that these mAbs recognize antigenic determinants contributed by both haloperidol and the carrier protein. Such mAbs are probably not appropriate for the subsequent generation of internal image anti-idiotypes.

Three of these mAbs were selected for detailed analysis of their binding affinities and specificities. 185(2)-1 and 189(2)-6 were selected on the basis of their high affinity for free haloperidol. We originally proposed that these two antibodies differed slightly in their selectivity of binding to analogues with substitutions in the chlorophenyl ring (13). Subsequent binding studies on protein A-purified mAb and detailed amino acid sequence determination has revealed that these mAbs, which were derived from two different fusions, are identical (41). Thus, we will concentrate only on 185(2)-1 for the remainder of this chapter. The selection of 190(2)-6, with only intermediate affinity for haloperidol, was based on the curious observation that it demonstrated a higher affinity for the related butyrophenone spiperone than it did for haloperidol. This difference in affinity was proportional to that observed for the actual D-2 receptor. These two mAbs, which were eventually sequenced and modeled using computer graphics, were renamed mAbs A and C, respectively, in order to simplify the presentation of data.

Out of 25 mAbs which have been produced against the OX-CONJ, two were selected for detailed study of molecular binding characteristics. These were originally assigned the clone numbers 258(2)-1 and 258(2)-11. These two mAbs were subsequently renamed mAbs D and E, respectively, and will be thus referred to throughout the remainder of this chapter.

Having selected four anti-haloperidol mAbs with moderate to high affinity for haloperidol, the next step became to determine their precise affinities and specificities for haloperidol and a series of closely related analogues. This permits an assessment of the similarities of these mAb binding sites with that of the actual D-2 dopaminergic receptor.

Data Analysis

Early methods for the analysis of equilibrium pharmacodynamic data in the absence of computing facilities attempted to render the basically nonlinear behavior of enzymes, drug receptors, and immunoglobulins mathematically tractable by performing a linearization. Scatchard analysis is one example of a linearization which has been extensively criticized on grounds of statistical inaccuracy and misinterpretation of results (20–23). It is now generally agreed that weighted nonlinear least-squares regression analysis is the method of choice for the analysis of such data because of the improved accuracy of analysis as well as the fact that simple models of ligand association can be easily compared to more complex models with two or more sites of interaction (24). The most simple equilibrium binding scheme, Eq. (1), assumes that the two Fab combining sites of an IgG each bind only one hapten and have no allosteric interactions and can account for most of the equilibrium binding interactions of haptens (H) with immunoglobulins (I).

$$H + I \underset{k_{-1}}{\overset{k_1}{\rightleftharpoons}} HI \tag{1}$$

The binding scheme depicted in Eq. (1) can be rearranged to allow for a more statistically appropriate model than the linearization provided by Scatchard (25) as follows:

$$HI = I_t \cdot \frac{1}{1 + K_d/H} \tag{2}$$

where K_d = equilibrium dissociation constant, I_t = IgG maximal binding capacity, H = free hapten at equilibrium, and HI = bound hapten at equilibrium.

In contrast to Scatchard analysis, in which there is highly correlated variance in both the dependent and independent variables, Eq. (2) can be solved for the parameters K_d and I_t as a function of H, and the primary source of variance is in the dependent variable HI. In addition, the data can be plotted in a semilog fashion, which avoids compression of the dependent variable at high levels of free hapten.

A program which we developed for the IBM PC called ANALAB has the following built-in functions for equilibrium association and dissociation (i.e., competition) binding which can be modeled for systems of one or two independent sites (26):

Association of a Radioligand

$$HI = I_{t1} \cdot \frac{1}{1 + K_{d1}/H} + I_{t2} \cdot \frac{1}{1 + K_{d2}/H} \tag{3}$$

Dissociation of a Radioligand by Competiton

$$HI = I_{t1} \cdot \frac{1}{1 + H/IC_{50}1} + I_{t2} \cdot \frac{1}{1 + H/IC_{50}2} \tag{4}$$

The significance of adding extra terms to the simple model of Eq. 2 can be determined by comparing the residual sum of squares for the partial and the full models by calculation of an F statistic according to the following equation (27):

$$\frac{SS_p - SS_f}{SS_f} \times \frac{(N - p)}{k} = F[k, N - p, \alpha (1)] \qquad (5)$$

where SS_p = sum of squares of partial model, SS_f = sum of squares of full model, N = number of data points, p = number of parameters in full model, $p-k$ = number of parameters in partial model, and $\alpha(1)$ indicates a one-tailed F test.

Another type of model which is commonly used to determine the extent of cooperativity is simply an extension of Eq. (2) in which the ratio of K_d/H is raised to the nth power (28). The value of n, determined by nonlinear fitting of equilibrium association binding data, is called the Hill slope. For an independent class of binding site the Hill slope would be expected to be 1.0. However, if positive or negative cooperativity is present the Hill slope will be greater than or less than 1.0. This model is included in ANALAB, even though the physical interpretation of data with a Hill slope which deviates from 1.0 cannot be determined without extensive biochemical and kinetic data to support the contention of allosteric interactions between subunits.

Application of this type of data analysis to equilibrium saturation binding data for mAbs A–E is shown in Fig. 9.3. Monoclonal antibody A has very

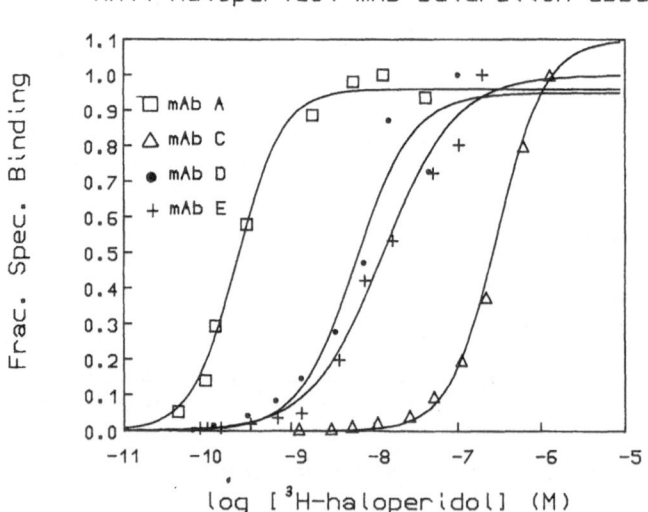

FIGURE 9.3. Saturation binding plot of mAbs A–E.

high affinity for haloperidol [K_d = 0.2($\pm$0.02) nM]. Monoclonal antibodies D and E bind with moderate affinity [(mAb D, K_d = 6.1($\pm$1.81) nM; mAb E, K_d = 16($\pm$4.9) nM], and mAb C binds haloperidol to a lesser degree [K_d = 398($\pm$0.32) nM]. In addition to high-affinity binding, it should be noted that the binding curves rise more sharply than would be expected for interaction of a small ligand with a single independent binding site. The Hill slopes of mAbs A, C, D, and E are 1.64$\pm$0.21, 1.28$\pm$0.06, 1.27$\pm$0.37, and 0.9$\pm$0.15, respectively. This observation would indicate possible positive cooperativity between binding sites. Although cooperativity between antigen binding sites on a single IgG molecule is not commonly reported, several laboratories have noted the contrary (28). However, the source of this apparent cooperativity is not clear at this time. The low affinity of mAb C for haloperidol greatly increases the cost and decreases the accuracy of radioligand binding assays with [^{3}H]haloperidol. Subsequent competition assays on mAb C were therefore performed using [^{3}H]spiperone (K_d = 200 nM).

Qualitative Models of mAb Combining Sites Based on Competition Binding Analysis and QSAR

In order to probe the various hydrophobic, electronic, and steric constraints present at the antibody binding site, 16 analogues of haloperidol bearing simple ring or chain substitutions were analyzed in competitive binding assays. A representative plot of competition binding data for haloperidol analogue R29808 is shown in Fig. 9.4. The structures of these analogues, their physicochemical properties, and their ability to displace [^{3}H]haloperidol, as evaluated by their respective IC_{50} values plus standard errors, have been reported previously (13). Quantitative structure-activity relationships (QSAR) can be used to develop a qualitative picture of molecular interactions between haloperidol and the combining sites of each mAb. This type of analysis involves application of multiple linear regression to correlate changes in the physicochemical properties of the haloperidol analogues with changes in binding affinity.

For this study, the Hansch approach was taken (29), whose model takes the form:

$$\text{Log } 1/IC_{50} = a(\text{hydrophobic}) + b(\text{electronic}) + c(\text{steric}) + d$$

where the terms in parentheses refer to any number of parameters designed to evaluate the given interaction. The relative magnitude and sign of the coefficients a, b, c, etc., indicate the degree to which the given interaction influences activity and the nature of the correlation (positively or negatively correlated). The parameters used to evaluate hydrophobicity, the electronic properties (charge, dipole, etc.), or the steric properties (molar volume, physical length, etc.) are obtained by recording free energy changes

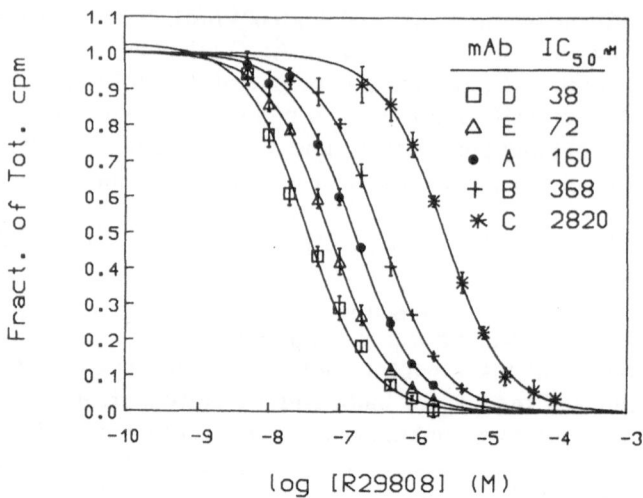

FIGURE 9.4. Competition binding for mAbs A–E. M. Sherman, D. Scott Linthicum, M. Bolger, Mol. Pharm. *29*:589–598, 1987, © by the American Society for Pharmacology and Experimental Therapeutics.

that occur as a result of modifying a reference compound. It is therefore necessary to equate the log of the IC_{50} value to these types of parameters, since equilibrium constants (such as IC_{50}) are related to free energy changes by the following equation:

$$\Delta G \cong -RT \text{ in } (I/IC_{50})$$

The experimental procedures used to arrive at QSAR parameters are often quite simple and have been reviewed extensively (30). Extensive tables listing physicochemical parameters for various chemical substituents have been compiled (31). A complete listing of the parameters used to evaluate the hydrophobic, electronic, and steric properties of the chemical substituents found in the structures of the 16 haloperidol analogues have been previously reported (13). These data were then analyzed by a multiple linear regression analysis program (Stat-System software package, Psychological Assessment Research, Inc., Odessa, Fla.) on an IBM XT personal computer. Table 9.1 contains the best-fit QSAR equations for each mAb and Figs. 9.5–9.7 show a schematic description of the type of molecular interactions which might be important in haloperidol recognition by these monoclonal antibodies.

Examination of the most significant QSAR equation for mAb A reveals the types of forces involved in the binding site interaction. The large positive coefficient associated with the length of the inter-ring alkyl chain

TABLE 9.1. Significant QSAR equations.

mAb	QSAR equation[a]	Statistical analysis[b]			
		n	r	s	F
A	$\log 1/IC_{50} = 1.97(\text{chain-X}) - 1.71(\text{min. width}) - 1.53(\text{L-R1}) - 0.65(\mu\text{-R2}) + 16.3$	16	0.93	0.47	18.68
C	$\log 1/IC_{50} = 2.05(\sigma_p - C=O) + 0.65(\text{chain}-Y) + 0.23(\mu-R2) + 0.51(\Delta\pi-R1) + 4.6$	16	0.92	0.30	14.9
D	$\log 1/IC_{50} = 3.02(\sigma_p - R2) - 0.59(\text{min. width}) + 0.56(\text{chain}-Y) + 9.39$	16	0.88	0.32	14.33
E	$\log 1/IC_{50} = 2.77(B1 - R2) + 2.54$	16	0.85	0.26	39.5

[a]Parameters:
Chain-X, indicator variable describing chain length: if chain is composed of 4 atoms, then chain-X = 1; if not 4, then chain-X = 0.
Min. width, the smallest dimension, in Å, of the first two chain members attached to ring 1, measured in the plane perpendicular to the ring.
L − R1, the length of the substituent attached to ring 1.
μ − R2, the diple moment, in debyes, of the substituent attached to ring 2.
σ_p − C = 0, electron-withdrawing or -donating capacity of the first two chain members attached to ring 1.
σ_p − R2, electron withdrawing or donating capacity of the substituent attached to ring 2.
Chain-Y, an indicator variable describing the length of the chain connecting ring 1 and the piperidine nitrogen: if 4 atoms or greater, then chain-Y = 1; if less than 4, chain-Y = 0.
$\Delta\pi$ − R1, hydrophobicity relative to fluorine of the substituent attached to ring 1.
B1 − R2, the smallest radius of the substituent attached to ring 2, as measured in the plane perpendicular to the ring plane.
[b]n, Number of analogues tested; r, correlation coefficient; s, standard deviation; F, F test value.

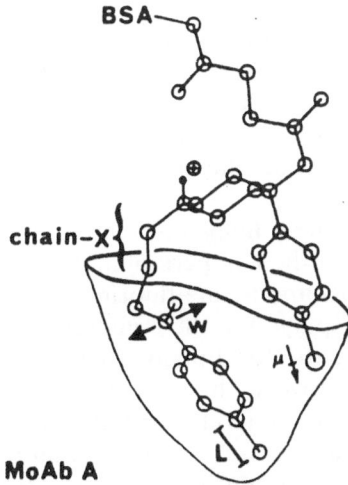

FIGURE 9.5. Qualitative schematic models for mAbs A and B.

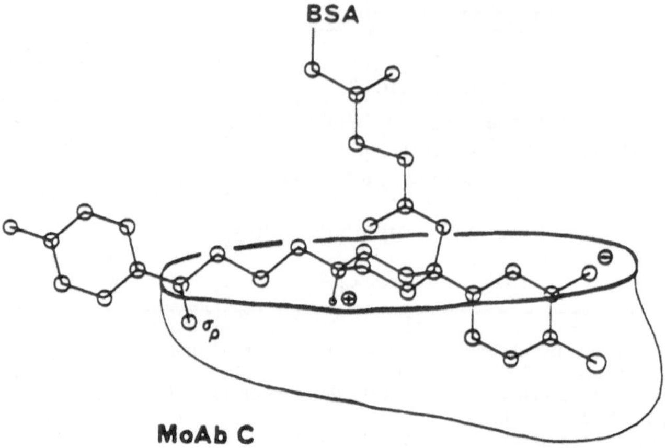

FIGURE 9.6. Qualitative schematic model for mAb C. M. Sherman, D. Scott Linthicum, M. Bolger, Mol. Pharm. *29*:589–598, 1987, © by the American Society for Pharmacology and Experimental Therapeutics.

indicates that analogues with a chain length that differs from the 4-atom length bind very poorly, especially the 3-atom-chain analogue. Also the large negative coefficient associated with the smallest physical dimension (width) of the substituent occupying the keto position in haloperidol indicates that analogues bearing planar sp^2 substituents in the keto position are highly favored by the antibody, whereas those bearing bulky sp^3 (tetrahedral) substituents are not able to bind. We have previously postulated that a folded conformation of haloperidol is preferentially recognized by mAb A (13). It is probable that a tetrahedral substituent in place of the keto group sterically prevents the analogue from assuming such a folded conformation in the binding site.

Another parameter which helps correlate binding affinity with physicochemical properties of the haloperidol molecule is the actual length, in Å, of the substituent occupying the fluorine position in haloperidol. The negative correlation displayed in the equation indicates that long substituents decrease binding affinity. Finally, there appears to be an electronic interaction occurring between the chlorine atom of haloperidol and an amino acid residue in the antibody binding site, as indicated by the selection of a dipole parameter for this position in the final equation. Substituents with high electron density at this position (Cl, Br, and OCH_3) are favored at the binding site over those with low electron density (Me, *i*-Pr). One would predict that the amino acid which interacts with this part of haloperidol would have an electron cloud which is highly polarizable in order to form dipole–induced dipole van der Waals contacts. The final

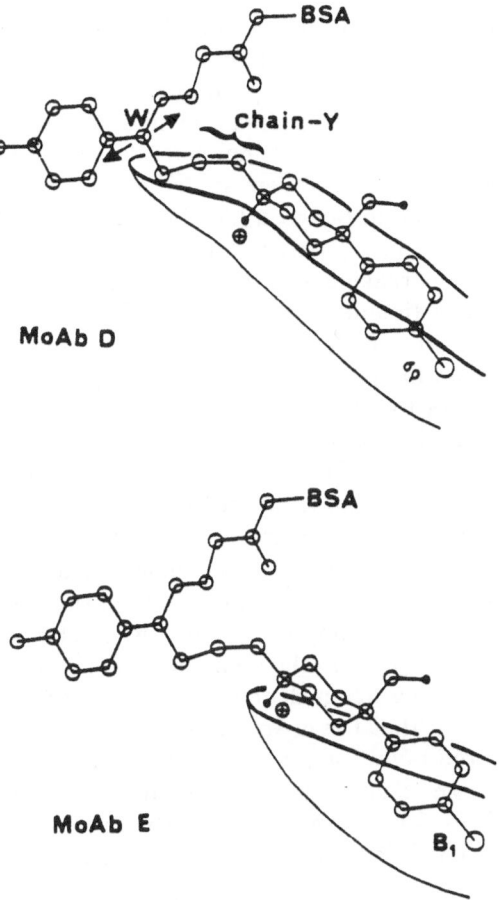

FIGURE 9.7. Qualitative schematic models for mAbs D and E. M. Sherman, D. Scott Linthicum, M. Bolger, Mol. Pharm. 29:589–598, 1987, © by the American Society for Pharmacology and Experimental Therapeutics.

equation displays a correlation coefficient of 0.93 and is therefore capable of explaining 86% of the variation in the data. The probability that this equation arose due to chance correlation is 0.1%.

Examination of the final QSAR equation for mAb C reveals that the most significant predictor of antibody binding activity is a parameter which describes the electron distribution between the substituent in the keto position and the adjacent aromatic ring. Compounds bearing substituents which withdraw electrons from the fluorophenyl ring are highly favored (carbonyl and amide functions), whereas neutral or electron-donating groups (sulfide, *sec*-alcohol, alkyl, cyclic ether) are not. This observation can be interpreted as meaning that the fluorophenyl ring is interacting with an electron-rich binding site residue or, alternatively, that the keto group is involved in an electron-donating interaction with an antibody amino acid. Sequence data and computer modeling facilitate this assignment. Several other parameters were selected by the computer program as increasing the correlation between binding data and molecular features. However, none of the added parameters are statistically significant, as demonstrated by a decreasing F-test value (13). The equation with the largest correlation coefficient, $r = 0.92$, indicates that short inter-ring alkyl chains are not favored and that once again the dipole properties of the substituent in the chlorine position may be contributing to binding affinity. Finally, there appears to be a small positive dependence on the hydrophobicity of the substituent in the fluorine position. These parameters, along with the observed preference for an extended conformer of haloperidol, are illustrated schematically in Fig. 9.6.

The equation describing the chemical factors responsible for moderate affinity binding of haloperidol to mAb D indicates that the most significant factor is the electron density of the chlorophenyl ring, with low electron density being favored. This implies a stacking mechanism between this ring and an electron-rich area in the combining site. Inter-ring chain length and keto width are once again deemed important for binding, but in a fashion different from that for mAb A. In the case of A, only a 4-atom chain was considered favorable, whereas in D, a chain length of 4 atoms or longer is acceptable. This implies that intramolecular folding is probably not occurring as in A. This is supported by the binding data for semirigid isomers (Fig. 9.8), which indicate a preference for the extended conformer (13). The dependences on chain length and on narrowness of the keto function are therefore interpreted as reflecting a rather narrow, groovelike binding cleft. Such grooves have been crystallographically observed in other antibody systems (32). The final equation provides a correlation coefficient of $r = 0.88$ and is significant at the 99.9% level of probability. A schematic model of mAb D's binding site is shown in Fig. 9.7.

The final equation for mAb E reflects the lack of specificity and low affinity of its combining site in that only one parameter is selected as significantly influencing binding affinity. This parameter, the minimum radius of the substituent occupying the chlorine position of haloperidol, actually has little value in terms of describing a ligand-receptor interaction. Qualitative inspection of the data indicates that a combination of two parameters, one describing the hydrophobicity or perhaps molar volume, and a second describing the electron-donating capacity of the substituent, explains the poor binding affinity of the methoxy analogue (R2842) much better. The final equation for mAb E provides a correlation coefficient of $r = 0.85$ and is once again highly significant (99.9%). The schematic model for mAb E in Fig. 9.7 illustrates recognition of the chlorophenyl ring.

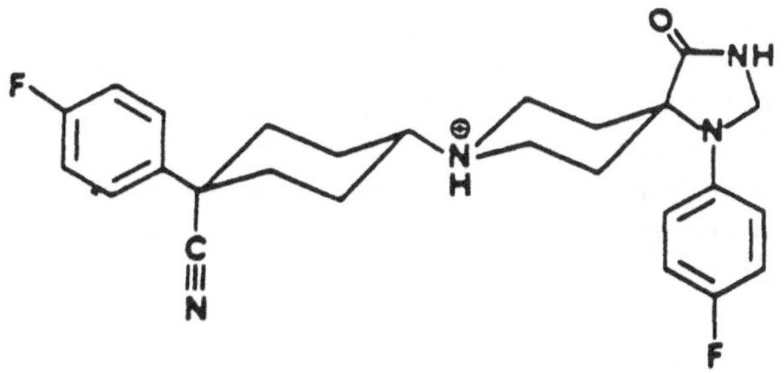

R 48 455

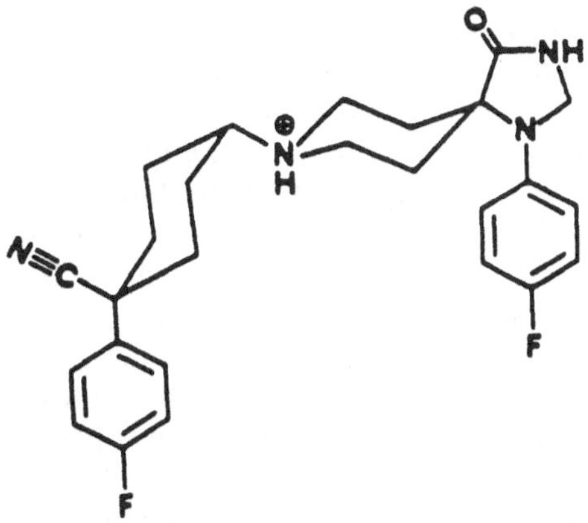

R 49 399

FIGURE 9.8. Semirigid isomers R 48 455, the equatorial or "extended" isomer, and R 49 399, the axial or "folded" isomer. M. Sherman, D. Scott Linthicum, M. Bolger, Mol. Pharm. 29:589–598, 1987, © by the American Society for Pharmacology and Experimental Therapeutics.

Correlations with D-2 Receptor Binding

If mAbs A through E are to be useful in preparation of $Ab_2\beta$, then they should have binding properties which are similar to that of the D-2 receptor. Having described antagonist binding properties, it became necessary to investigate agonist binding and then to try to determine the statistical correlations. Only mAbs C, D, and E bind dopamine to a significant extent, although the concentrations needed are rather high (mM range). Since all three of these antibodies appear to recognize haloperidol's chlorophenyl ring, this might imply that the chlorophenyl-piperidine region of haloperidol is the primary determinant of D-2 receptor recognition. However, this type of data must be interpreted cautiously in light of the fact that the neurotransmitters serotonin and epinephrine bind with equal or higher affinity (13). This finding would imply one of two things: mAbs C, D, and E possess binding sites which contain the basic elements of a prototype neurotransmitter receptor or, alternatively, these binding sites will accept any small molecule possessing a phenyl ring and a protonated nitrogen provided the concentration is high enough. Our data do not permit discrimination of these two possibilities at the present time. In order to assess similarities between the dopamine receptor and the four antibody binding sites, a more detailed comparison was needed.

Sixteen haloperidol analogues plus the semirigid isomers were synthesized by Janssen Pharmaceutica and kindly supplied for this study. The relative affinity of these analogues in binding to the D-2 receptor was determined using a standard homogenized brain membrane competition binding assay (13). Homologies between the D-2 receptor binding site and the antibody binding sites were then assessed by correlating the log of receptor IC_{50} values versus the log of antibody IC_{50} values for each mAb. Only data on the primary 16 haloperidol analogues were included in the correlation. These data indicate that the binding site of mAb A is more similar to the D-2 receptor's than is that of C, D, or E (correlation coefficient r = 0.82, 0.30, 0.15, and 0.05, respectively).

However, data on the semirigid isomers indicate that the receptor prefers the equatorial or "extended" isomer over the "folded" isomer by a factor of 100 to 1, which is exactly opposite to the preference of mAb A (folded preferred 30:1). Hence, it can be concluded, based on this collection of data, that the binding site of mAb A, although possessing regions which may resemble the actual dopamine receptor binding site, would most likely not prove to be a good template for the production of $Ab_2\beta$ internal image antibodies which cross-react with the receptor, due to recognition of an altered conformer of haloperidol. This type of study illustrates the power of combining extensive radioimmunoassay data with QSAR methodology in illustrating the likely types of physical interactions occurring on a molecular level at the antibody binding site. In the next section we provide physical evidence for the identity of amino acid residues within the antibody binding site which would give rise to these sorts of physicochemical interactions. Determination of the amino acid residues which form the binding sites of the mAbs A through E provides an opportunity to examine some interesting questions regarding mechanisms of immune response to a single, precisely defined hapten.

Amino Acid Sequence Determination of Fv Regions

The variable domains of each polypeptide chain in an immunoglobulin associate noncovalently to form the antigen binding site. This portion of an immunoglobulin is known as the Fv region, for "fragment variable," and maintains its antigen binding properties even when cleaved from the remainder of the molecule (33). McPC603 is a phosphorylcholine-binding immunoglobulin which has been studied crystallographically (34). The antigen binding site of McPC603 has been determined to consist of a pocket located near the top of the Fv and is formed by six adjacent polypeptide segments or "loops," three contributed by the light chain (L1, L2, and L3) and three by the heavy chain (H1, H2, and H3). The amino acid sequence of these loop segments varies extensively from one antibody to another, and they are therefore named "hypervariable" (HV) or "complementarity determining regions" (CDRs). The amino acid sequence of the remainder of the Fv is relatively conserved to ensure proper folding and association of the light and heavy domains. This portion is termed the "framework" region (FR). An unfolded, linearized chain would therefore be composed of three HV segments separated by four intervening FR segments, the entire segment being referred to as the variable region. This is connected at the C-terminus to the constant-region sequence.

The variable region of a light chain consists of 107 amino acids, whereas that of a heavy chain is composed of 113. Determining the amino acid sequence of these regions by enzymatic cleavage and Edman degradation is therefore a formidable task. Repeating the process on several antibodies is further complicated by the fact that each antibody has its own preferred set of cleavage conditions and patterns, due to its unique sequence. This greatly discourages the use of amino acid sequencing methodology when conducting comparative studies on several antibodies at once.

A more efficient and cost-effective approach is to sequence the mRNA coding for the immunoglobulin of interest. An immunoglobulin mRNA is schematically illustrated in Fig. 9.9 (36). The amino acid sequence encoded by the mRNA is identical to that of the final immunoglobulin chain, with the exception of a leader sequence which is post-translationally cleaved. The mRNA also has a polyadenylate tail characteristic of all mRNAs. The traditional method of obtaining the sequence of such a species would be to first isolate the specific mRNA from a cell lysate and produce a cDNA copy of the mRNA using the enzyme reverse transcriptase after first priming at the poly(A) tail with an oligo(dT). A double-stranded cDNA would then be produced and inserted into a plasmid such as M13 for cloning and sequencing (37). However, this approach requires the difficult task of isolation of a specific mRNA from the many thousands which exist in an antibody-secreting cell. Fortunately, one can take advantage of the fact that all immunoglobulins of a given class possess identical constant-region sequences (35). This allows one to design an oligomer which will hybridize specifically to any immunoglobulin mRNA, with little or no cross-reactivity to unrelated mRNAs. Since a monoclonal hybridoma cell by definition manufactures and secretes a single immunoglobulin species, such an oligomer will bind only to the desired species. Once hybridized, reverse transcriptase is then used to make a radiolabeled cDNA copy of

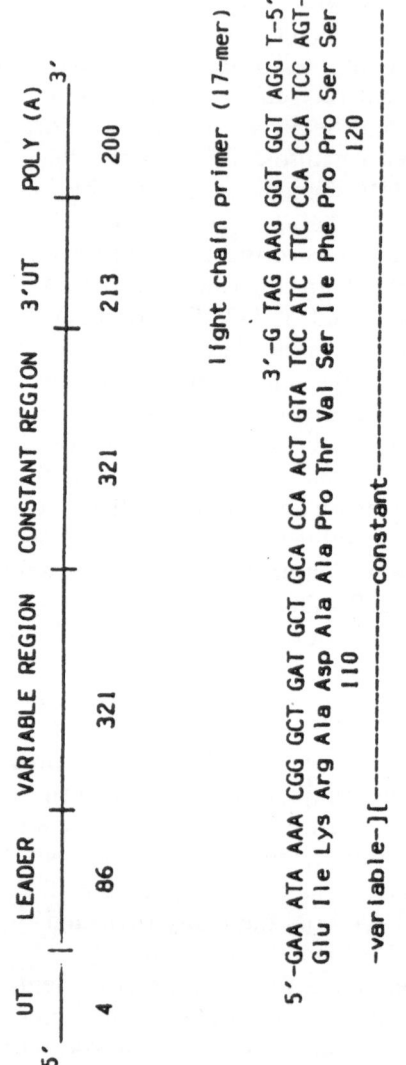

FIGURE 9.9. Schematic diagram of IgG mRNA. Sites of oligo binding to mRNA.

the message, which can be sequenced directly using Maxam and Gilbert cleavage reactions (38) or inserted into M13 for sequencing using the Sanger dideoxy approach (39).

Hamlyn and co-workers have discovered an even shorter method of sequencing immunoglobulin mRNA (40). Rather than isolating full-length cDNA for cloning and sequencing, dideoxynucleotides are added directly to the primer extension reaction such that a nested set of labeled fragments are produced with an identical 5' end as the mRNA is being copied. By separating these fragments on a polyacrylamide sequencing gel, the nucleotide sequence, and thus the amino acid sequence, can be obtained. Hamlyn designed two oligonucleotide primers, a 17-mer and a 15-mer, which will specifically bind to immunoglobulin G (IgG) light and heavy chains, respectively. These primers hybridize to a region near the variable–constant region junction of immunoglobulin mRNA. In the presence of reverse transcriptase, the 3' end of the primer is extended, producing a copy of a small portion of the constant-region sequence and the entire variable-region sequence. The nucleotide sequence of these oligonucleotide primers and the exact site of hybridization to IgG mRNA is illustrated in Fig. 9.9.

The Hamlyn method was therefore selected as the method of choice for rapidly generating amino acid sequences for the variable regions of all five anti-haloperidol mAbs. Hybridoma cells were grown in culture, harvested, and lysed, and total cytoplasmic RNA was isolated by centrifugation through cesium chloride. The mRNA fraction was obtained by selecting those species bearing a poly(A) tail. This was accomplished by affinity chromatography using cellulose oligo(dT). Total mRNA was then incubated with one of the IgG specific primers to select the immunoglobulin mRNA. Once hybridized, the mixture was divided among four tubes, each containing a different dideoxynucleotide and a mixture of all four deoxynucleotides (one being labeled with ^{32}P). Reverse transcriptase was then added, and an assortment of cDNA fragments was produced in each tube. Such fragments are randomly terminated at a specified base due to the presence of the nonextendable dideoxy analogue at the 3' end of the growing chain. Aliquots from each tube were then loaded onto four adjacent lanes of an ultrathin (0.3 mm) polyacrylamide gel, and the radiolabeled fragments were separated by size. Visualization of the bands via autoradiography produced a sequence "ladder" which permitted determination of the nucleotide sequence of the immunoglobulin variable region. A detailed account of these procedures is beyond the scope of this chapter and will appear elsewhere (41).

Sequencing Results

A representative sequencing gel is illustrated in Fig. 9.10. This gel records the nucleotide sequence coding for amino acids 60 to 115 of mAb A and mAb B heavy chains, and was obtained by electrophoresing [^{32}P]dTTP-

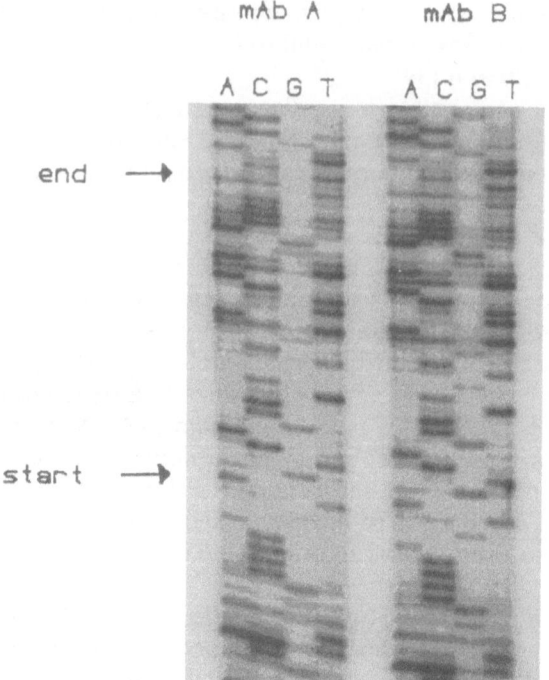

FIGURE 9.10. Representative sequencing Gel.

labeled transcripts on a 6% gel for 1.5 h. The sequence is that of the cDNA copy; the actual mRNA sequence is obtained by deducing the complementary strand. A total of 53 gels were utilized to deduce amino acid sequences for the light and heavy chains of mAbs A through E. Each chain was sequenced a minimum of two times using a different radiolabeled nucleotide so that the true identity of nucleotides obscured by a band running across all four lanes could be ascertained. These artifacts sometimes arise due to the fact that the deoxynucleotide used as the radiolabel is present in very low concentration compared to the other three (2 μM versus 50 μM) (42). This causes reverse transcriptase to pause or disassociate from the template when this base is required for chain elongation. To alleviate the problem, a high concentration of all four nucleotides was added as a "chase" to extend all chains which have terminated for reasons other than incorporation of a dideoxynucleotide. This eliminated the majority of sequence ambiguities. However, some ambiguities were persistent and remained despite repeated sequencing attempts. These premature terminations were most likely due to RNA secondary structure which prevents reverse transcriptase from reading through the sequence (43). Occasionally, running the sequencing reaction at a higher temperature (50°C) alleviated the problem. These sequence ambiguities constitute the major flaw of the Hamlyn approach to immunoglobulin sequencing as compared to methods such as M13 cloning which permit confirmation via sequencing of the complementary strand. However, the Hamlyn method saves a considerable amount of time and effort by avoiding cloning altogether.

This method frequently permitted the determination of the entire variable region, which requires reading 350 bases. This is very close to the practical limit for the number of bases which can be read accurately from priming at a single site (43). Fortunately, the last 75 to 100 bases of each chain code for highly conserved framework regions whose sequences are largely known, making it possible to assign the correct base to bands in this region whose identity would otherwise remain doubtful. In general, the ability to read to the end of the variable region was controlled by factors such as purity of the RNA preparation (presence of DNA or protein), use of fresh radiolabeled nucleotide stocks (<2 weeks old), and the stability of the gel under the high-temperature conditions of electrophoresis. All of these factors can help to eliminate a high background level of radioactivity, which tends to obscure faint bands in the upper regions of the sequencing ladder.

Another problem, unique to light chains, often made correct assignment of a given nucleotide difficult. This problem consisted of two bands appearing at a single position in the sequencing ladder, and arises due to the presence of one or more nonfunctionally rearranged light-chain gene segments which are contributed by the myeloma fusion partner (44). Fortunately, however, this sort of contamination was present in less than 20%, resulting in the darker of the two bands being the correct band. Other investigators have chosen to circumvent this problem by preparing full-length cDNA copies of both RNAs, excising the resulting bands from the gel, and sequencing each by Maxam and Gilbert cleavage reactions (45). This approach will probably replace the Hamlyn technique, at least in the sequencing of light chains. Methods capable of selectively isolating cytoplasmic RNA, as opposed to cytoplasmic plus nuclear RNA, are now available and significantly reduce the signal due to aberrant transcripts (42).

A complete listing of the nucleotide and deduced amino acid sequences of all five mAb variable regions is presented elsewhere, along with a discussion of comparisons to previously published germline sequences and a brief description of mechanisms by which sequence diversity is maintained in the immunoglobulin family. The sequences of mAbs A through E are then used to specifically illustrate these mechanisms (41).

Computer Modeling of the Hypervariable Region

Our current understanding of the structural basis for immunoglobulin function (antigen binding ability) is largely derived from X-ray crystallographic data. Unfortunately, however, the data base is small, consisting of three Fab structures (32,34,35) and three Bence-Jones protein structures (46–48) (light-chain dimers). Although other immunoglobulins have been crystallized (49), few of the resultant crystals diffract sufficiently well to permit resolution of individual amino acid side chains (less than 3.0-Å resolution needed). The structures of two new Fab fragments have been solved recently (50,51). However, this only emphasizes that the number of crystal structures solved each year is likely to remain rather constant, given current technology.

In order to circumvent difficulties in crystallization of immunoglobulins, investigators have increasingly begun to employ computer-assisted model building approaches to examine their primary sequence data. This is possible due to the high degree of sequence and structural homology that exists among the framework regions of the immunoglobulins studied by X-ray crystallography thus far. This high degree of homology has been the subject of several recent papers (52,53) and can be summarized as follows:

1. The sequence homology shared by the frameworks of the six published light-chain structures ranges from 65% to 84%. The comparison includes both kappa light chains and lambda light chains, both of which share even greater intrasubtype homology.
2. The heavy-chain framework regions of the three studied heavy chains share approximately 60% sequence identity, despite their differing origins (mouse versus human) and their differing subclasses (λ versus α).
3. When framework regions from all studied structures are superimposed, the percentage of alpha carbon atoms which can be superimposed within 1.5 Å ranges from 50% to 75%, with a root-mean-square error of from 0.61 to 0.85 Å (52). This extraordinary complementarity, particularly among the framework amino acids, is not surprising in light of the fact that without this common foundation, the relative positioning of the hypervariable loops would be lost, along with antigen binding ability.
4. If one compares the above sequences after taking into account similarities in physicochemical properties shared by many amino acids residues (Leu is often substituted for Met based on similar size and hydrophobicity), one obtains framework sequence variability values equivalent to those obtained for the highly conserved constant regions (54). This demonstrates that in many cases a specific class of amino acid residue at each position (hydrophobic, hydrophilic, charged, bulky, flexible, etc.) is responsible for maintenance of the framework structure rather than a distinct amino acid. As long as this pattern is maintained, the three-dimensional structures are likely to be superimposable.

The secondary structure of the hypervariable loops can vary extensively from one immunoglobulin to another due to variability in length. However, in cases where the lengths of the compared loops are identical, the structural homology is often quite high (52). This arises from the fact that despite

the high degree of variability in sequence observed for a selected hypervariable loop, some residues are less variable than others and serve in a
structural capacity to ensure proper packing interactions with adjacent
hypervariable loops (55). Hence, it has become possible to predict hypervariable loop secondary structure with a reasonably high degree of
accuracy. Although several binding site modeling attempts have been
published over the past 10 years, until recently none had been verified
via X-ray crystallography. This, however, changed recently with the publication of X-ray data for the anti-lysozyme antibody D1.3 (51), whose
structure had been previously predicted using model building techniques
(56). The experimentally determined structure matches the modeled
structure quite well. Superpositioning of predicted and observed framework regions provides a root-mean-square deviation of 1.0 Å, whereas
this value ranges from 0.63 to 0.97 Å for hypervariable loops excluding
H1 (2.07 Å). This study strongly implies that the modeling of immunoglobulin sequences for purposes of examining gross binding site anatomy
is indeed a viable substitute for crystallographic structure determination.

Selection of a Crystal Structure to Serve as a Model

The X-ray coordinates of the McPC603 Fv region were selected as the
model for the current study. McPC603 is an IgA, kappa-chain secretor
which binds phosphorylcholine, and represents the only Fab in the data
bank of mouse origin (34). The remaining two Fab structures, NEW (32)
and KOL (35), are of human origin and have lambda light chains, chains
which differ significantly from kappa chains in the conformation of their
hypervariable loops. In addition, the hypervariable loops of McPC603 are
either longer or equal in length to those of mAbs A through E (except for
H1 or mAb D which possesses a single insertion) and can therefore be
shortened if necessary. The homology between total and framework-only
sequences of mAbs A through E and McPC603 can be found in Table 9.2.
All five mAbs possess frameworks identical in length to those of McPC603.
The McPC603 framework is 60–63% homologous with that of mAbs A
through E. Proteins which are greater than 50% homologous are considered
sufficiently similar to serve as models for one another (57). This is es-

TABLE 9.2. Homology of anti-haloperidol mAb amino acid
sequences with McPC603.

mAb	Light chain	Heavy chain	Total	Framework only
A	68	44	55	62
C	68	41	54	63
D	62	44	53	60
E	62	41	52	62

pecially true if large, highly homologous segments coexist with less homologous segments, as is the case with framework and hypervariable segments of immunoglobulins. If residues are first classified by subtype (hydrophobic, aromatic, negatively charged, etc.), the framework homology value rises considerably to 85%. In addition to primary sequence homology, it was established that certain key residues which participate in the noncovalent interaction between heavy and light chains at their interface are conserved in mAbs A through E (58).

Selection of Model Hypervariable Loops from Existing Structures

The length of hypervariable segments L2, L3, and H1 are the same for mAbs A through E and McPC603 (excluding the H1 of mAb D which bears an insertion). In evaluating the suitability of the McPC603 loops to serve as a skeleton for the loops to be modeled, it was therefore only necessary to compare sequences to show that residues which serve in a structural capacity are conserved (55) and that the phi-psi angles of the residues in the McPC603 loops are compatible with the amino acid to be substituted. Compatibility refers to the fact that some amino acids are unique with respect to the allowed values of their torsion angles, phi (N—Cα) and psi (Cα—C). Certain combinations of angles are forbidden due to steric interactions between adjacent peptide bond atoms or with side chain atoms. Allowed angles can be predicted from a Ramachandran diagram which plots phi angle versus psi angle (59). This type of analysis has confirmed that the angles present in McPC603 loops L2, L3, and H1 are compatible with sequences for the equivalent loops in mAbs A through E. Models for remaining loops, whose lengths differ from those in McPC603, were selected from other immunoglobulin crystal structures. In situations where the length of the loop was not matched by a loop of known crystal structure, carefully selected residues were deleted from the McPC603 loop and the free ends joined using interactive computer graphics. Modeling of each individual loop is discussed elsewhere (41). The restrictions inherent in this type of backbone manipulation are rather stringent, as discussed below.

Restrictions on Making Amino Acid Deletions

The shortening of existing McPC603 hypervariable loops to produce loops with no crystallographic equivalent was necessary in the modeling of each anti-haloperidol mAb. In order to generate a shortened loop, three general energy-related restrictions must be satisfied. First, the deleted residues must not be structural in nature, as demonstrated by a low sequence variability and by their location in crystal structure stereographs (55). Deletion or manipulation of these amino acid side chains would result in gross structural alterations in the loop being modeled, as well as alter its packing interactions with adjacent loops. Once the appropriate amino acid residues have been deleted, the second restriction is that the peptide bond formed by joining the free ends of the two peptide chains must be planar (omega = 180 ± 10°), with a C—N distance of 1.33 ± 0.2 Å and an alpha carbon–alpha carbon distance of 3.8 ± 0.1 Å, in accordance with standard peptide geometry (60). All phi-psi angles in the final loop are then measured to ensure that none are forbidden according to Ramachandran plots. Finally, the amino acid sequence of the segment formed by joining the free ends of the peptide chains must be consistent with the formation of a tight β-turn, if such a turn is appropriate (61).

Not all amino acids are favored in tight β-turns. For instance, Ile and Val are rarely seen in turns due to their highly restricted phi-psi angles. On the other hand, Pro, Gly, Ser, Asp, and Asn are highly favored in β-turns due to their relative flexibility. Chou and Fasman examined 459 turns in the crystal structures of 29 proteins and determined the frequency with which each amino acid appears at each of the four β-turn positions (61). They discovered that certain residues are highly favored at each of the four positions and that these frequency hierarchies can be used to predict whether or not a given tetrapeptide sequence is compatible with the formation of a tight β-turn. In selecting residues to delete in the formation of the shorter hypervariable loops, all sequential tetrapeptides which could result from a given deletion were scored for turn potential. Scores equal to or greater than twice the average turn frequency were then considered as significant (62). The tetrapeptide with the highest turn potential was then considered as a potential site for chain reversal. Since this type of analysis is only accurate roughly 70% of the time (62), this score was most often used merely to confirm a turn that was deemed necessary based on local geometry and sequence homology considerations.

Computer Software Used in Modeling

The X-ray crystallographic coordinates for McPC603, determined by D. Davies' group and kindly provided to us by Dr. Arieh Warshel, are the most highly refined set of coordinates available to date. All others coordinates were obtained from the Brookhaven Protein Data Bank (Cambridge, England). Graphics were performed on an IBM PC/XT obtained through a grant from IBM (Project Socrates of the University of Southern California), using a molecular graphics program entitled MOLDISP written by Dr. Michael B. Bolger. The program features 500-atom capacity, color coding of residues, wire-frame, ball and stick, or space filling representation, stereo vision, animation, x-y-z axis rotation and movement, bond twisting, distance calculation, and dihedral angle measurement capabilities. The amino acid numbering system of Kabat et al. (63) is used throughout the remainder of this chapter.

Orientation of Haloperidol in the Antibody Binding Site

Inspection of antibody binding sites, whether derived from crystallographic data or from modeling studies, indicates that most of the hypervariable loop amino acid side chains within the central core of the β-barrel structure point upward toward the mouth of the central cavity and are closely packed together, restricting mobility. This is also true of the binding site models proposed for mAbs A through E. Haloperidol, on the other hand, possesses a great deal of flexibility as a result of the propyl chain connecting the fluorophenone moiety and the piperidine ring. Thus, it is expected, within certain energy-defined limits, that the haloperidol molecule will be able to conform to the contours of the binding cavity. Structurally rigid antigens would be expected to have a much more limited set of complementary binding sites within the immune repertoire compared to molecules for which many conformations are possible. Indeed, a recent theory regarding the nature of antigenic regions on the surface of proteins holds that antigenicity and mobility are highly correlated (64). Antigen flexibility was therefore assumed in positioning haloperidol within the binding sites of mAbs A through E.

The piperidinyl nitrogen of haloperidol is a prominent structural feature, and is 99% protonated at physiological pH (haloperidol pK_a 8.6). It is therefore expected that this positively charged moiety will interact with a negatively charged residue within the binding site. Ionic interactions of this sort have been estimated to account for up to -11.5 kcal/mol of the binding energy involved in certain drug-receptor interactions (65). All five anti-haloperidol mAbs possess H3 loops rich in negatively charged glutamate and aspartate residues. In fact, all five mAbs possess one of these two residues at heavy-chain position 95, a side chain which was consistently found in the interior of the antibody binding cavity regardless of the overall length of H3.

The second prominent feature of haloperidol is the presence of aromatic rings. Fluorescence quenching data would suggest that one of these rings is intimately involved in a stacking interaction with a tryptophan residue within the binding site for mAbs A and C (41). The fluorophenone ring has already been implicated as being the ring involved in the stacking interaction based on its relative accessibility, as demonstrated by the binding data, and based on its electron-poor nature, which makes it ideally suited for interaction with the electron-rich tryptophan ring system. In the case of mAb D, the requirement for low electron density on the chlorophenyl ring, coupled with the observation that fluorescence quenching does not occur upon binding, suggests that this ring is involved in an interaction with an electron-donating system, possibly a tyrosine or phenylalanine ring. Recognition of haloperidol by mAb E may involve a similar interaction, but in a less specific fashion.

Examination of the sequence data for mAb A has indicated a single hypervariable loop Trp residue, located at position 50 of the heavy chain in loop H2. Inspection of the computer model clearly demonstrates that this residue is located within the central binding pocket and that it is relatively exposed to the aqueous environment. π-π Aromatic ring stacking requires an inter-ring distance of 3.3 Å (66). After stacking the fluorophenone ring of haloperidol the required 3.3 Å from the surface of this residue, it is readily apparent that heavy-chain residues Glu 95 (Glu H-95) and Asp 100A (Asp H-100A) are within hydrogen bonding distance of the protonated piperidine nitrogen atom (60). This positioning requires a folding of the butyrophenone chain such that it resembles half of a cyclohexyl ring. This conformation therefore accounts for the relatively good binding affinity of the semirigid isomer R 49 399, the "folded" analogue of haloperidol which actually has a cyclohexyl ring in place of the butyrophenone chain.

The exact orientation of the chlorophenyl ring within the binding site of mAb A cannot be ascertained due to sequence ambiguities. In the stereo model for mAb A pictured in Fig. 9.11, the chlorine atom is tentatively placed in a pocket formed by parts of L1, L3, and H3. Prior to sequence determination, it was expected that the amino acid which interacts with this part of haloperidol would have an electron cloud which is highly polarizable, such as Tyr or Phe, in order to form dipole-induced dipole van

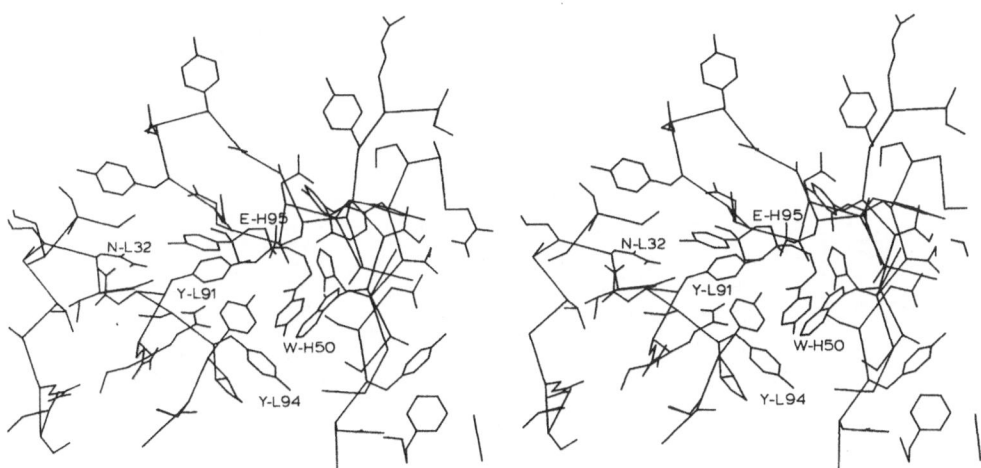

FIGURE 9.11. Stereo plot of mAb A/B.

der Waals contacts with the polarizable chlorophenyl ring. Residues in the vicinity of the chlorine atom include Tyr H-100, Asn L-32, and Tyr L-91. Thus, it is most likely that Tyr L-91 is serving as an electron acceptor, since tyrosine residues are well known for their bifunctional ring systems which are capable of donating and accepting electrons (67). Fortunately, the exact placement of the chlorine atom is irrelevant with respect to evaluating these two binding sites for dopamine receptor similarity: the relative orientation of the fluorophenone ring system and the piperidine ring alone are incompatible with orientations believed to be essential for dopamine receptor similarity, as discussed in the next section.

mAb C is a rather low affinity antibody (K_d for haloperidol = 398 nM) with an odd preference for both spiperone and for a hydroxylated metabolite of haloperidol. Binding data also indicated a preference for an extended haloperidol conformer. Architecturally speaking, mAb C differs significantly from A in that three residues are excised from the tip of the H3 loop. However, it resembles A in that tryptophanyl fluorescence quenching occurs upon haloperidol binding. As in A, a single Trp residue is found in an exposed position within the combining site of C, but at position 33 of the heavy chain. The residue at position H-50, the site of Trp in A, has mutated to a His in mAb C. However, examination of the stereo model indicates that this Trp is spatially adjacent to position H-50. Haloperidol's fluorophenone π-system was therefore positioned in a stacking interaction with this residue as with mAb A. Although Glu H-95 is conserved in mAb C, this cannot be the residue involved in the ionic interaction with haloperidol's protonated tertiary nitrogen due to its close proximity to the Trp residue (4 Å). Rather, Asp H96 is better positioned for such an interaction (8 Å). This residue lies outside the actual central cavity, but is accessible to haloperidol as a result of the shortened H3 loop. Haloperidol can be seen oriented in the binding site of mAb C in Fig. 9.12.

Placement of haloperidol in this orientation is consistent with other fine

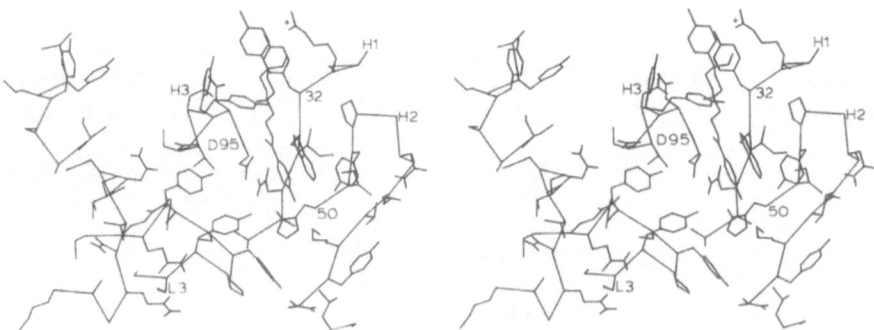

FIGURE 9.12. Stereo plot of mAb C.

specificity observations for mAb C. For example, this mAb has a twofold higher affinity for a haloperidol analogue which possesses an extra methylene group in the inter-ring chain (R8573). Examination of the model indicates that this extra chain length brings the piperidine nitrogen much closer to Asp H-96 (2.8 Å). The mAb's preference for spiperone is explained by the existence of a Tyr residue at position H-32 which is ideally oriented for a perpendicular ring stacking interaction with spiperone's phenyl ring. Unlike Trp ring stacking, tyrosines and phenylalanines are frequently observed to stack perpendicularly in protein structures, rather than in the parallel fashion (67). The overall low affinity of mAb C for haloperidol is explained by the repulsive interaction that occurs between Glu H-95 and haloperidol's electronegative keto group. This type of observation can be directly tested by performing site-directed mutagensis experiments (68) in which the Glu is replaced by a smaller, hydrogen-donating residue such as Ser. Also contributing to the overall low affinity of this mAb for haloperidol is the fact that the binding site is in the form of a shallow trough or groove with few van der Waals interactions possible.

mAb D has been characterized as binding an extended conformer of haloperidol with moderate affinity (K_d = 6.1 nM) through recognition of the chlorophenyl and piperidine ring systems. Narrow substituents in the keto position and long inter-ring chains are also favored. Lack of fluorescence quenching upon complex formation coincides well with the lack of an exposed Trp residue within the central cavity. The single most important determinant of high-affinity binding appears to be low electron density on the chlorophenyl ring, suggesting an interaction with an electronegative moiety. Since the stacking of aromatic rings is an energetically favorable interaction (67), the binding cleft of mAb D was examined for exposed Tyr or Phe residues adjacent to a negatively charged group for interaction with the piperidine nitrogen. Tyr H-50 satisfies these requirements and allows placement of the chlorine atom in a relatively hydrophobic pocket formed by residues Val L-94, Tyr L-96, and Trp H-47, a framework residue. Asp H-95 provides the required negatively charged group. The shortened H3 loop allows an extended conformer of haloperidol to occupy this orientation. Since this mAb was selected in response to the oxime conjugate (attachment through haloperidol's keto group with a short linker arm), it is logical that the bulky protein carrier approaches the binding groove from the least sterically hindered side (H3).

The preference for analogues with narrow substituents in the keto position is explained by the steric interaction with residues contributed by H3 and H1 which serve to form a channel at the interface. Formation of this channel is a direct consequence of inserting a residue into H1 and thus provides a rationale for selection of this unusual mAb. Existence of this channel also explains the poor binding of analogue R8479, haloperidol with a shortened inter-ring alkyl chain. This would place the fluorophenyl ring directly in this narrow channel, thus imitating the other poor-affinity analogues with wide tetrahedral groups in the keto position [aromatic rings are as wide as they are broad due to the π electron cloud (67)]. A stereo view of haloperidol in the binding groove of mAb D can be found in Fig. 9.13. Although the neurotransmitters dopamine, serotonin, and epinephrine do bind to mAb D, the very low affinity (mM range) is most likely a result of placing polar phenolic groups in the hydrophobic pocket which accepts the chlorine atom. Site-directed mutagenesis of one of the residues in the vicinity might produce an antibody with greater affinity for dopamine. Since it is still unclear whether dopamine's hydroxyl group serves as a hydrogen bond donor or acceptor at the receptor site, replacement of residue H-58 with a small proton acceptor (Asp) or a small proton donor (His) would be logical first choices.

As previously described, mAb E was characterized as being similar to mAb D, but rather insensitive to the electron density of the chlorophenyl ring, inter-ring chain length, or keto group width. The poorest-affinity analogue is haloperidol in which the chlorine atom has been replaced by a methoxy group. Substitution by the isosteric bromine and isopropyl moieties provides compounds with affinities equal to that of haloperidol. Clearly, a hydrophobic pocket accompanied by an adjacent negatively charged contact residue for the piperidine nitrogen is most likely involved. These requirements are satisfied by placing haloperidol within the binding groove in essentially the same orientation as in mAb D. Once again, Glu H-95, located at the mouth of the open-ended groove, provides the negative charge. The hydrophobic pocket is formed at the H2/L3 interface by His H-35, Tyr H-33, Val L-94, Tyr L-96, and the alkyl chain portions of Arg H-50 and Lys H-58. The lack of the additional inserted residue in H1 and the removal of a fourth residue from H3 effectively destroy the narrow channel which provided contact residues for the keto function in mAb D. A stereo view of haloperidol occupying the binding groove of mAb E can be found in Fig. 9.14.

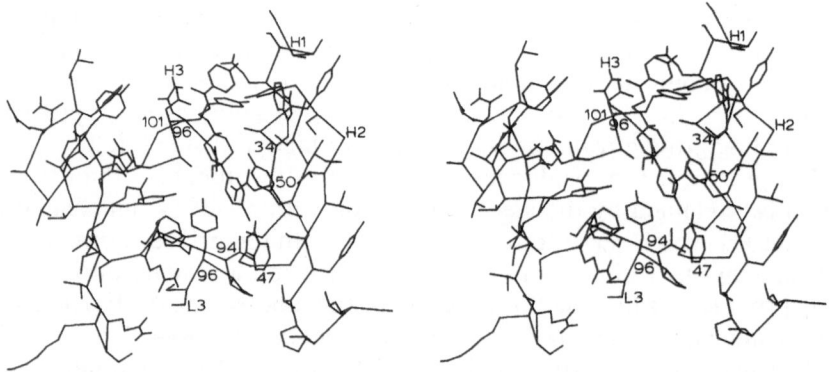

FIGURE 9.13. Stereo plot of mAb D.

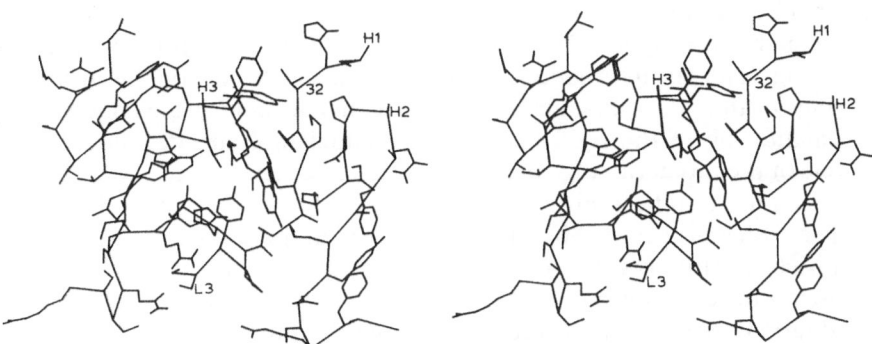

FIGURE 9.14. Stereo plot of mAb E.

Similarities of Anti-Haloperidol Binding Sites to Proposed Dopamine Receptor Maps

Mapping of the dopamine receptor (69–71) has been accomplished largely by determining the pharmacological potency of structurally rigid neuroleptics and their analogues. All such maps have several basic features as illustrated by the diagram in Fig. 9.15:

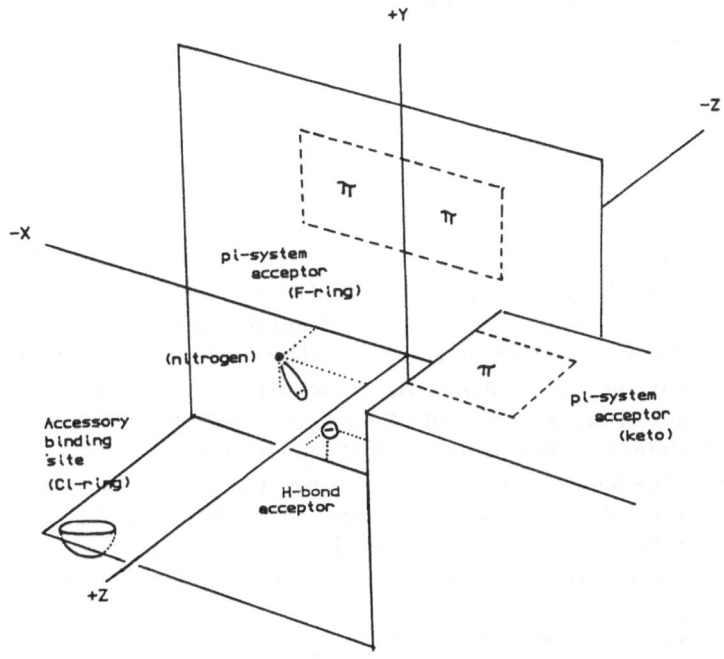

FIGURE 9.15. Model of dopamine receptor.

1. There exists a region, equivalent in size to two adjacent aromatic rings, which binds drug moieties with extended π-systems, probably through a ring stacking interaction. This region accepts one of the aromatic rings of the phenothiazine neuroleptics such as chlorpromazine, an aromatic ring of the rigid neuroleptics butaclamol and dexclamol, and one of the aromatic rings of agonists such as apomorphine. A π-binding region has also been proposed to be involved in antagonist binding only. The moieties accepted by this region are a bit more flexible and include π-rich substituents such as the keto group of butyrophenones. Replacement of the keto group with a phenyl ring provides the active diphenylbutyl derivatives (72), thus confirming the theory.

2. All active neuroleptics possess a basic nitrogen atom located 5 to 7 Å from the aromatic rings, which most likely interacts with a negatively charged residue roughly 2.6 Å below the normal plane defined by the center of the aromatic binding region and the basic nitrogen position. The variability in nitrogen-ring distance is assumed to reflect a somewhat flexible receptor surface residue (69). This is supported by the observation that the directional vector defined by the nitrogen lone pair electrons is critical to activity and not the absolute nitrogen-ring distance (73).

3. There exists an "accessory binding site" not involved in pharmacophore recognition located 9.6 Å from the primary aromatic binding region. This region accommodates bulky hydrophobic groups such as the *tert*-butyl group of butaclamol and the chlorophenyl ring of haloperidol, and is believed to be utilized only for high-binding-affinity compounds such as spiperone (74).

It is interesting to consider how the semirigid isomers R 49 399 and R 48 455 used in our binding studies would conform to the receptor model. The fluorophenyl ring attached to the five-membered oxazalone ring occupies the accessory binding site. The nitrogen lone pair would point downward in the same general direction as the above-mentioned ring to the negatively charged receptor moiety. However, only the extended isomer R 48 455 is able to then place its second fluorophenyl ring and cyano group in the π-system binding regions. The location of these groups in the folded isomer is far below the general plane of the piperidine ring rather than slightly above it and thus is not favored by the receptor. A similar situation occurs in the binding site of mAb A. Trp H-50, which would correspond to one of the π-system acceptor regions, is located below the general plane of the piperidine ring. Also, the basic nitrogen is only 3.8 Å away from the general center of the fluorophenone π-system (carbon 1 of the fluorophenyl ring) and therefore is not complementary to the proposed receptor-binding conformation.

The conformation of haloperidol adopted for binding to mAb C is actually quite close to that of the hypothesized receptor-active conformation for the closely related butyrophenone benperidol (75). Unfortunately, the relative positions occupied by the keto group and the fluorophenyl ring are transposed compared to the active benperidol structure. Thus the antibody binding site differs from the receptor binding site in that π-binding sites lie in the $+x$ direction rather than the $-x$ direction. mAb C also lacks a well-defined pocket equivalent to the receptor accessory binding site, a feature believed to contribute significantly to receptor affinity.

The binding sites of mAbs D and E would be equivalent to a dopamine receptor containing the accessory binding site and the negatively charged nitrogen binding site, but no π-system site. The π-system site is believed to be the region which binds dopamine's ring, based on the conclusion that the pharmacophoric group in apomorphine, a dopamine agonist, is also best accommodated in this site (76). However, this conclusion may be erroneous based on the observation that mAbs D and E bind dopamine. Tollenaere et al. have also postulated that the chlorophenyl ring is the true pharmacophore (73). It must be kept in mind, however, that the affinities of mAbs D and E for dopamine are quite low and may simply represent a rather nonspecific binding of small molecules with hydrophobic groups adjacent to protonated nitrogens.

Thus, the binding site of mAb C appears to resemble the actual dopamine receptor binding site more closely than does that of any of the other mAbs studied here. However, even though the distances between the three essential receptorlike regions appear to be correct, the relative orientation of these regions may not be. Also, there remains the problem of the repulsive interaction occurring between haloperidol's keto group and residue Glu H-95, which may be contributing to this antibody's low affinity for haloperidol. An internal image anti-idiotypic antibody to this region would most likely contain a positively charged residue in order to increase binding affinity with the idiotype. This would most likely not be well accommodated by the dopamine receptor.

Several questions have been posed here regarding the properties of a ligand which are necessary for generation of receptorlike antibody binding sites. With regard to the haloperidol/anti-haloperidol system, the data presented suggest that haloperidol itself is not an appropriate ligand for generation of receptorlike Ab_1 idiotypic antibodies. Based on the proposed overlap of the butyrophenone neuroleptics with the dopamine receptor map, the SUCC-CONJ appears to preserve features most critical to high-affinity receptor binding in that the fluorophenone and piperidinyl nitrogen remain exposed. The physicochemical forces responsible for high-affinity binding to the mAb, namely, ring stacking and ion-ion interactions, appear to mimic those employed by the dopamine receptor. It is important to note that all the mAbs binding the SUCC-CONJ (A and C) probably accommodate the fluorophenone moiety through a ring stacking interaction with a tryptophan residue. The receptor map for this region is characterized as a region the size of two adjacent aromatic rings with high affinity for π-systems (77). The antibody binding site models therefore suggest that this region in the receptor may also consist of a tryptophan ring. Likewise, the nitrogen binding region of the dopamine receptor is probably a glutamate or aspartate residue, as found in the antibody binding site.

Unfortunately, however, the high degree of flexibility present in the inter-ring alkyl chain of haloperidol results in the selection of antibody binding sites whose components are spatially unrelated to those of the actual receptor. The computer-generated models clearly indicate that the π-stacking region is incorrectly oriented with respect to the nitrogen binding region in all the SUCC-CONJ mAbs. For mAb A, this is a result of binding a folded haloperidol conformer. mAb C binds the required conformer, but the plane of the π-binding system (Trp) is incorrect, again due to chain flexibility. A more appropriate ligand for generation of receptorlike antibodies would be the semirigid extended isomer R 48 455, possibly conjugated through its amide nitrogen to a protein carrier. This would stimulate the production of mAbs with the proper arrangement of the three basic receptor features. It is also likely that the amino acid residues involved in the binding interaction would be similar to those used in mAbs A and C, since these mAbs were able to bind at least one of the isomers with high affinity compared to mAbs D and E.

A second question was posed regarding the effects of conjugation on the ability of a drug to maintain its receptor binding activity. Binding studies have demonstrated that the SUCC-CONJ was only 10-fold less active than haloperidol in displacing spiperone from the receptor binding site, whereas the OX-CONJ was 40-fold less active. Thus, coupling haloperidol through the keto group is more detrimental to receptor affinity than is coupling through the tertiary alcohol. This is supported by the mAb binding site models, which show that coupling through the alcohol gives rise to binding sites which maintain the important π-system binding region feature of the receptor, whereas the accessory binding site is mimicked in response to the OX-CONJ.

The Nature of the Ab_1 Antigenic Determinants (Idiotopes) Present in mAbs A through E

Several general questions have surfaced here regarding the exact nature of the Ab_1 idiotype/Ab_2 anti-idiotype interaction. Since no one has succeeded in obtaining X-ray diffraction data on an Ab_1/Ab_2 pair bound together, the existence of Ab_2 antibodies with fingerlike projections capable of inserting into an Ab_1 binding pocket is still unproven. This modeling study has indicated that mAb A possesses such a pocket-type binding site. In order to produce $Ab_2\beta$ mAbs (internal image anti-idiotypes), it would be necessary for the Ab_2 to have a long HV loop capable of extending down into the pocket. Such a loop may resemble the L1 loop found in mAbs D and E. This loop has been exposed by deletion of several residues from H3. Thus it appears that fingerlike projections are possible. Surprisingly, however, an Ab_2 antibody which mimics reovirus surface proteins was recently sequenced and shown not to contain fingerlike projections at all (78). Rather, the gross architecture of the HV region is identical to that found in mAb A, namely, a rather flat surface with a central pocket. Based on sequence homology with the original reovirus protein, the authors were able to localize the Ab_1 contacting residues as being in loop L2 of the anti-idiotype. This is curious since crystal data has demonstrated that L2 in often excluded from the central binding cavity by the L1 or H3 loop (34).

Our experiments to produce an anti-idiotypic antibody bearing a fingerlike projection complementary to the pocket of mAbs A and B have been unsuccessful. Several Ab_2s to mAb B have been made and have been found to cross-react with mAb A, which is not surprising in light of their matching sequences. However, these do not appear to bind to the dopamine receptor. Further characterization has indicated that these Ab_2s are of the $Ab_2\lambda$ type (79)—an antibody which binds determinants near the combining site but not actually within it. This assignment is based on the observation that the Ab_2, when bound to Ab_1, is capable of preventing haloperidol from binding to the Ab_1 but that haloperidol itself does not inhibit the binding of Ab_2 to Ab_1. Therefore, in this situation, the isolated Ab_2s were not of the internal image type.

A recent study has indicated that mouse autoantibodies use a limited set of Vh gene segments which are infrequently used in normal antibody responses (80). Thus autoimmune diseases might arise from defects in the control of Vh gene expression. The possibility of there being an autoimmune component to neurogenic diseases is still a fascinating and largely unexplored field. Elucidation of the conditions which allow production of internal image $Ab_2\beta$ antibodies may suggest possible mechanisms by which autoimmune diseases are triggered.

Conclusions

Based on the results of this study, it appears that the successful production of receptorlike antibodies is governed by four basic principles:

1. The hapten must be somewhat rigid to prevent the formation of antibodies which bind an inactive drug conformer.
2. The drug must be attached to the carrier protein in such a fashion that moieties believed to be directly involved in receptor binding are fully exposed and available for antibody recognition.
3. High-affinity antibodies do not necessarily possess the most receptorlike binding sites.

128   Michael B. Bolger, Mark A. Sherman, and D. Scott Linthicum

4. There appears to be a relatively limited set of amino acid residues used to bind a particular hapten moiety.

This study has demonstrated that in the absence of X-ray diffraction data, a combination of radioligand binding assays, quantitative structure-activity relationship analysis, sequencing technology, and computer graphics modeling can provide detailed, informative models of the molecular interactions which occur between drugs and protein receptors. In this study, the depth to which the chemical interactions responsible for neuroleptic-protein binding could be studied is viewed as being a direct consequence of advances in hybridoma technology. This technology has permitted the production of a single, homogeneous protein of predetermined binding specificity. This protein can be readily isolated, purified, and sequenced either by conventional methods or more rapidly via nucleotide sequencing. Linear sequences can then be transformed into informative three-dimensional structures using computer graphics as a modeling tool. This study has combined basic rules of protein anatomy to produce a hypervariable loop modeling protocol amenable to mechanization (41). The proposed antibody models are also directly testable. Our group is currently collaborating with crystallographers in an effort to cocrystallize mAb A with bound haloperidol. Verification of these models and others may eventually eliminate altogether the need to crystallize immunoglobulins when detailed information at the molecular level is desired.

Acknowledgments. We are grateful to the following for technical assistance: Sarah Combs, Doreen Yatko, John Lumb, Paul Kussie, Robert Deans, and JoAnne Goodnight. This work is supported by the following grants: BRSG S07 RR05792 (NIH), NS 22448 (NIH), 1761-A-4 from the National MS Society, and NS 00974 (NIH).

References

1. Nisonoff, A., and E. Lamoyi. 1981. Implications of the presence of an internal image of the antigen in anti-idiotypic antibodies: possible application to vaccine production. Clin. Immunol. Immunopathol. 21(3):397.
2. Gillet, J.G., S.V. Kaveri, O. Durieu, C. Delavier, J. Hoebeke, and A.D. Strosberg. 1985. β-Adrenergic agonist activity of a monoclonal anti-idiotypic antibody. Proc. Natl. Acad. Sci. USA 82:1781.
3. Chamat, S., J. Hoebeke, and A.D. Strosberg. 1984. Monoclonal antibodies specific for β-adrenergic ligands. J. Immunol. 133:1547.
4. Sawutz, D.G., D. Sylvestre, and C.J. Homcy. 1985. Characterization of monoclonal antibodies to the β-adrenergic antagonist alprenolol as models of the receptor binding site. J. Immunol. 135:2713.
5. Ng, D.S., and G.E. Isom. 1984. Binding of antimorphine anti-idiotypic antibodies to opiate receptors. Eur. J. Pharmacol. 102:187.

6. Glasel, J.A., and W.E. Myers. 1985. Rabbit anti-idiotypic antibodies raised against monoclonal anti-morphine IgG block Mu and Delta opiate receptor sites. Life Sci. **36**:2523.
7. Flurkey, K., M.B. Bolger, and D.S. Linthicum. 1985. Preparation and characterization of antisera and monoclonal antibodies to serotonergic and dopaminergic ligands. J. Neuroimmunol. **8**:115.
8. Diener, U., E. Knoll, and H. Wisser. 1981. Preparation of antibodies to catecholamines and metabolites—syntheses of various immunogens and characterization of the resulting antibodies. Clin. Chim. Acta. **109**:1.
9. Engback, F., and B. Voldby. 1982. Radioimmunoassay of serotonin (5-hydroxytryptamine) in cerebrospinal fluid, plasma, and serum. Clin. Chem. **28**:624.
10. Grota, L.J. and G.M. Brown. 1976. Antibodies to catecholamines. Endocrinology **90**:615.
11. Miwa, A.M. Yoshioka, A. Shirahata, and Z. Tamura. 1977. Preparation of specific antibodies to catecholamines and L-3,4-dihydroxyphenylalanine, Part I (Preparation of the conjugates). Chem. Pharm. Bull. **25**:1904.
12. Bolger, M.B., K. Flurkey, R.D. Simmons, D.S. Linthicum, P. Laduron, and M. Michiels. 1985. Preperation and characterization of antisera and monoclonal antibodies to haloperidol. Immunol. Invest. **14(6)**:523.
13. Sherman M.S., D.S. Linthicum, and M.B. Bolger. 1986. Haloperidol binding to monoclonal antibodies: conformational analysis and relationships to D-2 receptor binding. Mol. Pharmacol. **29**:589.
14. Creese, I., D.R. Burt, and S.H. Snyder. 1976. Dopamine receptor binding predicts clinical and pharmacological potencies of antischizophrenic drugs. Science **192**:481.
15. Poland, R.E., and R.T. Rubin. 1981. Radioimmunoassay of haloperidol in human serum: correlation of serum haloperidol with serum prolactin. Life Sci. **29**:1837.
16. Michiels, M., R. Hendriks, and J. Heykants. 1976. Antibodies to haloperidol: a very sensitive tool for the radioimmunologic determination of some butyrophenones. Preclinical research report (Janssen Pharmaceutica).
17. Olson, G.L., H.C. Cheung, K.D. Morgan, J.F. Blount, L. Todaro, and L. Berger. 1981. A dopamine receptor model and its application in the design of a new class of rigid pyrrolo[2,3-g]isoquinoline antipsychotics. J. Med. Chem. **24**:1026.
18. Seeman, P. 1981. Brain dopamine receptors. Pharm. Rev. **32**:230.
19. Tollenaere, J.P., H. Moereels, and M.H.J. Koch. 1977. On the conformation of neuroleptic drugs in the three aggregation states and their conformational resemblance to dopamine. Eur. J. Med. Chem. **12**:199.
20. Munson, P.J., and D. Rodbard. 1983. Number of receptor sites from Scatchard and Klotz graphs: a constructive critique. Science **220**:979.
21. Klotz, I.M. 1982. Numbers of receptor sites from Scatchard graphs: facts and fantasies. Science **217**:1247.
22. Klotz, I.M. 1983. Ligand-receptor interactions: what we can and cannot learn from binding measurements. Trends Pharmacol. Sci. **1983** (June):253.
23. Garfinkel, L., M.C. Kohn, and D. Garfinkel. 1977. Systems analysis in enzyme kinetics. Crit. Rev. Bioeng. **1977** (October):329.
24. Munson, P.J., and D. Rodbard. 1980. LIGAND: a versatile computerized approach for characterization of ligand-binding systems. Anal. Biochem. **107**:220.

25. Scatchard, G. The attraction of proteins for small molecules and ions. 1949. Ann. N.Y. Acad. Sci. **51**:660–672.
26. Bolger, M.B. 1986. PLOT4U: an interactive graphics and non-linear regression program for the analysis of experimental and simulated data, p. 49. *In* Proceedings of the 1986 Summer Computer Simulation Conference, Society for Computer Simulations.
27. Boxenbaum, H.G., S. Riegelman, and R.M. Elashoff. 1972. Statistical estimations in pharmacokinetics. J. Pharmacokin. Biopharm. **2(2)**:123.
28. Celis, E., R. Ridaura, and C. Larralde. 1977. Effects of the extent of DNP substitution on the apparent affinity constant and cooperation between sites. Immunochemistry **14**:553.
29. Hansch, C., and T. Fujita. 1964. ρ-σ-π Analysis. A method for the correlation of biological activity and chemical structure. J. Am. Chem. Soc, **86**:1616.
30. Fugita, T., J. Iwasa, and C. Hansch. 1964. A new substituent constant, π, derived from partition coefficients. J. Am. Chem. Soc. **86**:5175.
31. Hansch, C., and A. Leo. 1979. Substituent constants for correlation analysis in chemistry and biology. John Wiley & Sons, Inc., New York.
32. Saul, F.A., L.M. Amzel, and R.J. Poljak. 1978. Preliminary refinement and structural analysis of the Fab fragment from human immunoglobulin NEW at 2.0 Å resolution. J. Biol. chem. **253**:585.
33. Tonegawa, S. 1985. The molecules of the immune system. Scientific American **253(4)**:122.
34. Segal, D.M., E.A. Padlan, G.H. Cohen, S. Rudikoff, M. Potter, and D.R. Davies. 1974. The three-dimensional structure of a phosphorylcholine-binding mouse immunoglobulin Fab and the nature of the antigen binding site. Proc. Natl. Acad. Sci. USA **71**:4298.
35. Marquart, M., J. Deisenhofer, and R. Huber. 1980. Crystallographic refinement and atomic models of the intact immunoglobulin molecule KOL and its antigen-binding fragment at 3.0 Å and 1.9 Å resolution. J. Mol. biol. **141**:369.
36. Hamlyn, P. H., M.J. Gait, and C. Milstein. 1981. Complete sequence of an immunoglobulin mRNA using specific priming and the dideoxynucleotide method of RNA sequencing. Nucleic Acids Res. **9**:4485.
37. Maniatis, T., E.F. Fritsch, and J. Sambrook. 1982. Molecular cloning, a laboratory manual. Cold Spring Harbor Laboratory, New York.
38. Maxam A.L., and W. Gilbert. 1980. Sequencing end-labeled DNA with base-specific chemical cleavages. Methods Enzymol. **65**:499.
39. Sanger, F., S. Nicklen, and A.R. Coulson. 1977. DNA sequencing with chain-terminating inhibitors. Proc. Natl. Acad. Sci. USA **74**:5463.
40. Hamlyn, P.H., G.G. Brownlee, C.C. Cheng, M.J. Gait, and C. Milstein. 1978. Complete sequence of constant and 3′ noncoding regions of an immunoglobulin mRNA using the dideoxynucleotide method of RNA sequencing. Cell **15**:1067.
41. Sherman, M.S., R.L. Deans, and M.B. Bolger. 1986. Haloperidol binding to monoclonal antibodies: amino acid sequences and predictions of combining site structure via computer modeling (In press)
42. Griffiths, G.M., and C. Milstein. 1985. The analysis of structural diversity in the antibody response, p. 103. *In* T. A. Springer, (ed.), Hybridoma technology in the biosciences and medicine, Plenum Press, New York.
43. Smith, A.J.H. 1980. DNA sequence analysis by primed synthesis, Methods Enzymol. **65**:560.

44. Sood, A.K., H.L. Cheng, and H. Kohler. 1986. A highly efficient method of sequencing immunoglobulin mRNAs. 6th Intl. Congress Immunol., Toronto, Abstract **2.14.6**:102.
45. Shlomchik, M.J., D.A. Nemazee, V.L. Sato, J. van Snick, D.A. Carson, and M.G. Weigert. 1986. Variable region sequences of murine IgM anti-IgG monoclonal autoantibodies. J. Exp. Med. **164**:407.
46. Epp, O., P. Colman, H. Fehlhammer, W. Bode, M. Schiffer, R. Huber, and W. Palm. 1974. Crystal and molecular structure of a dimer composed of the variable portions of the Bence-Jones protein REI. Eur. J. Biochem. **45**:513.
47. Furey, W., Jr., B.C. Wang, C.S. Yoo, and M. Sax. 1983. Structure of a novel Bence-Jones protein (RHE) fragment at 1.6 Å resolution. J. Mol. biol. **167**:661.
48. Edmundson, A.B., K.R. Ely, R.L. Girling, E.E. Abola, M. Schiffer, F.A. Westholm, M.D. Fausch, H.F. Deutsch. 1974. Binding of 2,4-dinitrophenyl compounds and other small molecules to a crystalline lambda-type Bence-Jones dimer. Biochemistry **13**:3816.
49. Poljak, R.J. 1975. X-ray diffraction studies of immunoglobulins. Adv. Immunol. **21**:1.
50. Suh, S.W., T.N. Bhat, M.A. Navia, G.H. Cohen, D.N. Rao, S. Rudikoff, and D.R. Davies. 1986. The galactan-binding immunoglobulin Fab J539; an X-ray diffraction study at 2.6 Å resolution. (In Press).
51. Amit, A.G., R.A. Mariuzza, S.E.V. Phillips, and R.J. Poljak. 1986. Three-dimensional structure of an antigen-antibody complex at 2.8 Å resolution. Science **233**:747.
52. de la Paz, P., B.J. Sutton, M.J. Darsley, and A.R. Rees. 1986. Modeling of the combining sites of three anti-lysozyme monoclonal antibodies and of the complex between one of the antibodies and its epitope. EMBO J. **5**:415.
53. Padlan, E.A., and D.R. Davies. 1975. Variability of the three-dimensional structures in immunoglobulins. Proc. Natl. Acad. Sci. USA. **72**:819.
54. Padlan, E.A. 1977. Structural implications of sequence variability in immunoglobulins. Proc. Natl. Acad. Sci. USA **74**:2551.
55. Kabat, E.A., T.T. Wu, and H. Bilofsky. 1977. Unusual distributions of amino acids in complementary-determining segments of heavy and light chains of immunoglobulins and their possible roles. J. Biol. Chem. **252**:6609.
56. Chothia, C., A.M. Lesk, M. Levitt, A.G. Amit, R.A. Mariuzza, S.E.V. Philips, and R.J. Pol. 1986. The predicted structure of immunoglobulin D1.3 and its comparison with the crystal structure. Science **233**:755.
57. Chothia, C., and A.M. Lesk. 1986. The relation between the divergence of sequence and structure in proteins. EMBO J. **5**:823.
58. Chothia, C., J. Novotny, R. Bruccoleri, and M. Karplus. 1985. Domain association in immunoglobulin molecules, the packing of variable domains. J. Mol. Biol. **186**:651.
59. Ramachandran, G.N., and V. Sasisekharan. 1968. Conformation of polypeptides and proteins. Adv. Protein Chem. **23**:283.
60. Creigton, T.E. 1983. Proteins: structures and molecular principles. W. H. Freeman and Co. New York.
61. Chou, P.Y., and G.D. Fasman. 1977. Beta turns in proteins. J. Mol. Biol. **115**:135.
62. Chou, P.Y., and G.D. Fasman. 1978. Prediction of the secondary structure of proteins from their amino acid sequence. Adv. Enzymol. **47**:45.

63. Kabat, E.A., T.T. Wu, H. Bilofsky, M. Reid-Miller, and H. Perry. 1983. Sequences of proteins of immunological interest. U.S. Dept. of Health and Human Services, Public Health Service, NIH.

64. Tainer, J.A., E.D. Getzoff, H. Alexander, R.A. Houghten, A.J. Olson, R.A. Lerner et al. 1984. The reactivity of anti-peptide antibodies is a function of the atomic mobility of sites in protein. Nature (London) 312:127.

65. Andrews, P. 1986. Functional groups, drug-receptor interactions and drug design. Trends Pharm. Sci. 7(4):148.

66. Dower, S.K., S. Wain-Hobson, P. Gettins, D. Givol, W. R. Jackson, S.J. Perkins, C.A. Sunderland, B.J. Sutton, C. Wright, R. Dwek, et al. 1977. The combining site of the dinitrophenyl-binding immunoglobulin A myeloma protein MOPC 315. Biochem. J. 165:207.

67. Burley, S.K., and G.A. Petsko. 1985. Aromatic-aromatic interaction: a mechanism of protein structure stabilization. Science 229:23.

68. Sharon, J., M.L. Gefter, T. Manser, and M. Ptashne. 1986. Site-directed mutagenesis of an invariant amino acid residue at the variable-diversity segments junction of an antibody. Proc. Natl. Acad. Sci. USA 83:2628.

69. Olson, G.L., H.C. Cheung, K.D. Morgan, J.F. Blount, L. Todaro, and L. Berger. 1981. A dopamine receptor model and its application in the design of a new class of rigid pyrrolo[2,3-g]isoquinoline antipsychotics. J. Med. Chem. 24:1026.

70. Humber, L G., F.T. Bruderlein, A.H. Philipp, M. Gotz, and K. Voith. 1979. Mapping the dopamine receptor. 1. Features derived from modifications in ring E of the neuroleptic butaclamol. J. Med. Chem. 22:761.

71. Philipp A.H., L.G. Humber, and K. Voith. 1979. Mapping the dopamine receptor. 2. Features derived from modifications in the rings A/B of the neuroleptic butaclamol. J. Med. Chem. 22:768.

72. Janssen, P.A.J., and W.F.M. van Bever. 1978. Structure-activity relationships of the butyrophenones and diphenylbutylpiperidines, p. 1. D. Iversen, (ed.), Handbook of Psychopharmacology, Vol. 10. Plenum Press, New York.

73. Tollenaere, J.P., H. Moereels, and M.H.J. Koch. 1977. On the conformation of neuroleptic drugs in the three aggregation states and their conformational resemblance to dopamine. Eur. J. Med. Chem. 12:199.

74. Seeman, P., M. Watanabe, D. Grigoriadis, J.L. Tedesco, S.R. George, U. Svensson, J.L. Nilsson, J.L. Neumeyer 1985. Dopamine D-2 receptor binding sites for agonists: a tetrahedral model. Mol. Pharmacol. 28:391.

75. Moereels, H., and J.P. Tollenaere. 1978. A comparison between the conformation of dexclamol and the tricyclic and butyrophenone type dopamine antagonists. Life Sci. 23:459.

76. Humber, L.G., F.T. Bruderlein, and K. Voith. 1975. Neuroleptic agents of the benzocycloheptapyridoisoquinoline series. Mol. Pharmacol. 11: 833.

77. Humber, L.G., F.T. Bruderlein, A.H. Philipp, M. Gotz, and K. Voith. 1979. Mapping the dopamine receptor. 1. Features derived from modifications in ring E of the neuroleptic butaclamol. J. Med. Chem. 22(7):761.

78. Bruck, C., M.S. Co, M. Slaoui, G.N. Gaulton, T. Smith, B N. Fields, M.I. Greene et al. 1986. Nucleic acid sequence of an internal image-bearing mon-

oclonal anti-idiotype and its comparison to the sequence of the external antigen. Proc. Natl. Acad. Sci. USA **83:**6578.

79. Bona, C.A., and H. Kohler. 1984. Anti-idiotypic antibodies and internal images, p. 141. *In* J. C. Venter et al. (ed.), Monoclonal and anti-idiotypic antibodies: probes for receptor structure. Alan R Liss, New York.

80. Stevenson, F.K. 1986. Idiotypes and diseases. Immunol. Today. **7(10):**287.

10
Opiate Receptors and Molecular Shapes

JAY A. GLASEL

Introduction

The landscape forming the background of this paper is that of the present physicochemical understanding of the molecular interactions between ligands and receptors and the subsequent cellular events that accompany binding. Our laboratory has been active in studies of peptide conformations in solution (1–3) with the object of connecting these with conformations of the same molecules at macromolecular binding sites. Recently, advances in methods have made it possible to study the more biologically meaningful problem of conformations of ligands *at* specific macromolecular binding sites, especially by nuclear magnetic resonance spectroscopy (4), and we have turned to developing an exemplary system which would enable us to approach this problem. It is via this route that we became interested in the interactions of haptens with antibodies and of Ab_2 with anti-hapten Ab_1. In particular, the study of opiate ligands (or haptens)[1] and their interactions with opiate receptors has some real technical advantages, and these will be emphasized in this paper.

Opiates

Ligands which interact specifically with neuronal opiate receptors appear to be of two types. Those of type I are quite rigid polycyclic organic structures based upon the archetypal one derived from plants; that is, morphine (Fig. 10.1). Innumerable structure-activity studies have been done on synthetic analogues of this structure (5). All of the biologically active compounds of this type contain the tertiary nitrogen and the aromatic ring. The absolute stereochemistry of the asymmetric carbon atoms in the

---

[1]For the rest of this paper the words ligand and hapten will be used synonymously in order to bring immunological and receptor terminology into closer alignment with respect to interactions of small molecules with macromolecular binding sites.

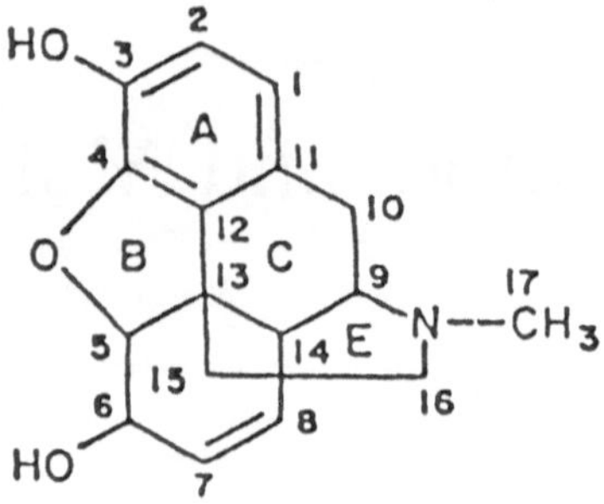

FIGURE 10.1. Covalent structure of morphine.

structures[2] is very important for both binding and biological action. In addition, certain simple substitutions on the nitrogen atom (e.g., an N-allyl group for the N-methyl) usually result in conversion of the molecule from an agonist (binding plus biological action) to an antagonist (binding with no biological action). The result is the availability of pairs of structurally similar molecules with similar binding behavior but differing effects upon the functioning of the receptor. The type II opiate structures consist of single-chain polypeptides which have lengths of from 5 to more than 30 residues (6). Invariably, the N-terminus is a tyrosyl residue. Initially, this led to conformational similarities being perceived to exist between the two types of opiates since, in a two-dimensional representation, the positions of aromatic groups with respect to the nitrogens bear some superficial resemblance to each other for a naive observer. This grouping is called a tyrammine moiety. In fact, close examination of the stereochemistry of the two types[3] shows that these sections of the molecules can be considered pseudo-enantiomeric, and it is easy to see from correctly constructed space filling or computer models that the spatial positions of the important groups are not similar at all, no matter how the peptide is folded (7). This fact has been almost universally ignored in the opiate-related literature but has experimental consequences which will be discussed in this paper. The peptide opiates bind to opiate receptors and have biological actions. The binding and actions of these type II ligands can be specifically blocked by type I antagonists. Type II antagonists have not been found. The biological actions of both types include analgesic effects on the central and peripheral nervous systems as well as various integrative effects on the central nervous system including narcosis, pleasure, and addictive liability (8). For these reasons, as well as due to the desire to develop analogues which have one of these effects to the exclusion of others, opiates have been the subject of great attention for many

_____

[2]For morphine: 5R, 6S, 9R, 13S, 14R
[3]For the opiate peptides: all S

years. One of the practical outcomes of this attention has been the recent commercial availability of many of these compounds with radioactive labels at very high specific activities along with a wide variety of nonlabeled congeners.

Opiate Receptors

One of the outcomes of this has been the awareness, essentially within the past decade, that neuronal receptors exist which bind these ligands specifically (with dissociation constants on the order of nanomolar) (9). The receptors are nonuniformly distributed throughout the central and peripheral nervous systems (10). They are known to be glycosylated, membrane-bound proteins and are present in total amounts on the order of a few micrograms of protein per laboratory animal brain. There are efficient bioassay systems available for opiate receptors, and the direct relationship between bioassayable potency and binding dissociation constants has led to the idea that binding affinity is directly correlated with pharmacological potency (10). In common with other receptor systems, biochemists, neurochemists, and pharmacologists have used the tool of binding isotherms to try to understand the interactions of opiates with receptors. In common with all macroscopic thermodynamic measurements, there is no direct connection between binding parameters (that is, K_d and B_{max}) and the molecular details of the binding (11). All that one can hope for are molecular schemes which are consistent with the observables.

Not in common with many other receptor systems are the readily studied behavioral effects on opiate-treated animals. Even a nominal critical review of this body of work would occupy considerable space. However, in summary, the results are interpreted by a majority of the workers in the field to indicate that there are multiple opiate receptors (either separate macromolecules, separate sites on one macromolecule, or separate states of a macromolecule) which are somewhat specific for the different types of opiates (12). In fact, type II has been broken down into two classes so that there are now commonly believed to be three classes of receptors (6). An important point for this discussion is that the natural opiate agonists of the two types isolated from plants or animal tissue all cross-react with the various receptors to a great extent as do the synthetic organic antagonists. In terms of this discussion, a characteristic difference between these putative receptors is often quoted to be the differences between the dissociation constants for the common antagonist, naloxone (*N*-allyloxymorphone) (Fig. 10.2). The maximum difference between the putative receptor classes corresponds to a change in binding free energy of 2.1 kcal/Mol (that is, a difference of 20%). A controversial question in this field is: Must there be more than one receptor if a ligand (such as naloxone) binds with more than one K_d?

As can be imagined, there has been a considerable effort to characterize

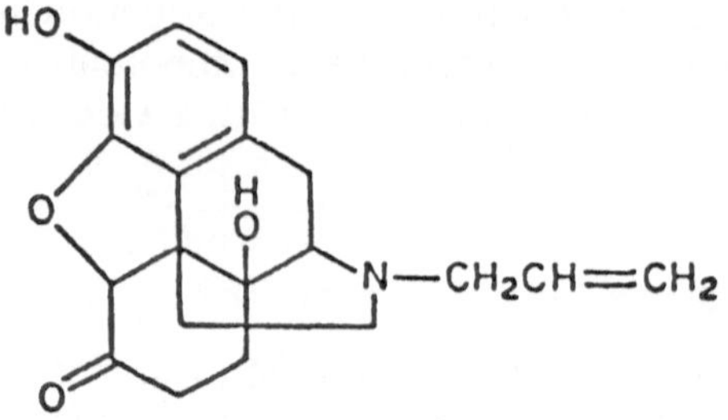

FIGURE 10.2. Covalent structure of naloxone.

and isolate the receptor macromolecules. With respect to physical characterization, there has been a certain amount of controversy. One of the contemporary methods of choice for initial characterization of membrane-bound proteins whose activity may be measured by some assay (such as ligand binding) is radiation inactivation (13). This is a deceptively simple technique which enables the investigator to determine the size of a membrane-bound macromolecule relative to known sizes of macromolecular markers both endogenous and exogenous to the membrane. The tricky part is ferreting out artifacts which change the thermodynamics of binding. One of these turns out to be, in the case of the opiate receptor, changing the membrane state during irradiation from that which exists in frozen homogenates to that which exists in lyophilized preparations. Early work which was done a few years ago using this technique (14) involved samples prepared in the latter way and, depending upon the study, resulted in molecular weights on the order of 200 to 400 kilodaltons (kDa) for the peptide opiate receptor. In collaboration with Dr. Ellis Kempner of the NIH, we carefully investigated the molecular weights of opiate receptors for a variety of agonists and antagonists in both frozen and freeze-dried homogenate states. We quickly found that the freeze-dried materials gave inconsistent results. The results of our work (Fig. 10.3, Table 10.1) show that the molecular weights of all the membrane-bound receptors we investigated are 100 kDa within a ±10% experimental error. Very recently, this result has been confirmed by another group using a different set of ligands (15). Radiation inactivation gives the *minimum* molecular weight of the entity which binds the ligand in the membrane-bound form of the receptor. If there are subunits, or other proteins, bound to the receptor, they will contribute to the observed molecular weight only if they are required for high-affinity binding. This point is important because there have been several reports of receptor solubilization studies which result in polypeptides of molecular weight of ~60 kDa that have been identified with the opiate receptor binding unit (16,17). The discrepancy is not a trivial point since, with other evidence, it points towards the involvement of a GTP binding protein closely associated with the native form of opiate

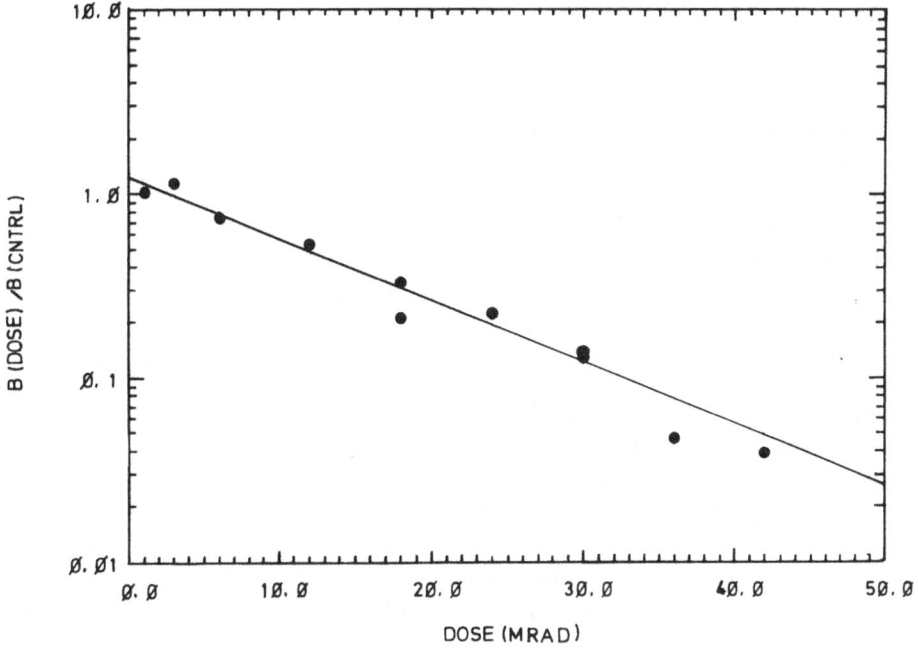

FIGURE 10.3. Example of radiation inactivation derived from binding isotherms of tritiated ligands to rat brain opiate receptor preparations. Binding of [³H]-etorphine to rat brain P2 homogenate monitored at a free ligand concentration of 0.2 nM. Irradiation for the doses indicated was at 13 Mev in the linear accelerator at the Armed Forces Radiobiology Research Institute (Bethesda, Md.). Samples were maintained frozen at − 135°C during irradiation and at − 70°C during shipment and storage.

TABLE 10.1. Radiation inactivation target sizes for opiate ligand receptors derived from nonlinear least-squares analysis of binding versus dose data at free ligand concentrations indicated.

Ligand	F(nM)	Slope	MW (kDa)
Etorphine	0.2	0.055 ± 0.008	99 ± 13
Morphine	1.5	0.057 ± 0.006	102 ± 10
Naloxone	1.0	0.067 ± 0.008	120 ± 15
DADLE	3.0	0.051 ± 0.008	91 ± 14

receptors in the neuronal membrane. In turn, biochemical work (18) indicates that receptor-ligand affinity, and therefore receptor binding site conformation, depends upon the existence of this moiety.

So far, attempts at sequencing the receptor via conventional methods have not been at all successful, and most of the groups involved with this effort have turned to molecular biological techniques—that is, indirect sequencing of the protein by use of cDNA sequencing. In one popular strategy for accomplishing this, a specific probe for translational product (the receptor binding unit) is needed. That is, in this case a macromolecular probe which binds to receptor with very high affinity is required. Attempts to raise such antibodies to the solubilized preparations which have thus far been obtained have not been satisfying from the standpoint of specificity and reversibility. Possible reasons are the impurity or lability of the available preparations, lack of immunogenicity of the receptor binding site, or changes in conformation of the receptor binding site when it is removed from its native membrane environment. It is in the area of development of specific probes for translational products, that a major interest in anti-idiotypic antibodies lies. The other interest in our laboratory concerns the meaning, in molecular spatial terms, of "images" of organic hapten geometries translated into those of peptides.

Immunological Work

Our work, up to now, has dealt strictly with opiates of type I. The reasons are as follows. The chemistry of these compounds has been well studied, a wide variety of them are available in macroscopic amounts, many are available labeled to theoretical values with tritium, and they are not subject to enzymatic degradation as is the case for peptide opiates. Attention to the extensive structure-activity work done on opiates shows that the only macromolecular binding sites of interest are those which recognize the tertiary nitrogen/aromatic ring system discussed at the beginning of this paper. No immunological work appears to have been done on the possible existence of a general relationship between the hapten-Ab_1 dissociation constant and that for Ab_2-Ab_1. However, it seems reasonable to require also that screening should be for hapten-Ab_1 binding that is at least as strong as that of hapten-receptor. Furthermore, screening for binding of solution conformations of hapten to antibody using nonsolution assay methods involving surface-adsorbed haptens or hapten conjugates seems inherently illogical. Hence, we had to develop a simple way of screening for high-affinity binding in solution (19).

Id$^+$ Anti-Opiate Antibodies

The monoclonal anti-morphine antibodies are rather interesting in themselves. We have concentrated upon four lines which show distinct patterns of cross-reactivities to opiate congeners (Fig. 10.4), along with dissociation constants in the low nanomolar range to several of them (20) (Table 10.2). One significant fact is that they do not show detectable cross-reactivities with opiate peptides. Another worker has independently found this to be true for other anti-morphine mAbs (21). This is consistent with the stereochemical, and therefore spatial, dissimilarity of the two types of opiates that have been mentioned. The mAbs are all IgG1s. Briefly, some of the chemical properties of these antibodies are as follows. Their binding sites are all blocked by a single phenoxybenzamine alkylation per Fab fragment. This reagent, incidentally, bears no resemblance to morphine. Chloramine-T or Bolton-Hunter iodination also blocks opiates from binding. These two facts may be interpreted to indicate that tyrosine residues are closely involved in binding opiates to both antibodies and receptors. Completely reduced and alkylated heavy and light chains recover binding activity when renatured. Finally, the light chains are all of the lambda class.

As a first effort we hope to derive, using a technique called the transferred nuclear Overhauser effect (22), the *bound* conformations of the opiates. Again, these particular molecules have some definite advantages for this type of study. First of all, the crystal structures of many of the opiate congeners are known to high precision. Secondly, we have shown that the molecules are quite rigid in solution with the exception of a simple chair/chair equilibrium in the important piperidine ring (23). Thus, we hope to see if the bound conformation is that which has lowest free energy in solution.

Anti-Idiotypic Antibodies

Our most extensive experiments are the result of immunization of rabbits with affinity-purified Fab fragments of our anti-morphine mAbs (24–26). Our assay is a solution competition one in which disruption of equilibrated binding between tritium-labeled morphine (or other hapten) and mAb by crude or purified serum fractions is observed. In common with published work by others, we find titers of individual animals are highly variable and aperiodic. However, we find that we can keep an animal producing titer over a long period of time. We have never observed any unusual physiological effects in the animals. This is not unexpected since, without any carrier mechanism, antibodies cannot cross the blood-brain barrier.

All of the experiments from this laboratory have been performed with antibodies purified in the following manner. It should first be noted that the reason opiate solutions are not used as eluents is that in receptor binding experiments even nanomolar concentrations of opiates will interfere with the assays due to the very small amounts of receptor present in neuronal membranes. Our purification scheme is based upon an affinity column of the appropriate mAb linked to Sepharose 4B-CL. Using a crude serum fraction resulting from "optimal" precipitation with saturated ammonium sulfate (27), approximately 99% of the protein comes through the

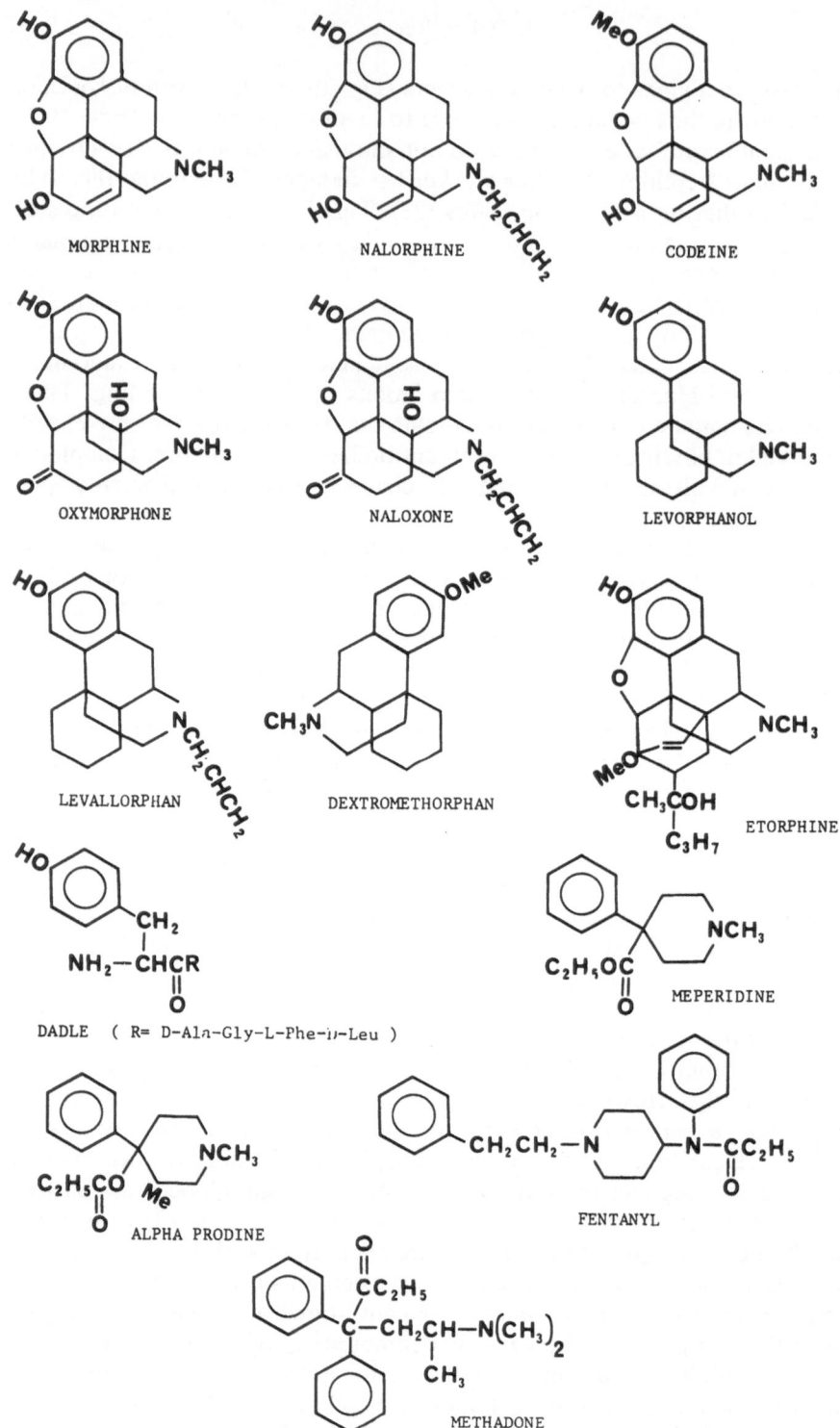

MORPHINE

NALORPHINE

CODEINE

OXYMORPHONE

NALOXONE

LEVORPHANOL

LEVALLORPHAN

DEXTROMETHORPHAN

ETORPHINE

DADLE (R= D-Ala-Gly-L-Phe-D-Leu)

MEPERIDINE

ALPHA PRODINE

FENTANYL

METHADONE

TABLE 10.2. IC_{50} values for displacement of tritiated morphine from anti-morphine mAbs by various drugs.

Drug	IC_{50} (nM) for immunoglobulin line:			
	12D4	11C7	10C3	3B9
Morphine	46	54	173	13
Codeine	250	70	200	10
Nalorphine	1,000	32	96	34
Oxymorphone	>> 10,000	741	20,000	8,100
Naloxone	>> 10,000	182	6,000	5,000
Levorphanol	200	447	4,600	199
Levallorphan	>> 10,000	166	11,500	500
Dextromethorphan	>> 10,000	>> 10,000	>> 10,000	>> 10,000
Etorphine	457	>> 10,000	>> 10,000	>> 10,000
DADLE	>> 10,000	>> 10,000	>> 1,000	>> 10,000
Meperidine	>> 10,000	>> 10,000	500	3,500
alpha-prodine	>> 10,000	>> 10,000	>> 10,000	>> 10,000
Fentanyl	>> 10,000	500	>> 10,000	>> 10,000
Methadone	>> 10,000	>> 10,000	>> 10,000	>> 10,000

column with the PBS loading buffer. We then elute the small amount of active material using 3 M sodium thiocyanate (pH = 8.1) or 6 M guanidine (pH = 7). The word active is used to mean that the material competes with labeled morphine reversibly in our solution assay procedure. Solid-state assays show that the materials we isolate are rabbit IgG with no detectable mouse IgG contamination. Preadsorption on a normal mouse IgG column still results in an active fraction, but we do not normally use this step in purification since we get more recovery by using the optimum ammonium sulfate precipitation (26) scheme for rabbit IgG and putting this crude material on our monoclonal anti-morphine IgG columns.

Quite early in our work we found that activity as just defined does not necessarily mean activity in competing with morphine at the opiate receptor. As has been pointed out, very small amounts of any entity which happens to bind either to the receptor (is a structural analogue of morphine) or to morphine (an anti-opiate analogue) will interfere with the receptor binding assay. However, we found that passing the material, obtained as just described, through an affinity column of morphine-Sepharose 4B-CL in order to remove any anti-opiate antibodies (putatively Ab_3) usually results in material which reversibly competes with opiate congeners at their receptors. In subsequent work, we have found that some preparations of crude rabbit antibodies (that is, the ammonium sulfate fraction) will com-

◁————————————————————————

FIGURE 10.4. Two-dimensional representations of covalent structures of opiate congeners. The uncharged forms are shown. In the physiological pH range, the tertiary amine groups will be protonated and DADLE ([D-Ala^2,D-Leu^5]enkephalin) will be zwitterionic. Taken from Ref. 20.

pete with opiates at the receptors. It therefore appears that the rabbit anti-mAb titer varies in its distribution of anti-idiotypic antibodies and may, at times, contain unknown amounts of Ab_3. Even assuming that the Ab_1 binding site physically resembles that of the receptor, Ab_2 competition with ligand at the Ab_1 binding site does not necessarily imply competition at the receptor binding site. That is, the population of anti-paratypic antibodies which binds specifically to receptor must be equal to or smaller than that binding to Ab_1. This would clearly have to be worked into any definition of Ab_2 mimicry of ligand structures. This point is discussed below.

The following discussion is intended to integrate two of our published results into the main theme of this paper. Since the work is in print, procedures will not be presented and concentration will be on a subset of the studies. One experiment involves the competition of our preparations with so-called opiate receptor-selective drugs for binding to receptors found in membrane preparations from rat brain (24). The other data involves the cross-reactivities of the anti-idiotypic fractions raised against each of the four lines of mAbs (25). That is, treating the anti-idiotypic fractions as drugs we asked the question: How good are they, compared to morphine and its congeners, in competition at the binding sites available in anti-morphine mAbs?

Our older results (23) should be kept in mind. That is, our anti-idiotypic antibody preparations block morphine binding to rat brain membranes extremely well. They also block enkephalin and naloxone binding to the same membranes. The line of mAb to which the anti-idiotypic antibody preparation being referred to now was raised does not bind enkephalin at all (in common with all of our anti-morphine mAbs) and naloxone very poorly.

Opiate receptor-"selective" drugs are ones which pharmacologists have developed that are supposed to bind very selectively to individual members of the putative classes of receptors. For example, the highly substituted peptide with the acronym DAMGE[4] (28) is thought to bind to opiate receptors of the morphine class almost exclusively while enkephalin is thought to bind almost exclusively to the opiate peptide receptor. This enables one to determine the binding constants of a drug to the *mu* and *delta* classes of receptor. By blocking the other classes with large amounts of both enkephalin and DAMGE, binding to the *kappa* receptor can be determined. The results of these experiments (Fig. 10.5) are that our preparation is more type I than type II receptor active, but not overwhelmingly so. These anti-idiotypic antibodies do not bind in detectable amounts to the ethylketocyclazocine (EKC) receptor (which is supposed to bind the

---

[4][D-Ala^2,MePhe^4,Gly-ol^5]enkephalin.

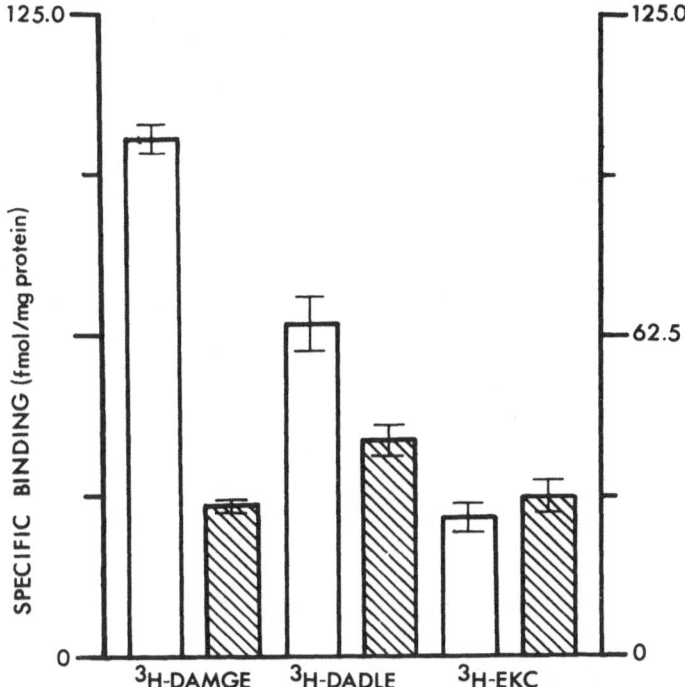

FIGURE 10.5. Inhibition of ³H-DAMGE, 3H-DADLE , and ³H-EKC (ethylketo-cyclazocine) binding to rat brain P2 homogenates by rabbit polyclonal anti-idiotypic antibodies raised in response to anti-morphine mAb (line 10C3). Radioligand incubation concentrations are 2, 4, and 4 nM, respectively. Open bars: controls (radioligand only); hatched bars: radioligand plus anti-idiotypic antibody. Ab_2 concentrations were those which inhibited the binding of morphine to Ab_1 by 50%. Taken from Ref. 25.

second, non-enkephalin/endorphin, opiate peptides). The next experiment is the cross-reactivity one (Fig. 10.6). This shows that all of the anti-idiotypic antibody preparations are able to compete for morphine with all of the other anti-morphine mAbs. Furthermore, anti-idiotypic antibodies raised against mAbs which *do not bind,* for example, naloxone or etorphine (a powerful opiate agonist with a very small K_d for the receptor) will compete for the naloxone or etorphine binding sites of mAbs which do (Table 10.3).

We have also found, in unpublished work done in collaboration with Prof. Rabi Simantov, The Weizmann Institute of Science, that our purified material blocks opiate binding to aggregating embryonic rat brain cells grown in culture.

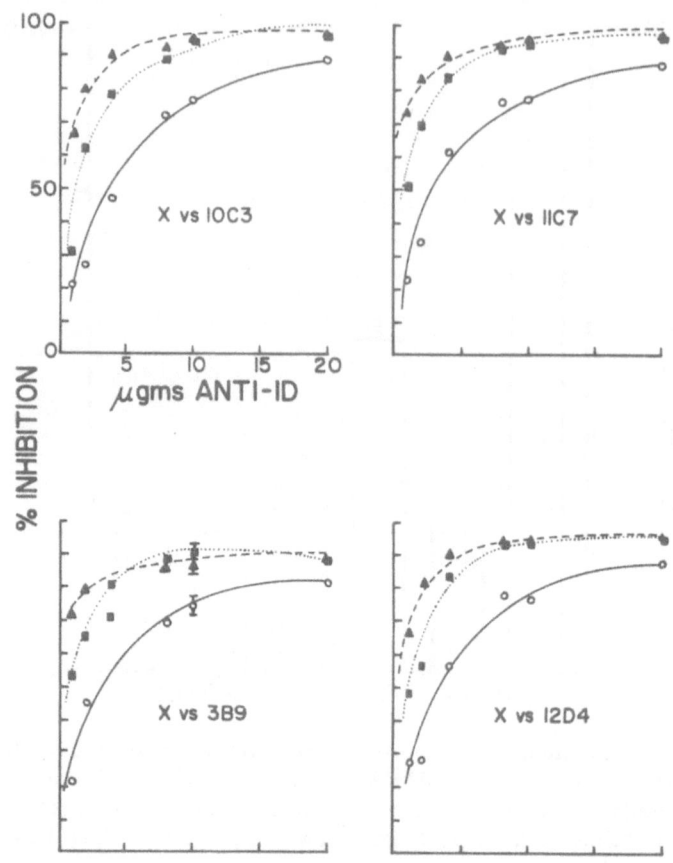

FIGURE 10.6. Displacement curves for the competition of affinity-purified anti-10C3, anti-11C7, and anti-3B9 lines of anti-morphine mAbs, for the morphine binding sites on affinity-purified 10C3, 11C7, 12D4, and 3B9. The curves were determined using incubation mixtures containing 200 nM morphine, 270 nM mAb IgG, and the indicated amounts of affinity-purified anti-idiotypic antibodies. Open circles: anti-10C3; filled squares: anti-11C7; filled triangles: anti-3B9. Curves are not theoretical fits. Typical error bars (one sigma) are shown for one set of points in the lower left-hand panel. Taken from Ref. 26.

The Physical Chemistry of Molecular Shapes and Images

The question is, obviously, what are we to make of this? In order to give some rational basis for discussion, some known features of the physical chemistry of ligand binding to macromolecules must be pointed out. The theme in this part of the paper is a remark that Richard Feynman, the

TABLE 10.3. Inhibition of drug-mAb binding by affinity-purified anti-idiotypic antibody.[a]

mAb	Drug	Anti-Id	%Inhibition
11C7	Naloxone	—	—
11C7	Naloxone	Anti-10C3	94 ± 1
11C7	Naloxone	Anti-11C7	98 ± 5
11C7	Naloxone	Anti-3B9	70 ± 10
3B9	Etorphine	—	—
3B9	Etorphine	Anti-10C3	93 ± 2
3B9	Etorphine	Anti-11C7	87 ± 8
3B9	Etorphine	Anti-3B9	94 ± 7

[a]Cold drugs were at concentrations of 100 μM. Tritiated drugs were at concentrations of 200 nM. The results represent the difference in specific binding of drug.

physicist, once made to the effect, "you can't disprove a vague theory." The concept of images of haptens present in anti-Ids is a vivid one, but almost entirely verbal and therefore vague in the molecular conformational sense. The purpose of this section is to examine what is definitely known about binding interactions between molecules and what this has to say about the image idea.

Considering just opiate binding to our Ab_1s and to receptor, there is no indication, on the basis of the comparison of binding constants, that the ligand atomic groups involved in binding with opiate receptors and those involved in binding to antibodies are interacting in the same manner. For example, cross-reactivity studies show that the rank order of binding constants to the receptor are different than to the mAbs. In addition, there is the problem that while opiate peptides cross-react with receptors, they do not bind to the anti-type I mAbs which have been selected to recognize the tyrammine portion of the molecule. This seems to be incontrovertible evidence that the binding site "images" of the anti-morphine mAbs do not complement that of the opiate peptides.

The experimental facts are that a population of polyclonal anti-idiotypic antibodies we have purified from rabbits compete with type I opiates for binding at the monoclonal anti-opiate antibody binding site and at the opiate receptor sites. Thus, while the mAb binding site has no affinity for opiate peptides, the receptor sites do, and here the opiate peptides compete with the rabbit antibodies in experiments designed to examine binding to each class of receptor.

For binding of opiate haptens to our antibodies and to receptors, we must account for ~12 kcal/mol free energy. Allowing for the loss of rotational and translational entropy upon binding, this energy can easily be explained on the basis of dispersion forces, superimposed upon a charged interaction involving the protonated nitrogen. This is in agreement with the structure-activity studies which have been done on opiate congeners. One potential objection which might arise to this rationalization of small molecule binding is that it does not take into account entropic changes due to water displacement from either binding site or small molecule during binding. However, because of the size of the small molecule (the solvent-accessible surface area of morphine is ~300 Å, and it is not highly hydrated) this should not be an important effect either at the antibody or receptor binding sites.

In trying to define what the term "image" of a hapten is in geometric terms, we are faced with a very difficult problem right at the beginning. That is, the reaction of antibody with antibody, just as that of antigen with antibody, must be described in a very different way than for haptens to antibody. First of all, entropic *losses* on binding resulting from degree of freedom loss in side chain movement as well as entropic *gains* due to water displacement are extremely important in protein-protein binding, but largely incalculable. The best current thinking would explain any high-affinity protein-protein binding on the basis of sterically available close complementarity over large areas of the macromolecules, resulting in entropic gain by displacement of bound water which is combined with electrostatic, hydrogen bonding, and dispersion interactions to overcome the negative entropy change due to loss of degrees of freedom. Certainly, the only pieces of experimental evidence so far forthcoming [reference is made to recent crystallographic work on lysozyme-anti-lysozyme (29)] seem to support this molecular interpretation of the thermodynamics observed. These results indicate that indeed the recognition site of antibody for antigen is disperse and that close interactions of the two surfaces extend to areas outside of the immunoglobulin fold.

The question is: if this picture holds for Ab_2-Ab_1 interactions, where does this leave the concept of hapten "images"? Care must be taken to draw a distinction between an *operational* concept and a molecular one. There is no doubt, on the basis of a large body of work, that, operationally, anti-idiotypic antibodies will compete with haptens for Ab_1 binding sites. There are actually two questions involved. First, can the geometric pictures of Ab_2-Ab_1 and hapten-Ab_1 binding be the same? There is an immediate problem with respect to both binding and immunogenicity. If Ab_2 is structurally similar to Ab_1, how do binding recognition sites both buried in immunoglobulin folds have access to each other? Although there is no published evidence that demonstrates the structural similarity of Ab_2 with what is accepted to be the general case for Ab_1, it seems a reasonable assumption that they are roughly similar, and we will have to be content with this until definitive structural work is done. However, the fact that haptens will compete with anti-idiotypic antibodies at the Ab_1 site seems to be clear evidence that parts of the Ab_2 peptide chains can come into close contact with the Ab_1 binding site. The only realistic molecular picture of this is to give up the generalized picture of the hapten being buried within a cleft and assume instead that the binding site is a surface patch on Ab_1. This also solves the problem of immunogenicity of the paratopes of Ab_1. There has been a great deal of controversy among immunologists concerning what portions of an antigen lead to immunogenicity. Various concepts have been introduced including flexibility of peptide chains in immunogenic regions (30,31). However, recently a very satisfying picture of antigenic structures in macromolecules has been presented (32,33) which fits the known facts without any ad hoc assumptions. The basic concept is that solvent-accessible surfaces of antigens which protrude above the mean macromolecular altitude are those which are immunogenic. Since $Id^+ Ab_1$ is immunogenic by definition, it follows that the paratope is a protuberant feature on Ab_1. This brings up the second question. Can the section of Ab_2 which recognizes this site be a spatial translation of the hapten structure into peptide residue terms?

The mode of interaction of haptens with Ab_1, just as the interaction of drugs with receptors, is presently regarded by theorists as being fundamentally different from that of macromolecule with macromolecule. Small molecule binding to macromolecules is characterized by negative free energies which are the result of subtracting a relatively small entropy loss from a relatively small positive enthalpy of interaction. The entropy term comes from losses of degrees of freedom and amounts (when translated into energy terms) to some 14 kcal/mol. The interaction enthalpy changes are on the order of 26 kcal/mol, thus yielding the 12 kcal/mol for binding which has been mentioned. In direct contrast, when macromolecules bind to each other, surface area patches on the order of 1,000 to 2,000 $Å^2$ are involved. These are accessible to solvent before binding, but become buried during the binding. The entropy gain contribution to binding free energies due to this is on the order of -25 to -50 kcal/mol. The entropy losses due to changes in degrees of freedom have been variously estimated but a rough order of magnitude is 20 to 30 kcal/mol. Hence, without reference to any other details of specific interactions, hydrophobic interactions can account for antibody-antibody binding with dissociation constants of nanomolar or less by subtracting these two *large* terms from one another. An attempt at quantitatively explaining any additional stabilization of Ab_2-Ab_1 binding on the basis of a more or less exact replication of the electron density pattern between hapten and the projecting portion of Ab_2 is going to be very hard to do, and perhaps is not required. Provided that the peptide extension of Ab_2 fills the surface binding site of Ab_1 sufficiently, hapten and the Ab_2 binding surface cannot coexist in the site. Depending upon relative free energies of binding (remembering they are of different origins), either hapten or Ab_2 will predominantly populate the site. This explanation requires that the Fab portions of Ab_1 and Ab_2 be rigid enough so that breaking apart the close approach of the antibodies to each other (due to the disruption in complementarity when the small ligand occupies its binding site) is sufficient to cause a large shift of entropic contributions away from hydrophobic bonding.

The theory of drug-receptor interactions is a subject of much contemporary interest. As discussed above, binding of drug to receptor must, in general, be similar physically to binding of hapten to Ab_1. To be succinct, drug designers are interested in explaining (and then predicting) "goodness of fit" of drug to receptor (34). One appealing picture which has emerged is that of receptor binding sites as areas in the proteins which exhibit high hydrophobicity and flexibility. There is also obvious interest in explaining why, for two molecules with approximately equal binding affinities, one may be an antagonist and one an agonist. Without any evidence to the contrary, the physical basis for the interaction of Ab_2 with receptor must also be the same as that for the interaction of Ab_2 with Ab_1. That is, two macromolecules are binding with consequent loss of degrees of freedom entropy which must be compensated for by gain in entropy due to displaced solvent over a large surface area. Therefore, the basic question is: where does specific binding affinity of Ab_2 to receptors arise? To repeat, it can't be of strictly "imaging" origin because the binding of macromolecules with the affinities we are talking about is fundamentally different than binding of haptens to macromolecules. That is, a large part of the binding free energy comes from subtraction of entropic contributions which are relatively nonlocal and therefore cannot be a replica of the original hapten's spatial distribution.

On the basis of the above, the following may be suggested as a testable hypothesis. The anti-idiotypic antibodies we are interested in, and are selecting, are those which have polypeptide domains which extend far enough into a surface antihapten-Ab_1 binding site to be interfered with by the hapten. The Ab_2-Ab_1 interaction is stabilized by complementarity between these extended domains within and other complementarity outside the binding site. Anti-idiotypic antibodies which have domains either too large or too small do not bind to Ab_1 with sufficient affinity for us to detect. Turning to the interaction of Ab_2 with receptors for the same small molecule, the same sort of effect can be invoked. The generalized receptor binding site is quite hydrophobic, and there are doubtless a number of nonspecific complementarity arrangements between Ab_2 and receptor. Having selected Ab_2s with an extending peptide domain which is long enough to conflict with hapten binding at the Ab_1 site, it is not too hard to believe that a similar but less specific thing could happen at the receptor site. Moreover, it is quite possible that some of these Ab_2s will *not* interfere with ligand binding at the receptor whereas they do at the Ab_1 site because the orientation of the extended domains is different in the two cases. In summary, the proper introduction of the domains into the receptor binding site in such a manner that they reversibly compete with the "normal" ligand must be determined by factors other than structural imaging.

Using this model explains, without torturing the data and in accord with the principle of Occam's razor, what we observe in the case of the opiate receptor. Our Ab_1s do not bind opiate peptides at all, which is not surprising since the two types of opiates do not have any real spatial resemblance to one another. The Ab₂s which we have produced all interfere with hapten binding to anti-type I opiate antibodies. That is, the putative domains are all large enough to fit into anti-type I antibody binding sites even if the site is a different paratope from the one to which Ab₂ was raised. When Ab₂ interacts with receptor, on the other hand, both type I and type II opiates interfere with binding. The explanation is that the opiate receptor binding site is actually polyfunctional insofar as it is large enough to bind both types of opiate ligands in what must be different spatial arrangements. Our present Ab₂s have large enough peptide extension to interfere with the binding of both types of opiates to the receptor. A further question then arises as to what happens with respect to function of the receptor when Ab₂ binds to receptor alone? The only way to test this in any such system is via bioassay. In the opiate system, the answer, on the basis of preliminary work done in collaboration with Prof. Charles Chavkin, Washington State University, is that at least one of our Ab₂s acts as a classical antagonist. That is, it binds to receptor and blocks agonist binding but does not cause receptor functioning. If this is so, it is interesting that this would be the first polypeptide antagonist for opiate receptors. On the basis of what has been proposed here, this should be the effect of most Ab₂-receptor binding. Ab₂ agonist effects would be due to binding of the extending peptide domain to the same residues on the receptor to which the original drugs bind, although productive reactions which produce agonist behavior must involve transition states.

If this model is correct, it may possibly lead to the tailoring of Ab₂s with agonist or antagonist activity by changing the length of the intra-binding site peptide domains using molecular biological means and/or the composition of the side chains by chemical means. We hope that our chemical and physical work, now in progress, will enlighten us on some of these points.

Acknowledgments. This work was supported by NIH grant NS15700 and NSF grant DMB 85-10693.

References

1. Becker, E.L., H.E. Bleich, A.R. Day, R.J. Freer, J.A. Glasel, and J. Visintainer. 1979. Nuclear magnetic resonance conformational studies on the chemotactic tripeptide. Biochemistry 18:4656.
2. Bleich, H.E., A.R. Day, R.J. Freer, and J.A. Glasel. 1979. NMR rotating frame relaxation studies of intramolecular motion in peptides. Tyrosine ring motion in methionine-enkephalin. Biochem. Biophys. Res. Commun. 87:1146.
3. Bleich, H.E., J.D. Cutnell, A.R. Day, R.J. Freer, J.A. Glasel, and J.F. McKelvy. 1976. Preliminary analysis of 1-H and 13-C spectral and relaxation behavior in methionine-enkephalin. Proc. Natl. Acad. Sci. USA 73:2589.

4. Clore, G.M., and A.M. Gronenborn. 1982. Theory and applications of the transferred nuclear Overhauser effect to the study of the conformations of small ligands bound to proteins. J. Magn. Reson. **48**:402.

5. Casy, A.F. 1971. The structure of narcotic analgesic drugs. Biochem. Pharmacol. **1971**:1.

6. Hughes, J. 1983. Opiod Peptides. Br. Med. Bull. **39**:1.

7. Maryanoff, B.E., and M.J. Zelesko. 1978. Stereochemical considerations in structural comparison of enkephalins and endorphins with exogenous opiate agents. J. Pharm. Sci. **67**:590.

8. Palmer, G.C. 1983. Effects of psychoactive drugs and cyclic nucleotides in the central nervous system. Prog. Neurobiol. **21**:1.

9. Glasel, J.A., and R.F. Venn. 1984. Opiate and opioid peptide binding measurements in vertebrate CNS. Curr. Meth. Cell. Neurobiol. **2**:185.

10. Snyder, S.H., and S. Matthysse. 1975. Opiate receptor mechanisms. Neurosci. Res. Prog. Bull. **13**:1.

11. Scatchard, G. 1949. The attractions of proteins for small molecules and ions. Ann. N.Y. Acad. Sci. **51**:660.

12. Kosterliztz, H.W., and S.J. Paterson. 1980. Characterization of opioid receptors in nervous tissue. Proc. Roy. Soc. (London). **B210**:113.

13. Shorr, R.G.L., E.S. Kempner, M.W. Strohsacker, P. Nambi, R.J. Lefkowitz, and M.G. Caron. 1984. Determination of the molecular size of frog and turkey erythrocyte beta-adrenergic receptors by radiation inactivation. Biochemistry **23**:747.

14. McLawhon, R.W., J.C. Ellory, and G.J. Dawson. 1983. Molecular size of opiate enkephalin receptors in neuroblastoma-glioma hybrid cells as determined by radiation inactivation analysis. Biol. Chem. **258**:2102.

15. Ott, S., T. Costa, M. Wuster, B. Hietel, and A. Herz. 1986. Target size analysis of opioid receptors. No difference between receptor types, but discrimination between two receptor states. Eur. J. Biochem. **155**:621.

16. Simonds, W.F., T.R. Burke, Jr., K.C. Rice, A.E. Jacobson, and W.A. Klee. 1985. Purification of the opiate receptor of NG108-15 neuroblastoma-glioma hybrid cells. Proc. Natl. Acad. Sci. USA **82**:4974.

17. Demoliou-Mason, C.D., and E.A. Barnard. 1984. Solubilization in high yield of opioid receptors retaining high-affinity delata, mu and kappa binding sites. FEBS Lett. **170**:378.

18. Childers, S.W. 1984. Interaction of opiate receptor binding sites and guanine nucleotide regulatory sites: selective protection from Nethylmaleimide. J. Pharmacol. Exp. Ther. **230**:684.

19. Glasel, J.A., W.M. Bradbury, and R.F. Venn. 1983. A rapid binding assay for immunoglobulins which may be used as an alternative to ammonium sulfate precipitation methods. J. Immunol. Meth. **63**:291.

20. Glasel, J.A., W.M. Bradbury, and R.F. Venn. 1983. Properties of murine anti-morphine antibodies. Mol. Immunol. **20**:1419.

21. Goldstein, A., R.W. Barrett, I.F. James, L.I. Lowney, C.J. Weitz, L.L. Knipmeyer, and H. Rapoport. 1985. Morphine and other opiates from beef brain and adrenal. Proc. Natl. Acad. Sci. USA **82**:5203.

22. Bothner-By, A.A. 1979. Nuclear Overhauser effects on protons and their use in the investigation of structures of biomolecules, p. 177. *In* Biological Applications of Magnetic Resonance. R. Shulman (ed.), Academic Press, New York.

23. Glasel, J.A., and H.W. Reiher. 1985. Very high frequency proton NMR studies of the conformations of opiate agonists and antagonists. Magn. Reson. Chem. **23:**236.
24. Glasel, J.A., and W.E. Myers. 1985. Rabbit anti-idiotypic antibodies raised against monoclonal anti-morphine IgG block mu & delta opiate receptor sites. Life Sci. **36:**2523.
25. Myers, W.E., and J.A. Glasel. 1986. Subclass specificity of anti-idiotypic anti-opiate receptor antibodies in rat brain, guinea pig cerebellum, & neuroblastoma X glioma (NG108-15). Life Sci. **38:**1783.
26. Glasel, J.A., and L.A. Pelosi. 1986. Morphine-mimetic anti-paratypic antibodies: crossreactive properties. Biochem. Biophys. Res. Commun. **136:**1177.
27. Herbert, G.A. 1976. Improved salt fractionation of animal serums for immunofluorescence studies. J. Dent. Res. Suppl. **55:**A33.
28. Spain, J.W., B.L. Roth, and C.J. Coscia. 1985. Differential ontogeny of multiple opioid receptors (mu, delta, and kappa). J. Neurosci. **5:**584.
29. Amit, A.G., R.A. Mariuzza, E.V. Phillips, and R.J. Poljak. 1985. The three-dimensional structure of an antigen-antibody complex at 6 Å resolution. Nature (London) **313:**156.
30. Tainer, J.A., E.D. Getzoff, H. Alexander, R.A. Hougten, A.J. Olson, and R.A. Lerner. 1984. The reactivity of anti-peptide antibodies is a function of the atomic mobility of sites in a protein. Nature (London) **312:**127.
31. Westhof, E., D. Altschuh, D. Moras, A.C. Bloomer, A. Mondratgon, A. Klug, and M.H.V. Van Regenmortel. 1984. Correlation between segmental mobility and the location of antigenic determinants in proteins. Nature (London) **311:**123.
32. Fanning, D.W., J.A. Smith, and G.D. Rose. 1986. Molecular cartography of globular proteins with application to antigenic sites. Biopolymers **25:**863.
33. Novotny, J., M. Handschumacher, E. Haber, R.E. Bruccoleri, W.B. Carlson, D.W. Fanning, J.A. Smith, and G.D. Rose. 1986. Antigenic determinants in proteins coincide with surface regions accessible to large probes (antibody domains). Proc. Natl. Acad. Sci. USA **83:**226.
34. Andrews, P.R., D.J. Craik, and J.L. Martin. 1984. Functional group contributions to drug-receptor interactions. J. Med. Chem. **27:**1648.

11
Monoclonal Antibodies to Benzodiazepines

Their Application in Making Anti-Idiotypic Antibodies
to the Benzodiazepine Receptor and in Revealing the
Existence of Benzodiazepines and Benzodiazepine-like
Molecules in the Brain

ANGEL L. DE BLAS, DONGEUN PARK,
JAVIER VITORICA, AND LAKSHMI SANGAMESWARAN

Introduction

Our interest in the brain benzodiazepine receptor (BZDR) is derived from both the social and neurobiological importance of the benzodiazepines (BZDs). The benzodiazepines (like diazepam or Valium) are the most widely prescribed drugs (Fig. 11.1). Several thousand metric tons of BZDs are consumed in the United States every year (1). The BZDs are used for their anxiolytic, antiepileptic, and muscle-relaxing properties. Their psychotropic effects are due to the existence of benzodiazepine receptors in the neuronal surface membranes which have very high affinity (with K_d in the nanomolar range) and specificity for benzodiazepines (2, 3). The benzodiazepine receptor is part of a protein complex that includes the $GABA_A$ receptor (GABAR), the chloride channel, and a receptor for barbiturates (4–10). Benzodiazepines potentiate the effect of GABA on opening the chloride channel by increasing the total Cl^- current flowing through the channel (11–15). This mechanism explains the powerful effects of these drugs since GABA is the main inhibitory neurotransmitter in the brain [up to 40% of the brain synapses use GABA as a neurotransmitter (16)]. Therefore, benzodiazepines displace the balance between inhibitory and excitatory synapses towards the former. This effect is the base for their anxiolytic, anticonvulsant, and muscle-relaxing properties (the muscle relaxation is due to an effect on the CNS and not to a direct effect on the muscle).

The brain BZDR is a protein that not only binds the classical anxiolytic agonist benzodiazepines (like diazepam and flunitrazepam) but also binds

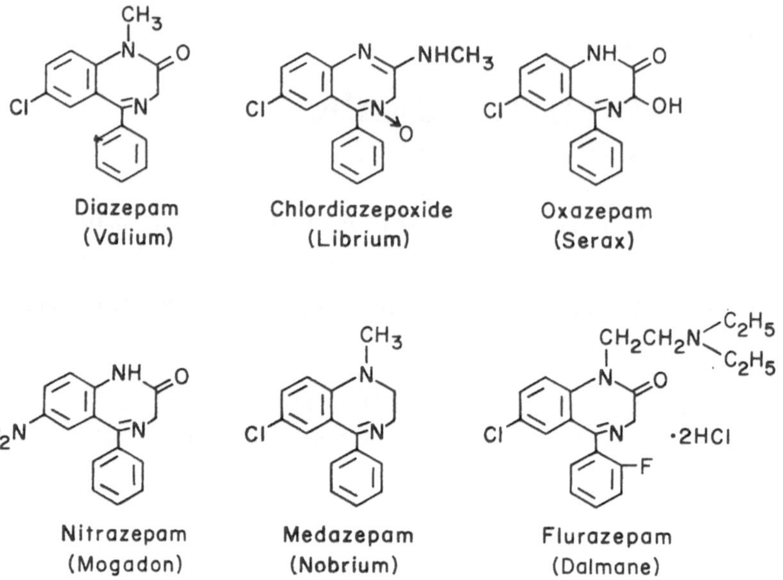

Diazepam
(Valium)

Chlordiazepoxide
(Librium)

Oxazepam
(Serax)

Nitrazepam
(Mogadon)

Medazepam
(Nobrium)

Flurazepam
(Dalmane)

FIGURE 11.1. Structures of some common prescription benzodiazepines.

with very high affinity (K_d in the nanomolar range) to both antagonists (like Ro15-1788) and inverse agonists [like β-carboline carboxylate ethyl ester, βCCE, and β-carboline carboxylate methyl ester, βCCM (Fig. 11.2)].

The functional interactions between the BZDR and the GABAR can be studied with in vitro binding assays because GABA increases the binding affinity of BZDR for the agonists while it decreases the affinity of the BZDR for the inverse agonists like βCCE and βCCM. However, it does not affect the affinity of the BZDR for the antagonists like Ro15-1788. These effects are the basis for a simple ligand binding test for determining if a BZDR ligand is an agonist, antagonist, or inverse agonist (17). This test involves assaying the inhibition of the binding of [³H]Ro15-1788 to the BZDR by the test substance both in the presence and absence of GABA. GABA will potentiate, decrease, or have no affect on the inhibition by the test substance of the [³H]Ro15-1788 binding to the BZDR if the substance is an agonist, inverse agonist, or antagonist, respectively (17). We have used this test to demonstrate that both the anti-idiotypic antibodies and the endogenous BZD-like substances bind to the BZDR with agonist properties (see below).

Several types of brain BZDRs have been identified based on their different binding, molecular, and detergent solubilization properties as well as on their different distribution in various brain regions (see Refs. 9 and 10 for a review). The molecular basis of this BZDR heterogeneity is not

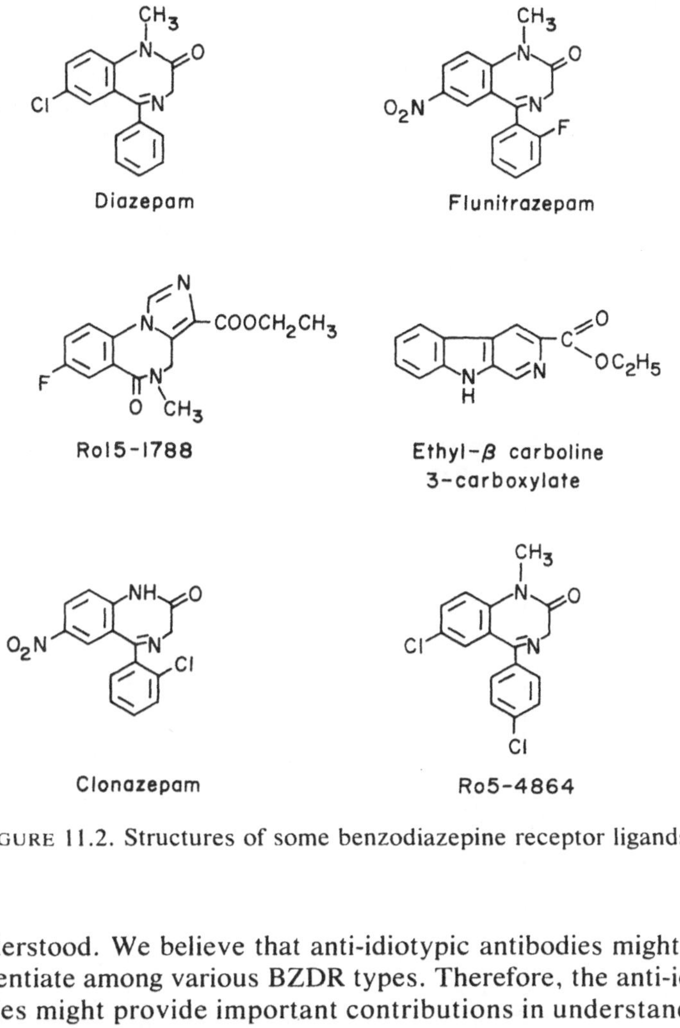

FIGURE 11.2. Structures of some benzodiazepine receptor ligands.

well understood. We believe that anti-idiotypic antibodies might be able to differentiate among various BZDR types. Therefore, the anti-idiotypic antibodies might provide important contributions in understanding the BZDR heterogeneity.

There are also "peripheral-type" benzodiazepine receptors which are abundant in kidney and other tissues as well as in the brain. These receptors can be differentiated from the aforementioned "central-type" BZDR by the binding of some benzodiazepines that distinguish between the two types. Thus clonazepam and Ro5-4864 (Fig. 11.2) are specific ligands for the "central type" and "peripheral type" BZDR, respectively (2,3,18). Both types of receptor, however, bind diazepam and flunitrazepam. The "peripheral type" BZDR is not associated to the GABAR nor to the chloride channel, and it is mainly localized in the external membrane of the mitochondria (19, 20). The function of this "peripheral-type" BZDR is still unknown.

Molecular Properties of the BZDR-GABAR Complex

The membrane-bound BZDR can be photoaffinity labeled with [^{3}H]flunitrazepam ([^{3}H]FNZ) or [^{3}H]Ro15-4513 (21, 22). Most of the labeling occurs ·in a 51,000-M_r peptide. Two other peptides with M_r of 54,000 and 57,000, respectively, are also labeled (Fig. 11.3).

The BZDRs have been solubilized with various detergents such as Triton X-100, Lubrol-PX, sodium deoxycholate, and digitonin (see Ref. 24 for a review) and purified by affinity chromatography on immobilized benzo-

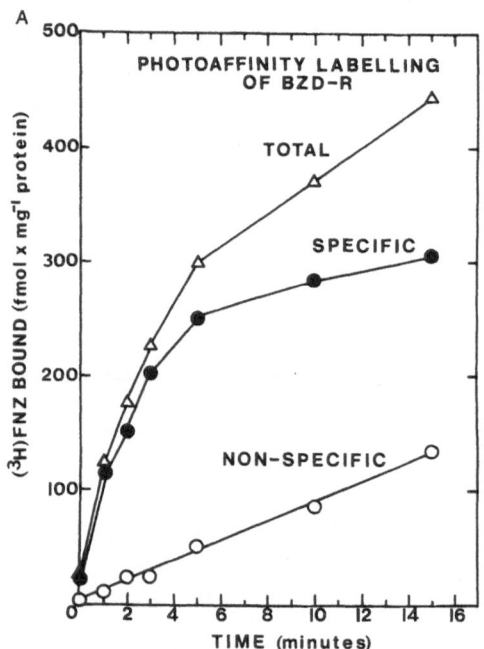

FIGURE 11.3. Photoaffinity labeling of benzodiazepine receptors from rat brain membranes. (A) 1.7 ml of brain membranes in 50 mM Tris hydrochloride, pH 7.4 (1 mg protein/ml) were incubated with 3 nM [^{3}H] FNZ in the presence or absence of 10^{-6} M clonazepam for 2 h at 0°C in the dark. The membranes were then irradiated with UV light (360 nm) for different lengths of time. Unbound [^{3}H] FNZ was eliminated after washing the membranes four times with 10 volumes of 50 mM Tris hydrochloride, pH 7.4. The [^{3}H] FNZ binding values obtained in the presence and absence of 10^{-6} M clonazepam were considered to represent the nonspecific and total binding, respectively. The specific binding is the difference

diazepines (Figs. 11.4–11.6 and Refs. 25–27). The functional interactions between the BZDR, the GABAR, and the chloride channel are lost after the solubilization of the BZDR with these detergents. However, all the activities and the functional interactions among the elements of the com-

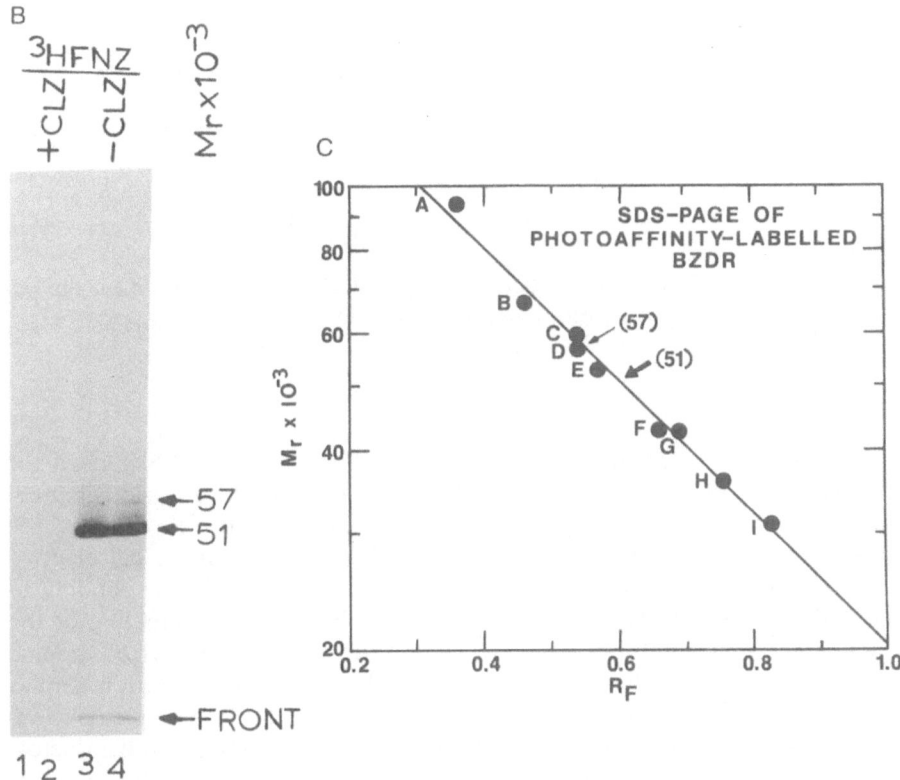

between total binding and nonspecific binding. (B) A fluorogram of the [³H] FNZ-labeled peptides after SDS-PAGE of the photoaffinity-labeled membranes. The membranes were photoaffinity labeled for 3 min. Lanes 3 and 4 show that most of the [³H] FNZ is bound to a peptide of 51,000 M_r. Minor bands corresponding to peptides of 57,000 and 54,000 M_r are also seen. Lanes 1 and 2 show that the [³H] FNZ binding to these peptides could be blocked by 10^{-6} M clonazepam. (C) M_r values of the [³H] FNZ binding peptides calculated from their mobilities in SDS-polyacrylamide gel electrophoresis relative to standard proteins of known M_r. The following proteins were used in the calibration: A = phosphorylase *b* (M_r 94,000); B = bovine serum albumin (67,000); C = catalase (60,000); D = α-tubulin (57,000); E = β-tubulin (53,000); F = actin (43,000); G = ovalbumin (43,000); H = lactate dehydrogenase (36,000); I = clathrin-associated proteins (31,000). Electrophoresis was in 10% acrylamide gels (23).

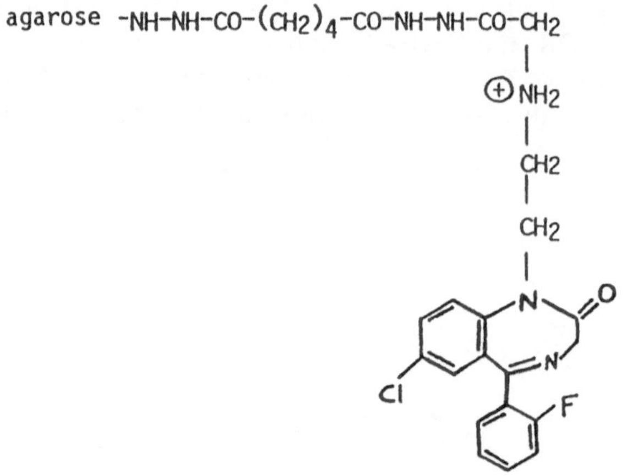

agarose -NH-NH-CO-(CH$_2$)$_4$-CO-NH-NH-CO-CH$_2$

FIGURE 11.4. The immobilized benzodiazepine Ro7-1986/1 was used for the purification of the BZDR by affinity chromatography.

plex are preserved when the zwitterionic detergent CHAPS is used for the solubilization (24, 28). The BZDR, GABAR, and the chloride channel activities copurify (27, 29), which demonstrates that the three activities are part of a molecular complex. The SDS-polyacrylamide gel electrophoresis of the purified BZDR shows two main bands corresponding to 51,000- and 57,000-M_r peptides (Fig. 11.5). Photoaffinity labeling of the purified receptor with [^{3}H]FNZ shows that the high-affinity benzodiazepine binding sites are on the 51,000-M_r peptide (Fig. 11.6). This result is similar to the one obtained with intact membranes (Fig. 11.3). Some radiation-inactivation experiments (31) and studies on the hydrodynamic parameters of the Triton X-100-solubilized receptor (32) suggest that the receptor complex has a M_r of 200,000 to 350,000. Thus the complex might be a tetramer protein. However, other radiation-inactivation (33) and hydrodynamic studies with CHAPS-solubilized receptor (34), where all the activities of the complex are present and the functional interactions between the subunits are preserved, indicate that the [receptor-Cl$^-$ channel] complex has an M_r of approximately 600,000. One explanation for the discrepancy between the M_r values obtained when using Triton-X 100 or CHAPS is the possible dimerization of the receptor in the latter (34). This dimerization might have a functional meaning.

At the present time we are immunizing rabbits and BALB/c mice with the purified BZDR to make conventional antisera and monoclonal antibodies (mAbs) to the receptor. Two other groups have recently reported success in making antibodies to the purified BZDR (35–37).

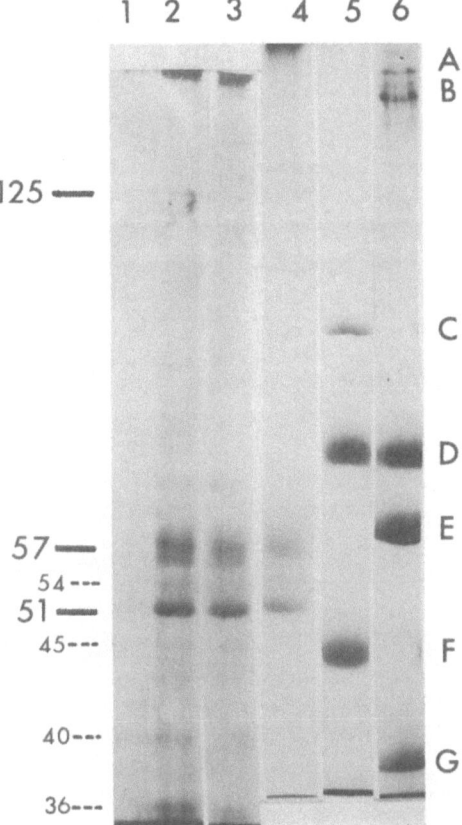

Figure 11.5. Purification of the BZDR by affinity chromatography on the immobilized benzodiazepine Ro7-1986/1 (Fig. 11.4). The procedure is based on that described by Sigel et al. (26). The membranes from bovine brain cortex were solubilized with 0.5% (w/v) sodium deoxycholate in 10 mM Tris hydrochloride, pH 7.4, and 150 mM KCl in the presence of protease inhibitors. The extract was passed through the affinity chromatography column and the retained material was eluted with 10 mM chlorazepate in 0.05% (v/v) Triton X-100 and 10 mM potassium phosphate buffer, pH 7.4. The proteins were precipitated with 10% TCA, washed with acetone, and subjected to SDS-PAGE in 10% acrylamide gels (23). Lanes 1–3 correspond to various fractions eluted from the affinity column and stained by a silver method (30). Lane 2 corresponds to the fraction with the highest receptor content as determined by [^{3}H] muscimol binding. Lane 4 corresponds to the same fraction as lane 2 but containing five fold more protein. Lanes 4–6 were stained with Coomassie brilliant blue. The purified receptor showed two main peptides with 51,000 M_r and 57,000 M_r, respectively. There are also minor peptides of 125,000, 54,000, 45,000, 40,000, and 36,000 M_r. Lanes 5 and 6 contain the following molecular weight markers: A = thyroglobulin (330,000); B = ferritin (220,000); C = phosphorylase b (94,000); D = bovine serum albumin (67,000); E = catalase (60,000); F = ovalbumin (43,000); G = lactate dehydrogenase (36,000). Lanes 1–3 and 4–6 are from different SDS-PAGE runs. The M_r values were determined as in Fig. 11.3.

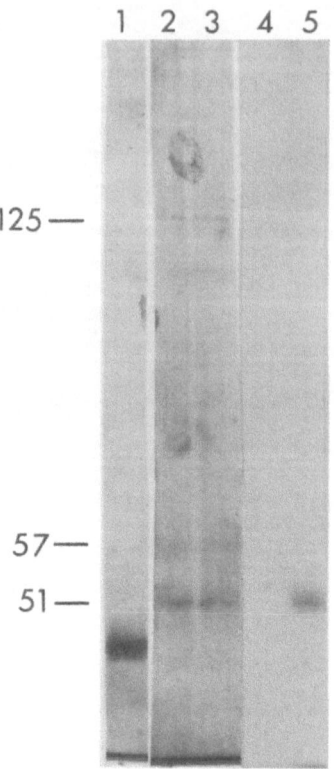

FIGURE 11.6. Photoaffinity labeling of the purified BZDR with [³H] FNZ. The chlorazepate was eliminated from the affinity-purified receptor of Fig. 11.5 by ion-exchange chromatography on DEAE-Sephacel. The purified BZDR was then photoaffinity labeled with 20 nM [³H] FNZ. Lanes 2 and 3 show the purified receptor stained by the silver method and lanes 4 and 5 are the fluorograms. In lane 4 the photoaffinity labeling was done in the presence of 20 μM unlabeled flunitrazepam. Lane 1 has the 43,000-molecular-weight marker ovalbumin. The SDS-PAGE was as in Fig. 11.5.

Monoclonal Antibodies to Benzodiazepines

Our initial attempts to prepare relatively large quantities of purified BZDR (which are needed for the immunizations) were frustrated by the difficulty in obtaining the BZD Ro7-1986/1 which is necessary for the purification of the BZDR by affinity chromatography (Fig. 11.4). We thought that one possible way to circumvent this obstacle was to prepare mAbs to the BZDR via the anti-idiotype antibody approach. Thus, BALB/c mice were immunized with 3-hemisuccinyloxyclonazepam conjugated to bovine serum albumin (BZD-BSA). The preparation and characterization of the mAbs to BZDs s has been reported elsewhere (38). A competition assay was used for detecting both the auto-anti-idiotypic mAbs to the BZDR and the mAbs to BZDs. The assay was based on the inhibition by the hybridoma culture media of the [^{3}H]FNZ binding to the membrane BZDR. Table 11.1 shows that four mAbs (out of the 591 generated) affected the binding of [^{3}H]FNZ to the membrane-bound BZDR. The stimulation of [^{3}H]FNZ binding induced by the mAb 21-7F9 was due to the binding of this mAb to both the brain membranes and [^{3}H]FNZ as shown below. Table 11.1 also shows that the four antibodies bound [^{3}H]FNZ, which indicates that they are anti-BZD mAbs and not auto-anti-idiotypic mAbs to the BZDR.

Table 11.2 shows that the anti-BZD mAbs bind [^{3}H]FNZ, [^{3}H]diazepam, and [^{3}H]Ro5-4864 with very high affinities. However, no binding of [^{3}H]Ro15-1788, [^{3}H]βCCE, and [^{3}H]muscimol to the mAbs was detected.

The binding affinities of the four anti-BZD mAbs for [^{3}H]FNZ were calculated by Scatchard analysis. The K_ds are 0.5×10^{-9} M for 21-7F9, 2.3×10^{-9} for 21-8C6, 4.1×10^{-9} M for 22-4G9, and 4.2×10^{-9} M for 26-5H12. These are comparable to the affinities of the brain membrane benzodiazepine receptors for the same ligand. A comparative study of the affinities of the mAbs for various benzodiazepine receptor ligands shows that the four mAbs have similar (although not identical) specificities for the BZDR ligands tested. Thus, all four mAbs bound with very high affinities to Ro5-3453, diazepam, Ro11-6896, flunitrazepam, Ro5-3438, Ro5-4864, Ro11-6893, and medazapam (the K_d values are in the nM range).

TABLE 11.1. Binding of [^{3}H] FNZ to anti-benzodiazepine mAbs and to brain membranes in the presence of the mAbs.[a]

mAb	Binding to membranes[b]		fmol bound to mAb[c]
	fmol	% of control	
P3X63Ag8.6.5.3	9.5 ± 0.1	100	0
21-7F9	20.7 ± 0.3	218	154 ± 2
21-8C6	2.2 ± 0.1	23	79 ± 10
22-4G9	2.7 ± 0.1	28	34 ± 3
26-5H12	6.8 ± 0.6	71	42 ± 2

[a]Values are the mean ± SD of duplicate samples of a representative experiment.
[b]The reaction mixture was 375 μl of 50 mM Tris hydrochloride, pH 7.4, 50 μl of hybridoma culture medium, and 50 μl of bovine brain membranes (approx. 250 μg protein), which was incubated at 37°C for 1 h. The mixture was cooled to 4°C and 25 μl of [^{3}H] FNZ was added to a final concentration of 0.1 nM and incubated for 1 h at 4°C. The membrane-bound [^{3}H] FNZ was determined after filtration through a Whatman GF/B filter as described below. Clonazepam (10^{-6}M) was used for determining the nonspecific binding. The binding of [^{3}H] FNZ to the membranes was not affected by control mAbs that did not recognize benzodiazepines nor by the culture medium of the parental myeloma P3X63Ag8.6.5.3.
[c]The reaction mixture was 50 μl of hybridoma culture medium plus 450 μl of 50 mM Tris hydrochloride, pH 7.4, containing 0.66 nM [^{3}H] FNZ. The reaction proceeded for 1 h at 4°C. The nonspecific binding was determined by using 10^{-6} M unlabeled FNZ. The antibody-benzodiazepine complexes were precipitated with 7.5% (wt/vol) polyethylene glycol 8,000 using 0.125% (wt/vol) bovine gamma-globulin as a carrier. The precipitated complexes were recovered by filtration through Whatman GF/B filters and the radioactivity determined by liquid scintillation counting (24).

TABLE 11.2. Binding of ^{3}H-labeled receptor ligands to anti-benzodiazepine mAbs.[a]

^{3}H-labeled ligand (nM)	Concn. nM	fmol bound to mAb:			
		21-7F9	21-8C6	22-4G9	26-5H12
[^{3}H]FNZ	0.66	154 ± 2	79 ± 10	34 ± 3	42 ± 2
[^{3}H]Diazepam	0.46	108 ± 2	51 ± 3	9 ± 3	7 ± 1
[^{3}H]Ro5-4864	0.86	176 ± 13	100 ± 8	2 ± 1	26 ± 2
[^{3}H]Ro15-1788	1.15	0	0	0	0
[^{3}H]βCCE	2.0	0	0	0	0
[^{3}H]Muscimol	2.45	0	0	0	0

[a]Data are mean ± SD values from duplicate samples of a representative experiment. The reaction mixture was 50 μl of hybridoma culture medium plus 450 μl of 50 mM Tris hydrochloride, pH 7.4, containing the radioligand. The reaction proceeded for 1 h at 4°C. The nonspecific binding was determined by using 10^{-6} M corresponding unlabeled ligand. The antibody-benzodiazepine complexes were precipitated with 7.5% (wt/vol) polyethylene glycol 8,000 using 0.125% (w/v) bovine γ-globulin as a carrier (24).

However, the antibodies have very low affinities for the receptor inverse agonists (βCCE and βCCM) and antagonists (Ro15-1788 and CGS 8216) and the triazolopyridazine agonist CL218,872. The high binding affinities for Ro5-4864 should be noted since this is a specific ligand of the "peripheral type" BZDR (see above). Thus, the anti-BZD mAbs bind benzodiazepine ligands for both the "peripheral type" and "central type" BZDRs.

Anti-Idiotypic Antibodies to the Benzodiazepine Receptor

Syngeneic immunizations of BALB/c mice with the anti-BZD mAbs 21-7F9, 26-5H12, 21-8C6, and 22-4G9 were done as described in Table 11.3. The sera were periodically tested for the presence of anti-idiotypic antibodies that blocked the binding of [^{3}H]FNZ to the BZDR. After six months of repetitive immunizations, the serum of one mouse (number 3, immunized with the mAb 26-5H12) had anti-idiotypic antibodies that bound to the BZDR with the properties of an agonist. The serum inhibited the binding of [^{3}H]FNZ to the BZDR, and GABA potentiated the inhibitory effect of the antiserum on the binding of [^{3}H]Ro15-1788 to the BZDR (Tables 11.3 and 11.4). As shown in tables 11.3 and 11.4, neither the serum from mouse number 3 nor the sera from mice 1, 2, and 4 contained antibodies that bound [^{3}H]FNZ. These four mice were immunized with either the mAb 21-7F9 or 26-5H12. These monoclonal antibodies are IgMs that were quickly eliminated from the host sera. This was not the case for the anti-BZD IgGs 21-8C6 and 22-4G9, which remained in the host's sera for many

TABLE 11.3. Inhibition of [³H] FNZ and [³H] Ro15-1788 binding to the rat brain membrane BZDR by the sera from mice immunized with anti-BZD mAbs.[a,b]

Sera from animal	mAb immunogen	[³H] FNZ binding to BZDR membrane		[³H] FNZ binding to the sera (cpm)	[³H] Ro15-1788 binding to BZDR			
					No GABA		10^{-4} M GABA	
		cpm	% of control		cpm	% of control	cpm	% of control
1	21-7F9	633 ± 17	81 ± 2	14 ± 1	980	86	985	86
2	21-7F9	629 ± 43	81 ± 6	14 ± 1	709	62	486	43
3	26-5H12	375 ± 12	45 ± 2	11 ± 1	1,054	92	1,058	93
4	26-5H12	637 ± 0	82 ± 0	11 ± 1	1,264	111	1,316	115
5	21-8C6	362 ± 18	46 ± 2	923 ± 214	1,107	97	1,171	103
6	21-8C6	567 ± 8	73 ± 1	391 ± 76	1,057	93	894	78
7	22-4G9	433 ± 3	56 ± 0	243 ± 17	1,041	91	981	86
8	22-4G9	504 ± 38	65 ± 5	175 ± 87	1,125	99	996	87
9	Control serum	777 ± 10	100 ± 1	12 ± 0	1,178	103	1,348	118
10	Control serum	769 ± 1	99 ± 0	15 ± 4				
11	No serum	780 ± 29	100 ± 4		1,141	100		

[a]BALB/c mice were injected intraperitoneally with 100 μg of purified anti-BZD mAb at days 0, 14, 49, 93, and 159. The first injection was given in complete Freunds adjuvant and boost injections were given in incomplete Freunds adjuvant. Sera were collected at day 169 (10 days after the last injection).

[b]The inhibition of the radioligand binding to the BZDR by 50 μl of the diluted sera (diluted 1:10 in 50 mM Tris hydrochloride, pH 7.4) and the binding of the radioligands to the Abs of the sera were assayed as in Table 11.1. [³H] FNZ (5×10^{-11} M) and [³H] Ro15-1788 (4×10^{-9} M) were used in the membrane binding assays and [³H] FNZ (5×10^{-11} M) in the antibody binding assay.

TABLE 11.4. Inhibition of [³H] FNZ and [³H] Ro15-1788 binding to the rat brain membrane BZDR by the sera from mice immunized with anti-BZD mAbs.ª

Sera from animal	mAb immunogen	[³H] FNZ binding to BZDR membrane		[³H] FNZ binding to the sera (cpm)	[³H] Ro15-1788 binding to BZDR			
					No GABA		10^{-4} M GABA	
		cpm	% of control		cpm	% of control	cpm	% of control
3	26-5H12	372 ± 52	40 ± 6	13 ± 1	796 ± 11	64 ± 1	519 ± 6	41 ± 0
4	26-5H12	896 ± 8	95 ± 1	11 ± 1	1,357 ± 55	109 ± 4	1,284 ± 51	103 ± 4
5	21-8C6	840 ± 6	89 ± 1	111 ± 1	1,379 ± 46	111 ± 4	1,341 ± 12	108 ± 1
6	21-8C6	893 ± 34	95 ± 4	27 ± 1	1,329 ± 60	107 ± 5	1,275 ± 47	102 ± 4
9	Control serum	906 ± 6	96 ± 1	14 ± 1	1,386 ± 38	111 ± 3	1,348 ± 16	108 ± 1
10	Control serum	946 ± 21	101 ± 2	15 ± 1	1,423 ± 30	114 ± 2	1,339 ± 31	108 ± 2
11	No serum	940 ± 98	100 ± 10		1,244 ± 29	100 ± 2	1,210 ± 132	97 ± 11

ªThe experimental procedure was the same as the one used to obtain the data in Table 11.3 except that the sera used in this experiment were collected at day 204 (45 days after the last injection).

TABLE 11.5. Inhibition of [^{3}H] FNZ binding to the mouse anti-BZD mAbs by sera from rabbits immunized with anti-BZD mAbs.[a]

Rabbit serum	mAb immunogen	[^{3}H] FNZ binding (cpm) to mAb:			
		21-7F9	26-5H12	21-8C6	22-4G9
1	Control serum	1,802 ± 103	952 ± 28	578 ± 104	169 ± 27
2	21-8C6	2,151 ± 87	833 ± 22	2,036 ± 171	
3	22-4G9	2,190 ± 71	878 ± 56		1,463 ± 24
4	21-7F9	1,990 ± 79	1,135 ± 6		
5	26-5H12	2,186 ± 20	413 ± 13		
—	No serum	1,785 ± 147	772 ± 18	549 ± 4	171 ± 5

[a]Rabbits were immunized with 1 mg of purified anti-BZD mAb at days 0, 30, 47, and 58. The first immunization was given intradermally (several injections) in complete Freunds adjuvant and the others in incomplete Freunds adjuvant. The sera were collected at day 68 (10 days after the last immunization). Fifty μl of diluted rabbit sera (diluted 1:5 in 50 mM Tris hydrochloride, pH 7.4) were used for inhibiting the binding of [^{3}H] FNZ to the anti-BZD mAbs. The binding assay was described in Table 1. The concentration of [^{3}H] FNZ used was 5×10^{-11} M.

days after immunization. The injected anti-BZD mAbs that remained in the sera (and not anti-idiotypic antibodies to the BZDR) were responsible for the inhibition of [^{3}H]FNZ binding to the BZDR (Tables 11.3 and 11.4). At the present time we are also attempting to make monoclonal anti-idiotypic antibodies to the BZDR.

In view of the difficulty in raising anti-idiotypic antibodies by syngeneic immunizations with the anti-BZD mAbs, we have also immunized rabbits with the mouse anti-BZD mAbs. After three immunizations, the serum from rabbit number 5 (immunized with the anti-BZD mAb 26-5H12) showed the presence of anti-idiotypic antibodies that blocked the binding of [^{3}H]FNZ to the mAb idiotype (Table 11.5). However, the antiserum did not block the binding of [^{3}H]FNZ to the BZDR (not shown). An unexpected result was obtained with the sera from the rabbits immunized with 21-8C6 and 22-4G9. The sera from these animals dramatically increased the binding of [^{3}H]FNZ to the mAb idiotype (Table 11.5). A similar phenomenon (potentiation of the binding of the antigen to the antibody idiotype by the anti-idiotypic antibody) has also been found in a different system and it is described in detail by Sawutz and Homcy in Chapter 12 of this book. We do not know yet the functional meaning of this observation. Nevertheless, it is tempting to speculate that the interactions between idiotypic and anti-idiotypic antibodies with hormones and neurotransmitters and their membrane receptors might be involved in the regulation of receptor function by the immune system. Alterations of the balance of the idiotype-anti-idiotype-receptor network might trigger some autoimmune diseases, as discussed in other chapters of this book.

Monoclonal Antibody 21-7F9 Recognizes Endogenous Brain Benzodiazepine-like Molecules

The binding of the anti-BZD mAb 21-7F9 to the rat brain membranes (Table 11.1) was confirmed by immunocytochemistry and immunoblotting, as reported elsewhere (39). The immunocytochemistry experiments showed that the anti-BZD mAb binds to BZD-like molecules distributed throughout the brain. Nevertheless, these are present in neurons but not in glial cells (Fig. 11.7). These molecules are in the perikarya and processes of many but not all neurons. The immunoblots show that the anti-BZD mAb 21-7F9 also binds to a number of peptides present in the homogenates and membranes of various rat brain regions. No qualitative differences between these regions were found. The molecular weights of the reactive peptides range between 20,000 and 300,000 M_r (Fig. 11.8). The most abundant peptide with BZD-like epitopes is the 100,000 M_r. Some peptides (including the 100,000-M_r species) are associated with a synaptosomal membrane fraction (39).

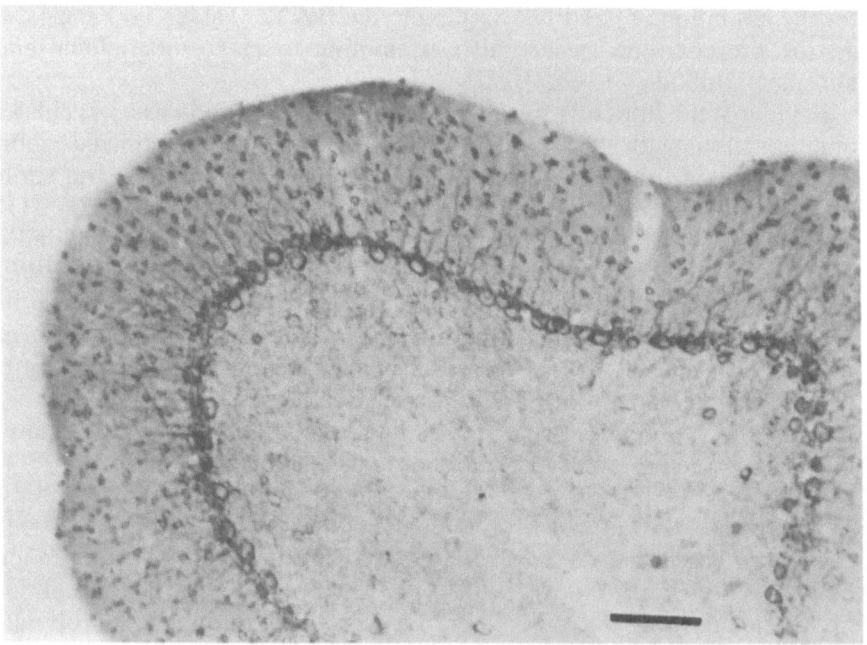

FIGURE 11.7. Immunocytochemistry with the anti-BZD mAb 21-7F9: Sprague-Dawley rat cerebellum. Differential interference-contrast optics. The peroxidase-antiperoxidase immunocytochemistry was done as explained elsewhere, using hybridoma culture medium (40, 41).

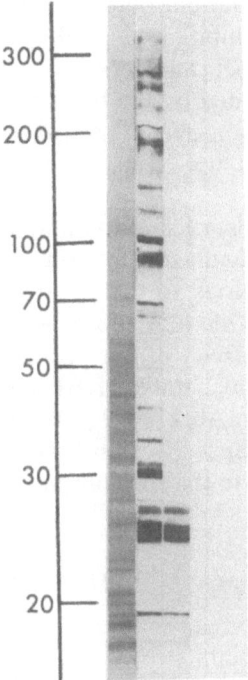

FIGURE 11.8. Immunoblots of brain membranes with mAb 21-7F9. The left strip shows the staining of the membrane proteins with amido black. The middle and right strips show the immunoblots with mAb 21-7F9 without and with 10^{-6} M FNZ. On the left side of the figure the apparent molecular weight of the peptides in kilodaltons is indicated. The immunoblotting procedure using hybridoma culture medium and based on a mouse peroxidase-antiperoxidase (PAP) reaction has been described elsewhere (42). Notice the differences between the results obtained with the crude membrane preparation used in this experiment and the results obtained when using the whole homogenate or purified synaptosomal membranes (39). The M_r 100,000 peptide is present in all the preparations.

The binding of the anti-BZD mAb 21-7F9 to the brain molecules (in the immunocytochemistry and immunoblot assays) as well as the binding of [^{3}H]FNZ to the mAb are blocked by FNZ and diazepam but not by the antagonist Ro15-1788. Other effective blockers of the binding of the mAb to the tissues are Ro11-6896, Ro11-6893, and Ro5-4864. In contrast, βCCE and βCCM were ineffective blockers. Thus, these results suggest that there are very close structural similarities between BZDs and the endogenous brain BZD-like epitopes recognized by 21-7F9.

The BZD-like molecules were purified from the rat brain homogenates by immunoaffinity chromatography on immobilized 21-7F9. The retained material was eluted with 0.2 M acetic acid, which was eliminated by drying under reduced pressure, and dissolved in 50 mM Tris hydrochloride, pH 7.4. The eluted material inhibited the binding of agonists, antagonists, and inverse agonists to the BZDR and behaved as a BZDR agonist in the presence of GABA, as determined by the [^{3}H]Ro15-1788 binding test explained above. The purified substance from the rat brain had a molecular weight smaller than 1,000, and it was resistant to proteases and heating for 5 minutes at 100°C.

The pharmacological properties of the immunoaffinity-purified molecules also differ from those of the diazepam binding inhibitor (DBI) peptide (43) in that (i) whereas DBI binds to the β-carboline site of the BZDR, the immunoaffinity-purified BZD-like substance binds to the BZD (agonist) binding site and (ii) the activity of our preparation, unlike that of DBI, is pronase resistant (44). Thus, unlike DBI and β-carbolines, the purified substance recognized by mAb 21-7F9 potentiates the inhibitory effect of GABA on the GABA synapses, which indicates that it is an agonist and, therefore, strongly suggests that it is an anxiolytic substance.

Purification of a Benzodiazepine from the Bovine Brain

The most active substance that blocked the [^{3}H]FNZ binding to the BZDR after the immunoaffinity chromatography purification was further purified to homogeneity by gel filtration through Sephadex G-25 followed by two sequential HPLC steps. In these studies we used bovine brains. The purified substance was characterized by HPLC and mass spectroscopy as *N*-desmethyldiazepam or nordiazepam (45).

The main questions that need to be answered are whether the brain benzodiazepines are endogenous or exogenous and whether they have biological significance. It is unlikely that a heterocyclic molecule containing a chlorine atom could be totally biosynthesized by mammals (Fig. 11.9).

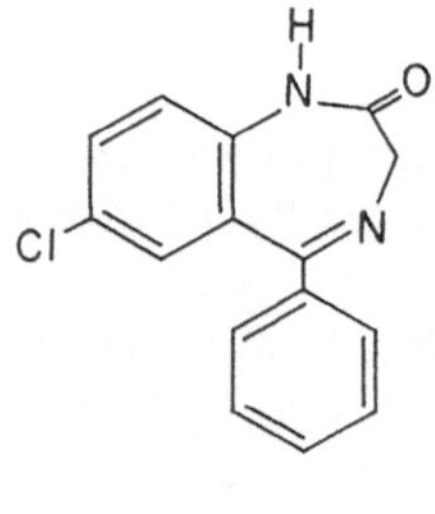

N-desmethyl
diazepam

FIGURE 11.9. Structure of *N*-desmethyldiazepam (nordiazepam).

A possibility worth considering is that N-desmethyldiazepam accumulates in the mammalian brain as a result of the intake of synthetic benzodiazepines that could be present in the drinking water or the animal food or may even be purposely given to the cattle. N-Desmethyldiazepam is a metabolite that is formed in the body during the degradation of several benzodiazepines such as diazepam, chlordiazepoxide, clorazepate, and prazepam (46). N-Desmethyldiazepam (nordiazepam) is also used as a prescription benzodiazepine. The elimination from the tissues of N-desmethyldiazepam is slow due to its low clearance rate. The elimination half-life of this substance in humans is between 50 and 100 h (46). Furthermore, N-desmethyldiazepam and other benzodiazepines accumulate in the body after prolonged use. The slow elimination rate and the accumulation in the tissues seem to be due to the tight binding of these molecules to serum albumin and other tissue proteins (46). The protein immunoreactivities detected with the anti-benzodiazepine mAb 21-7F9 in the immunoblotting experiments (Ref. 39 and Fig. 11.8) could result from BZDs that are tightly bound to proteins and not from the amino acid epitopes. If this is the explanation for the presence of N-desmethyldiazepam in the bovine brain, then serious toxicological implications would be derived from our observations. In this case the localization of the BZD-like brain molecules might reflect accumulation and detoxification processes.

The hypothesis of an industrial origin of the N-desmethyldiazepam seems unlikely, especially in the case of laboratory rats that have been fed with a controlled diet throughout their lives. (As described in the previous section, the initial experiments were done with rat brains.) Moreover, we have found BZD-like immunoreactivity in the five human cerebellum samples tested, which have been stored in paraffin since 1940. The distribution of the immunoreactivity (Fig. 11.10) is identical to that observed with fresh rat cerebellum fixed with 4% PLP (Fig. 11.7). The immunocytochemistry experiments with the 1940 human brain samples indicate that the brain benzodiazepine-like molecules that are recognized by the mAb 21-7F9 have a biological, not a chemical, origin since the first chemical synthesis of benzodiazepines was done in 1955 (47) while Librium appeared on the market in 1960 (47). Nevertheless, we are not sure that the BZD-like immunoreactivities detected by immunocytochemistry and immunoblotting in the brains correspond to the isolated N-desmethyldiazepam. Until this is established, we would like to leave open the possibility that the isolated benzodiazepine might have an industrial origin.

Regardless of the origin of the endogenous BZDs, they seem to be mostly in bound form, attached to proteins and probably to other biomolecules too. These might represent not only storage compartments for their use as neuromodulators of the GABA transmission but also compartments for their biotransformation, accumulation, and elimination. Benzodiazepines are lipophilic compounds that are converted to more hydrophilic metabolites and conjugated to other biomolecules by the cytochrome P-450 enzyme system. This system plays a major role in the biotransformation of

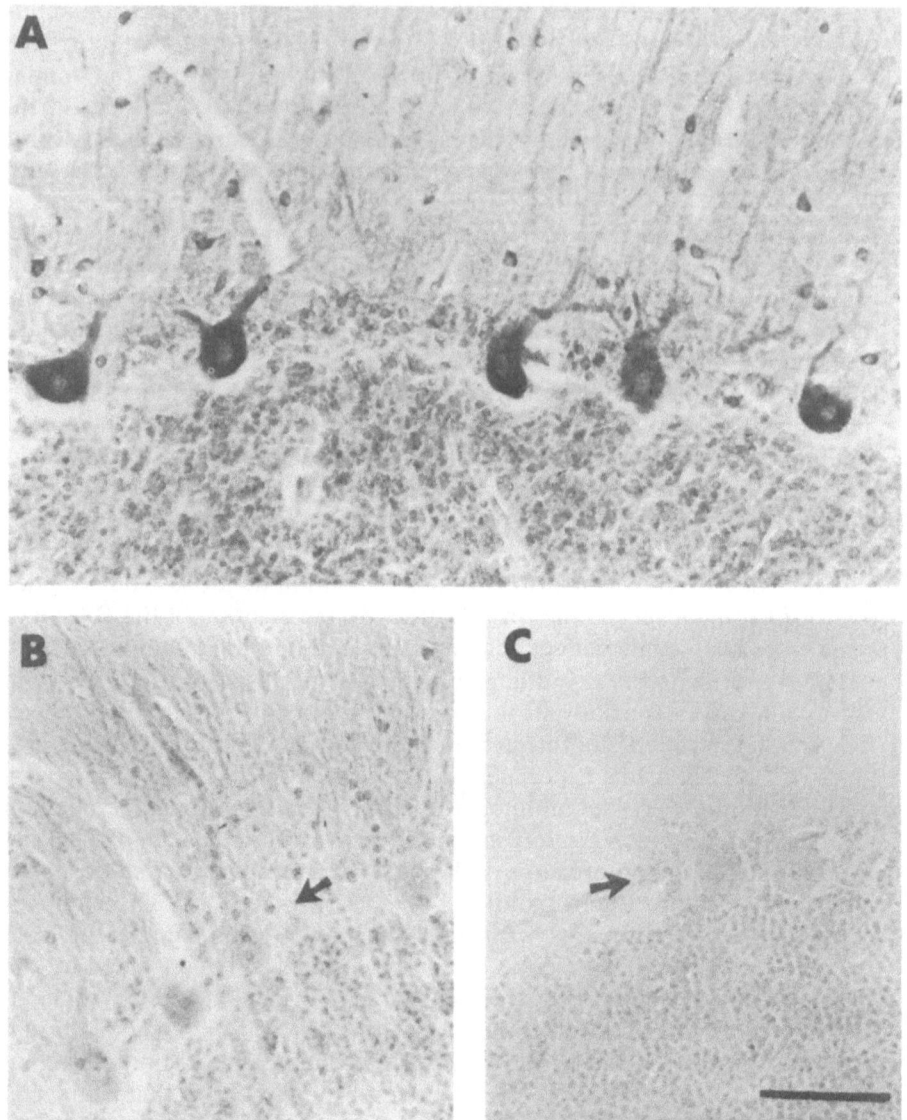

FIGURE 11.10. Human cerebellum stored in paraffin since 1940. Immunocyto-chemistry with the anti-BZD mAb 21-7F9 (A and B). Panel C shows a negative control using the same immunocytochemical procedure with the culture medium from the parental myeloma line P3X63Ag8.6.5.3. The binding of the mAb 21-7F9 to the cerebellum was blocked by 10^{-7} M diazepam (B). The arrows show the Purkinje cell layer. The bar represents 100 μm. From Ref. 45.

xenobiotics (drugs, mutagens, carcinogens, etc.) as well as endogenous substances such as steroids and prostaglandins (48).

Another possibility worth considering is that benzodiazepines could be biosynthesized by microorganisms and/or plants and that they might become components of our diet (like vitamines, for example). These substances could accumulate in the brain tissue as discussed above and they could affect GABA transmission. N-Desmethyldiazepam could be taken from the diet as such or in the form of a precursor that could be converted to N-desmethyldiazepam in the animal body. This interpretation is consistent with the results obtained with the 1940 human brains. Some support for this hypothesis also comes from the demonstration of benzodiazepine biosynthesis in the fungus *Penicillium cyclopium* (49, 50). This possibility would be an example of functional modulation of a neurotransmitter receptor (GABA receptor) by a dietary component (benzodiazepine), which in turn could affect brain function and behavior.

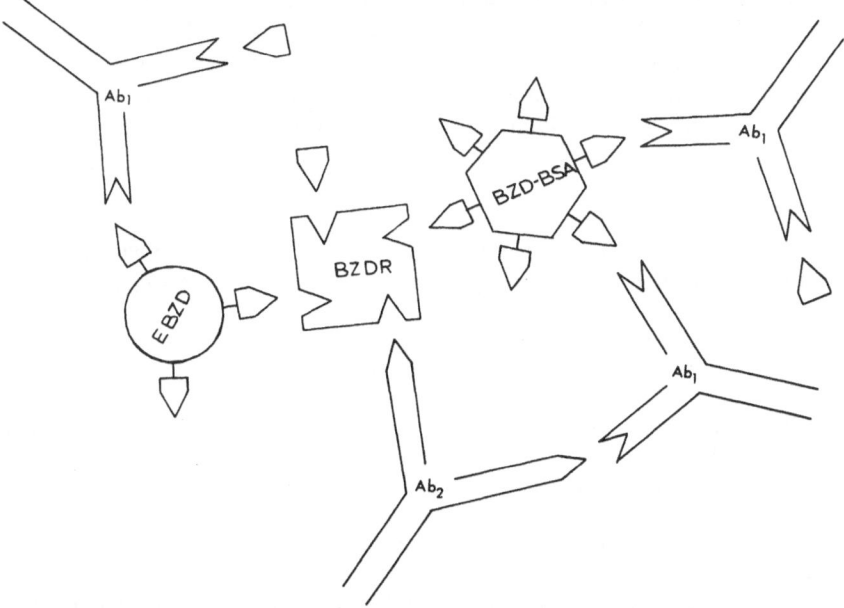

FIGURE 11.11. Schematic illustration of the approach followed in the study of both the BZDR and the endogenous ligands by using anti-benzodiazepine monoclonal antibodies. The benzodiazepines (arrowhead-shaped figures) were conjugated to BSA (BZD-BSA) and monoclonal antibodies to benzodiazepines (Ab₁) were made. We are using the Ab₁ antibodies to generate the anti-idiotypic antibodies Ab₂. The latter reacts with the benzodiazepine receptors at the benzodiazepine binding site (the agonist site). We have also used the antibody Ab₁ (21-7F9) for identifying endogenous brain molecules with benzodiazepine-like epitopes (EBZD). Free benzodiazepines react with both the anti-BZD monoclonal antibody Ab₁ (or 21-7F9) and the BZDR.

Concluding Remarks

We have shown that, although difficult, it is possible to make anti-idiotypic antibodies to the BZDR. During the course of these studies, we have also made monoclonal antibodies to BZDs. We are using these anti-BZD mAbs for (i) the development of benzodiazepine immunoassays of urine and serum samples, (ii) the demonstration of the existence of endogenous BZD-like molecules in the brains of humans, rats, and cows, and (iii) the isolation of endogenous BZDs from the bovine brain. Figure 11.11 shows a schematic illustration of our anti-BZD and anti-idiotypic antibody approach for the study of the benzodiazepine receptors and their putative endogenous ligands.

Acknowledgments. We especially thank Dr. Haruo Okazaki of the Mayo Clinic for the 1940 human brain samples. We also thank Diane Godden and Linda Cerracchio for typing the manuscript. This research was supported by Grant NS17708 from the National Institute of Neurological and Communicative Disorders and Stroke, by a grant from the Epilepsy Foundation of America, by an Esther A. and Joseph Klingenstein Fellowship to A.L.D., and by a Fundación Juan March fellowship to J.V.

References

1. Tallman, J.F., S.M. Paul, P. Skolnick, and D.W. Gallager. 1980. Receptors for the age of anxiety: molecular pharmacology of the benzodiazepines. Science **207**:274.
2. Bräestrup, C., and R. Squires. 1977. Specific benzodiazepine receptors in rat brain characterized by high-affinity [^{3}H]diazepam binding. Proc. Natl. Acad. Sci. USA **74**:3805.
3. Möhler, H., and T. Okada. 1977. Benzodiazepine receptor: demonstration in the central nervous system. Science **198**:849.
4. Olsen, R.W. 1981. GABA-benzodiazepine-barbiturate receptor interactions. J. Neurochem. **37**:1.
5. Richards, J.G., and H. Möhler. 1984. Benzodiazepine receptors. Neuropharmacology **23**:233.
6. Ticku, M.K. 1983. Benzodiazepine-GABA receptor-ionophore complex. Current concepts. Neuropharmacology **22**:1459.
7. Turner, A.J., and S.R. Whittle. 1983. Biochemical dissection of the γ-aminobutyrate synapses. Biochem. J. **209**:29.
8. Guidotti, A., M.G. Corda, B.C. Wise, F. Vaccarino, and E. Costa. 1983. GABAergic synapses. Supramolecular organization and biochemical regulation. Neuropharmacology **22**:1471.
9. Squires, R.F. 1984. Benzodiazepine receptors, pp. 261–306. *In* A. Lajtha (ed.), Handbook of neurochemistry, Vol. 6. Plenum Press, New York.
10. Tallman, J.F., and D.W. Gallager. 1985. The GABAergic system: a locus of benzodiazepine action. Annu. Rev. Neurosci. **8**:21.

11. Choi, D.W., D.H. Farb, and G.D. Fischbach. 1981. Chlordiazepoxide selectively potentiates GABA conductance of spinal cord and sensory neurons in cell cultures. J. Neurophysiol. **45**:621.
12. Study, R.E., and J.L. Barker. 1982. Cellular mechanisms of benzodiazepine action. J. Am. Med. Assoc. **247**:2147.
13. Bormann, J., and D.E. Clapham. 1985. γ-Aminobutyric acid receptor channels in adrenal chromaffin cells: a patch clamp study. Proc. Natl. Acad. Sci. USA **82**:2168.
14. Bormann, J., P. Ferrero, A. Guidotti, and E. Costa. 1985. Neuropeptide modulation of GABA receptor Cl⁻ channels. Regulatory Peptides **246(4)**:33.
15. Barker, J.L., and D.G. Owen. 1986. Electrophysiological pharmacology of GABA and diazepam in cultured CNS neurons, pp. 136–165. In R.W. Olsen and J.D. Venter (eds.), Receptor biochemistry and methodology, Vol. 5. Alan R. Liss, New York.
16. Bloom, F., and L. Iversen. 1971. Localizing ³H-GABA in nerve terminals or rat cerebral cortex by electron microscopic autoradiography. Nature **229**:628.
17. Möhler, H., and J.G. Richards. 1981. Agonist and antagonist benzodiazepine receptor interaction in vitro. Nature (London) **294**:763.
18. Syapin, P., and P. Skolnick. 1979. Characterization of benzodiazepine binding sites in cultured cells of neuronal origin. J. Neurochem. **32**:1047.
19. Anholt, R.R.H., P.L. Pedersen, E.B. De Souza, S.H. Synder. 1986. The peripheral-type benzodiazepine receptor. Localization to the mitochondrial outer membranes. J. Biol. Chem. **261**:576.
20. Basile, A.S., and P. Skolnick. 1986. Subcellular localization of "peripheral-type" binding sites for benzodiazepines in rat brain. J. Neurochem. **46**:305.
21. Möhler, H., M.K. Battersby, and J.G. Richards. 1980. Benzodiazepine receptor protein identified and visualized in brain tissue by a photoaffinity label. Proc. Natl. Acad. Sci. USA **77**:1666.
22. Möhler, H., W. Sieghart, J.G. Richards, and W. Hunkeler. 1984. Photoaffinity labelling of benzodiazepine receptors with a partial inverse agonist. Eur. J. Pharmacol. **102**:191.
23. Laemmli, U.K. 1979. Cleavage of structural proteins during the assembly of the head of bacteriophage T4. Nature (London) **227**:680.
24. Mernoff, S.T., H.M. Cherwinski, J.W. Becker, and A.L. De Blas. 1983. Solubilization of brain benzodiazepine receptors with a zwitterionic detergent. Optimal preservation of their functional interaction with the GABA receptors. J. Neurochem. **41**:752.
25. Martini, C., A. Lucacchini, G. Ronca, S. Hrelia, and C.A. Rossi. 1982. Isolation of putative benzodiazepine receptors from rat brain membranes by affinity chromatography. J. Neurochem. **38**:15.
26. Sigel, E., F.A. Stephenson, C. Mamalaki, and E.A. Barnard. 1983. A γ-aminobutyric acid/benzodiazepine receptor complex of bovine cerebral cortex. J. Biol. Chem. **258**:6965.
27. Kirkness, E.F., and A.J. Turner. 1986. The gamma-aminobutyric/benzodiazepine receptor from pig brain. Purification and characterization of the receptor complex from cerebral cortex and cerebellum. Biochem. J. **233**:265.
28. Stephenson, F.A., and R.W. Olsen. 1982. Solubilization by CHAPS detergent of barbiturate-enhanced benzodiazepine-GABA receptor complex. J. Neurochem. **39**:1579.

29. Sigel, E., and E.A. Barnard. 1984. A γ-aminobutyric acid/benzodiazepine receptor complex from bovine cerebral cortex. J. Biol. Chem. **259**:7219.
30. Merril, C.R., D. Goldman, S.A. Sedman, and M.H. Ebert. 1981. Ultrasensitive stain for proteins in polyacrylamide gels shows regional variations in cerebrospinal fluid proteins. Science **211**:1437.
31. Chang, L.-R., and E.A. Barnard. 1982. The benzodiazepine/GABA receptor complex: molecular size in brain synaptic membranes and in solution. J. Neurochem. **39**:1507.
32. Stephenson, T.A., E.A., Watkins, R.W. Olsen. 1982. Physiochemical characterization of detergent-solubilized γ-aminobutyric acid and benzodiazepine receptor proteins from bovine brain. Eur. J. Biochem. **123**:291.
33. Bräestrup, C., and M. Nielsen. 1986. Benzodiazepine receptor binding *in vivo* and efficacy, pp. 167–184. *In* R.W. Olsen and J.C. Venter (eds.), Receptor biochemistry and methodology, Vol. 5. Alan R. Liss, New York.
34. Ray, J.P., S.T. Mernoff, L. Sangameswaran, and A.L. De Blas. 1985. The Stokes radius of the CHAPS-solubilized benzodiazepine receptor complex. Neurochem. Res. **10**:1221.
35. Schoch, P., J.G. Richards, P. Haring, B. Takacs, C. Stahli, T. Staehelin, W. Haefely, and H. Möhler. 1985. Co-localization of GABA-A receptors and benzodiazepine receptors in the brain shown by monoclonal antibodies. Nature (London) **314**:168.
36. Haring, P., C. Stahli, B. Takacs, T. Staehelin, and H. Möhler. 1985. Monoclonal antibodies reveal structural homogeneity of gamma-aminobutyric acid/benzodiazepine receptors in different brain areas. Proc. Natl. Acad. Sci. USA **82**:4837.
37. Stephenson, F.A., S.O. Casalotti, C. Mamalaki, and E.A. Barnard. 1986. Antibodies recognizing the GABAa/benzodiazepine receptor including its regulatory sites. J. Neurochem. **46**:854.
38. De Blas, A.L., L. Sangameswaran, S.A. Haney, D. Park, C.J. Abraham, Jr., and C.A. Rayner. 1985. Monoclonal antibodies to benzodiazepines. J. Neurochem. **45**:1748.
39. Sangameswaran, L., and A.L. De Blas. 1985. Demonstration of benzodiazepine-like molecules in the mammalian brain with a monoclonal antibody to benzodiazepines. Proc. Natl. Acad. Sci. USA **82**:5560.
40. De Blas, A.L. 1984. Monoclonal antibodies to specific astroglial and neuronal antigens reveal the cytoarchitecture of the Bergmann glia fibers in the cerebellum. J. Neurosci. **4**:265.
41. De Blas, A.L., R.O. Kuljis, and H.M. Cherwinski. 1984. Mammalian brain antigens defined by monoclonal antibodies. Brain Res. **322**:277.
42. De Blas, A.L., and H.M. Cherwinski. 1983. Detection of antigens on nitrocellulose paper immunoblots with monoclonal antibodies. Anal. Biochem. **133**:214.
43. Guidotti, A., C.M. Forchetti, M.G. Corda, D. Konkel, C.D. Bennet, and E. Costa. 1983. Isolation, characterization, and purification to homogeneity of an endogenous polypeptide with agonist action on benzodiazepine receptors. Proc. Natl. Acad. Sci. USA **80**:3531.
44. Costa, E., M. Ferrari, P. Ferrero, and A. Guidotti. 1984. Multiple signals in GABAergic transmission: pharmacological consequences. Neuropharmacology **23**:989.

45. Sangameswaran, L., H. Fales, P. Friedrich, and A.L. De Blas. 1986. Purification of benzodiazepines from the bovine brain. Human brains contain natural substances with benzodiazepine-like immunoreactivity. Proc. Natl. Acad. Sci. USA **83**:9236.
46. Guenter, T.W. 1984. Pharmacokinetics of benzodiazepines and of their metabolites, pp. 241–386. In J.W. Bridges and L.F. Chasseaud (eds.), Progress in drug metabolism, Vol. 8. Taylor and Francis, Ltd.
47. Sternbach, L.H. 1983. The discovery of CNS active 1,4-benzodiazepines, pp. 1–6. In E. Costa (ed.), The benzodiazepines. From molecular biology to clinical practice. Raven Press, New York.
48. Lampe, J., G. Butschak, and W. Scheler. 1984. Cytochrome P-450 dependent biotransformation of drugs and other xenobiotics, pp. 277–336. In K. Ruckpaul and H. Rein (eds.), Cytochrome P-450. Akademie-Verlag, Berlin.
49. Luckner, M. 1984. Secondary metabolism in microorganisms, plants and animals, 2nd ed., pp. 272–276. Springer-Verlag, Berlin.
50. Waller, G.R., and O.C. Dermer. 1981. Enzymology of alkaloid metabolism in plants and microorganisms, pp. 317–401. In E. E. Conn (ed.), The biochemistry of plants, Academic Press, New York.

12
Idiotypy and the β-Adrenergic Receptor

DAVID G. SAWUTZ AND CHARLES J. HOMCY

Introduction

The β-adrenergic receptor has been one of the most thoroughly studied membrane proteins to date. Interest in the receptor stems primarily from its role in the regulation of cardiovascular function at all levels. As a result, there have been many approaches to the study and characterization of the receptor polypeptide and the elucidation of how the receptor interacts with other membrane components. A novel method, the use of anti-idiotypic antibodies raised against antibodies specific for β-adrenergic receptor ligands, affords an immunological approach to the problem. Here we describe some pharmacological and physiochemical features of the β-adrenergic receptor and discuss some of the relevant literature on applications of immunological techniques to the study of the receptor at the molecular level.

Historical Aspects of the β-Adrenergic Receptor

Adrenergic receptors were initially classified by Ahlquist (1) into two distinct receptor systems, termed α and β, based on the varied physiological responses of catecholamines. These hormones, termed epinephrine and norepinephrine, are synthesized in sympathetic nerve terminals and interact with membrane-bound β_1- and β_2-adrenergic receptors as well as α_1- and α_2-adrenergic receptors. The binding of a catecholamine to the β receptor leads to an increase in adenylate cyclase activity and produces a cascade of events resulting in a biological response (2,3). It was Sutherland and co-workers (4) who, in 1959, demonstrated that $3',5'$-cyclic adenosine monophosphate (cAMP) was the cytosolic entity that mediates epinephrine-stimulated increased glycogen phosphorylase activity in the liver. It is now well accepted that the catecholamine-stimulated increase in adenylate cyclase activity is mediated through a guanyl nucleotide regulatory binding protein, termed G_s (5). α-Adrenergic receptors, which bind endogenous catecholamines with comparable affinity, are readily distinguished by their effector-coupled responses (6). α_1-Adrenergic receptors are linked to Ca^{2+} entry and phosphatidylinositol turnover (7–9). There is evidence suggesting that α_2-adrenergic receptors are coupled to adenylate cyclase through a GTP binding protein (G_i) in an inhibitory manner (10–12).

The molecular characterization of the β-adrenergic receptor proceeded slowly until the early 1970s. By that time, the availability of a variety of radiolabeled hormone agonists and antagonists allowed the radioligand binding assay to become an invaluable tool in the pharmacological characterization of the receptor's hormone binding site (13,14). The combination of highly sensitive radioligand binding assays and the ability to monitor agonist-stimulated adenylate cyclase activity (15) has enabled researchers to study the physiological regulation of the β-adrenergic receptor at the molecular level. For example, an increase in β-adrenergic receptor number has been demonstrated in hyperthyroidism (16,17), while a decrease in receptor number has been observed under hypothyroid conditions (18,19). In addition, insights into the underlying mechanisms inherent in receptor regulation have been gained through receptor binding assays. Perhaps the best example is that of receptor desensitization or the attenuation of a physiological response resulting from continuous exposure to a hormone (20). Not only has a loss of receptors from the cell surface due to an internalization process been documented, but also an actual uncoupling of the receptor from components of the adenylate cyclase system has been observed as an early event in certain models.

An understanding of how various components of the β-adrenergic receptor-adenylate cyclase system interact to produce a transmembrane signal will come only after each component is purified and reconstituted within an artificial membrane system (21). With the development of reliable solubilization and binding assay techniques, purification of the receptor has become an attainable goal. The β-adrenergic receptor is an intrinsic membrane protein that requires detergents for its solubilization. When liberated into a detergent micelle, the receptor molecules can be characterized and purified by biochemical techniques, including gel filtration and affinity chromatography. Although high concentrations of disrupting agents such as urea have been shown to liberate the receptor (23), the most useful and effective agent for solubilization is digitonin, a sterol detergent analogous to the bile acids. Digitonin appears to be the only nonionic detergent that does not interfere to any great extent with ligand binding (24). The development of affinity supports consisting of agarose coupled to either alprenolol (25,26) or acebutolol (27), both receptor antagonists, has proven invaluable in isolating the solubilized receptor. This approach combined with high-performance gel filtration chromatography has recently enabled the entire purification protocol to be completed within a period of hours rather than days (22,28,31,32).

There are now several reports on the purification of the hormone-binding subunit of the β-adrenergic receptor, of both the β_1 and β_2 subtypes. Shorr et al. (26) initially purified the β_1-adrenergic receptor from frog erythrocyte membranes and reported a subunit molecular weight of 58,000 daltons (Da) for the receptor based on photoaffinity labeling of the receptor polypeptide. The same group then purified the β_1 receptor from turkey erythrocyte membranes and reported that the purified, iodinated receptor was composed of two subunits having molecular weights of 40,000 and 45,000 Da (22). These values were slightly higher than those previously reported by Atlas and Levitzki (41,000 and 37,000 Da), which were also based on photoaffinity labeling of the receptor (29). Cubero and Malbon subsequently purified the β_1 receptor from rat fat cells (30). Iodination and SDS-PAGE of the purified receptor protein revealed that it migrated with a molecular weight of 67,000 Da.

Our group initially purified the mammalian β_2-adrenergic receptor from canine lung to apparent homogeneity (27). A 53,000-Da protein was identified as the receptor peptide after SDS-PAGE and staining with Coomassie blue. Benovic et al. (31) subsequently purified the β_2-adrenergic receptor from hamster, guinea pig, and rat lungs and reported a molecular weight of 64,000 Da for the purified peptides from each tissue. Recently, Graziano et al. (32) purified the β_2 receptor from rat hepatic tissue and also reported a subunit molecular weight of 67,000 Da for the receptor. We have now completed a 34,000-fold purification of the β_2 receptor from calf lung (28). The purified receptor has a subunit molecular weight of 67,000 Da, retains ligand binding properties, and has an isoelectric point of 6.0 (Fig. 12.1).

Purification of the β-adrenergic receptor has led to the isolation of cDNA clones for both β receptor subtypes. The amino acid sequence of the hamster lung β_2-adrenergic receptor was first reported by Dixon et al. (33). The receptor polypeptide contains 418 amino acids, which corresponds to a molecular weight of approximately 46,000 Da and is in agreement with the molecular weight determined for the deglycosylated receptor (49,000 Da), reported by Stiles et al. (34). Based on the amino acid sequence of the polypeptide, seven relatively large hydrophobic domains have been proposed as possible membrane spanning regions. This would indicate that the receptor protein is highly intercalated within the plasma membrane.

More recently, Yarden et al. (35) reported the amino acid sequence for the turkey erythrocyte β_1-adrenergic receptor. The receptor polypeptide has 483 amino acids and a calculated molecular weight of 54,000 Da. This is slightly higher than the molecular weight of the native avian receptor, which was determined to be 45–48,000 Da (36). The β_1-adrenergic receptor from turkey and the β_2-adrenergic receptor from hamster have a 76% homology in their respective amino acid sequences within the proposed membrane spanning regions of the polypeptides. The homology between the two receptor subtypes decreases to only 34% when the hydro-

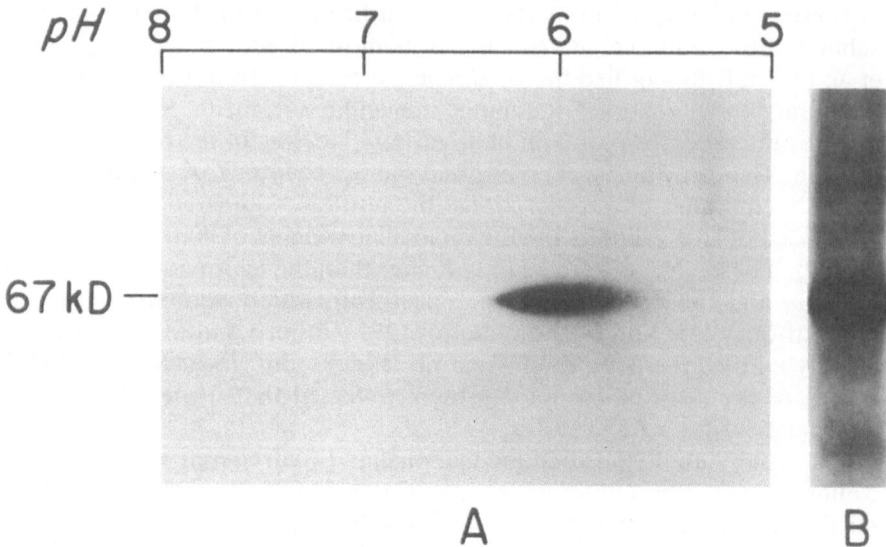

FIGURE 12.1. Two-dimensional gel pattern of the β₂-adrenergic receptor. (A) Purified calf lung β₂ receptor (28), radiolabeled with ¹²⁵I by a chloramine-T method, was electrophoresed by IEF in the first dimension followed by SDS-PAGE in the second dimension. A major protein is evident at a molecular weight of 67,000 Da and a pH value of approximately 6.0 +/− 0.2 pH units after autoradiography. (B) ¹²⁵I-labeled β₂-adrenergic receptor used for (A) was analyzed in one dimension by SDS-PAGE on a 10% polyacrylamide gel. From Ref. 28 with permission.

philic regions are compared. Significant homology in the membrane spanning regions of both β receptor subtypes and rhodopsin, another membrane receptor which interacts with a guanyl nucleotide binding protein in a manner analogous to that of the β-adrenergic receptor, has also been noted. This homology suggests that there is a conservation of amino acid sequence within membrane receptors that couple to GTP regulatory proteins, and this homology extends across species lines.

Antibodies to the β-Adrenergic Receptor

The development of antibodies to the β-adrenergic receptor, a major goal of several laboratories, would provide valuable tools for investigating receptor regulation and structure. In an analysis of desensitization that results in loss of ligand binding, anti-receptor antibodies not dependent on a functional binding site would allow an independent method of quantifying receptor turnover. Homology between receptor subtypes could be studied by using antibodies to identify specific epitopes that might differ between two related polypeptides, i.e., the β₁ and β₂ receptors. Although several groups have now successfully raised antibodies to the β receptor protein, little substantive information has yet come from studies with these probes.

Wrenn and Haber (37) initially reported an antibody raised to a partially purified cardiac β_1-receptor subtype. This antibody appeared, however, to be of generally low affinity. Couraud et al. (38) and Fraser and Venter (39) have both reported on the development of monoclonal antibodies to the β_1-adrenergic receptor of the turkey erythrocyte. In light of the antibody titers reported by these authors, however, the antibodies were of such low affinity that their sensitivity is probably insufficient to permit their successful use as probes of receptor structure or metabolism. In addition, Strader et al. (40) have also obtained antisera raised to a partially purified β receptor preparation and reported that these antisera noncompetitively inhibit adenylate cyclase activation.

Because of the inherent difficulties associated with the purification of significant quantities of intact receptor polypeptide for use as immunogens, generating antibodies against purified receptor peptide fragments or synthetic peptides whose sequence is based on the amino acid sequence deduced from the cDNA clone of the receptor becomes an attractive alternative. Such an approach was recently employed to generate a polyclonal anti-peptide antibody that could also bind to the β receptor (33). A fusion protein of a 34-amino-acid peptide purified from CNBr cleavage fragments of the hamster lung β_2-adrenergic receptor and the N-terminal domain of the yeast RAS^{sc1} protein SC1N was used as an immunogen (41). The fusion protein was expressed in *E. coli,* purified, and injected into rabbits. The resulting antisera immunoprecipitated ^{125}I-cyanopindolol-labeled, solubilized β receptor from hamster lung in addition to the ^{125}I-cyanopindolol-labeled receptor from human A431 epidermoid carcinoma cells and turkey erythrocytes. That the anti-peptide antibody was indeed recognizing the β receptor was confirmed by the use of the synthetic peptide to inhibit antibody immunoprecipitation of the receptor at a peptide concentration of 100 μM. Binding of the antibody to purified receptor after SDS-PAGE and nitrocellulose blotting could also be demonstrated, a further indication that the antibody was specific for the receptor polypeptide.

The production of monoclonal anti-peptide antibodies from peptides synthesized on the basis of the deduced sequence of the β receptor may allow the assignment of a functional role to a particular protein domain, such as the ligand binding site, within the receptor molecule.

Antibodies as Models of the Receptor Binding Site

β-Adrenergic receptors are present in the plasma membrane in a very small number of copies, generally in the range of 4,000 to 10,000 per cell (42). When partially purified membranes of a given tissue are used to generate anti-receptor antibodies, there is a high probability that antibodies to other membrane components will be developed as well. Anti-idiotypic antibodies raised to anti-ligand antibodies can be used to avoid this problem and to obtain antibodies directed against the receptor ligand binding site. If an antibody can be generated to a ligand that binds to the hormone receptor, then it may be possible to isolate an anti-idiotypic antibody specific for the first antibody's combining site. The anti-idiotypic antibody might then be predicted to recognize an epitope within the receptor's binding site that is similar to that found in the binding site of the anti-ligand antibody.

To achieve this result, the anti-ligand antibody must bind a spectrum of β-adrenergic receptor ligands in a way that mimics the receptor itself. The antibody then becomes a model of the receptor's ligand binding site. For the β-adrenergic receptor, a number of structurally different receptor antagonists are available, most of which share a common propanolamine side chain (see Fig. 12.2). One such ligand, alprenolol, can be coupled to carrier proteins with a high degree of substitution through the propylene side chain. Hoebeke et al. (43) demonstrated that unfractionated antiserum to an alprenolol-protein conjugate bound alprenolol in a manner analogous to that of the β receptor. Rockson et al. (44) reported the production of polyclonal anti-alprenolol antibodies that displayed ligand binding specificities similar to those of the receptor. They demonstrated that the relative binding potency of several β-adrenergic receptor agonists and antagonists paralleled the in vivo potency of the ligands for modulating β receptor-stimulated adenylate cyclase activity. In addition, the antibody specific for high-affinity ligand binding could be fractionated on an acebutolol-Sepharose resin and eluted with (−)-propranolol, an analogue of alprenolol. This class of antibody preferentially bound the (−)-enantiomer of propranolol, paralleling the stereospecific recognition properties of the β receptor and thus representing a more accurate model of the receptor's ligand binding site.

Monoclonal antibodies to alprenolol have also been isolated by immunizing mice with ligand-protein conjugates and produced according to standard hybridoma techniques. Chamat et al. (45) first described the production of four anti-alprenolol monoclonal antibodies after somatic cell fusion of NS-1 myeloma cells and splenocytes isolated from BALB/c mice that had been immunized with alprenolol coupled to bovine serum albumin. These antibodies had a range of alprenolol binding affinities (K_A) from 2 $\times$ 10^6 to 24 $\times$ 10^6 M^{-1} based on [^{3}H]dihydroalprenolol binding studies. Two of the antibodies, 37A4 and 10E4, exhibited higher affinity (2-fold and 20-fold, respectively) for the l-isomer of propranolol compared to the d-isomer. None of the antibodies, however, bound agonists with high affinity.

By using the β-adrenergic receptor antagonist alprenolol coupled to hemocyanin as an immunogen, we have successfully isolated four hybridomas (1B7, 5B7, 5D9, and 2G9) that secrete anti-alprenolol antibodies

(46). Although their binding properties differ, three of these antibodies are highly selective for antagonists and have poor affinity for the entire class of β receptor agonists (Table 12.1). Antibody 1B7 is an exception in that it binds isoproterenol with relatively high affinity, although at a level below its binding affinity for the antagonists cyanopindolol, alprenolol, and propranolol.

Antibody 1B7 is unique in that it demonstrates an entire range of binding specificites for both antagonists and agonists, which generally follows the ligand binding specificities of the β receptor itself (Table 12.2). It binds ^{125}I-cyanopindolol with an affinity approximately 10 times greater than that for alprenolol and mimics the receptor's binding characteristics in its recognition of agonists (isoproterenol > epinephrine > norepinephrine) as well. It is the only antibody that clearly recognizes agonist molecules with an affinity comparable to that of the β$_2$-adrenergic receptor. In all likelihood, 1B7 represents the class of antibody detectable in the antisera of mice that demonstrated agonist-inhibitable binding.

On the basis of their preference for antagonists and their marked stereoselectivity observed in ligand binding assays, it appears that 1B7, 5B7, and 5D9 recognize as their principal determinant the asymmetric carbon moiety in the propanolamine side chain common to almost all β receptor antagonists. In the case of 1B7, an increase in the number of substitutents on the terminal amino group seems to be a requirement for increased agonist binding affinity, since norepinephrine binds to the antibody with the lowest affinity and isoproterenol binds to it with the highest affinity. Changes in antagonist affinity generally appear to be more related to alterations in the aromatic ring substituents. For example, the significantly

TABLE 12.1. Ligand affinity binding constants of anti-alprenolol monoclonal antibodies.

	K_D (μM) for mAb[a]:			
Ligand	1B7	5B7	5D9	2G9[b]
Antagonists				
Cyanopindolol	0.00014			
Alprenolol	0.006	0.02	0.25	10.50
l-Propranolol	0.007	0.05	0.20	4.90
d-Propranolol	3.600	7.80	26.00	2.10
Acebutolol amine	0.024	19.00	38.00	
Pindolol		0.14		
Agonists				
Isoproterenol	0.03	—	—	—
Epinephrine	18.00	—	—	
Norepinephrine	2,900.00	—	—	

[a]Ligand-antibody binding constants were determined as previously described (46).
[b]IC$_{50}$ values from [^{3}H]-DHA competitive inhibition curves.
[c]Dose-dependent inhibition of [^{3}H]-DHA binding not observed.

TABLE 12.2. Comparison of ligand binding affinities for antibody 1B7 and the β-adrenergic receptor.

Ligand	K_D (μM) for:	
	1B7[a]	β Receptor[b]
Antagonists		
Cyanopindolol	0.00014	0.0001 (28)
Alprenolol	0.006	0.0034 (3)
l-Propranolol	0.007	0.0046 (3)
d-Propranolol	3.60	0.28 (3)
Acebutolol amine	0.024	
Agonists		
Isoproterenol	0.03	0.40 (3)
Epinephrine	18.00	4.60 (3)
Norepinephrine	2,900.00	49.00 (3)

[a]The ligand affinity binding constants for antibody 1B7 were determined from saturation binding data or from competitive inhibition curves of [^{3}H]-DHA binding as previously described (46).
[b]The ligand affinity binding constants for the β-adrenergic receptor were taken from the literature as cited in the table.

lower binding affinity of acebutolol amine for each antibody when compared with cyanopindolol, alprenolol, and propranolol might be due to decreased overall hydrophobicity of the molecule, because the aryl amine group would be largely protonated at physiological pH. These observations suggest that full antagonist potency is realized only when: (i) the oxymethylene group ($-OCH_2$) is inserted at the 1 position of the aromatic ring; (ii) the hydroxyl group at the asymmetric carbon is in the correct configuration; and (iii) the terminal amino group is substituted with an isopropyl group (Fig. 12.2). Although each of the agonist molecules fulfills at least one of these requirements, none has the oxymethylene substituent, and in addition to the presence of the two hydroxyl groups on the aromatic ring, this is capable of accounting for the significantly lower binding affinities of the agonists for the antibodies.

Our studies suggest that binding sites of anti-ligand antibodies possessing the required structure for high-affinity agonist binding will also be capable of binding antagonists with high affinity. In addition, such an antibody will also likely recognize the entire spectrum of adrenergic ligands stereoselectively and with a potency order similar to that of the β-adrenergic receptor. In our previous study (44), antibodies purified by agonist elution from an acebutolol-affinity resin also generally demonstrated the highest affinity for antagonists when compared to fractions eluted by antagonists or by salt elution. Antibody 1B7 was the only monoclonal antibody to demonstrate high-affinity agonist binding; therefore, when the antibody binding site is constructed in such a way as to recognize agonists with

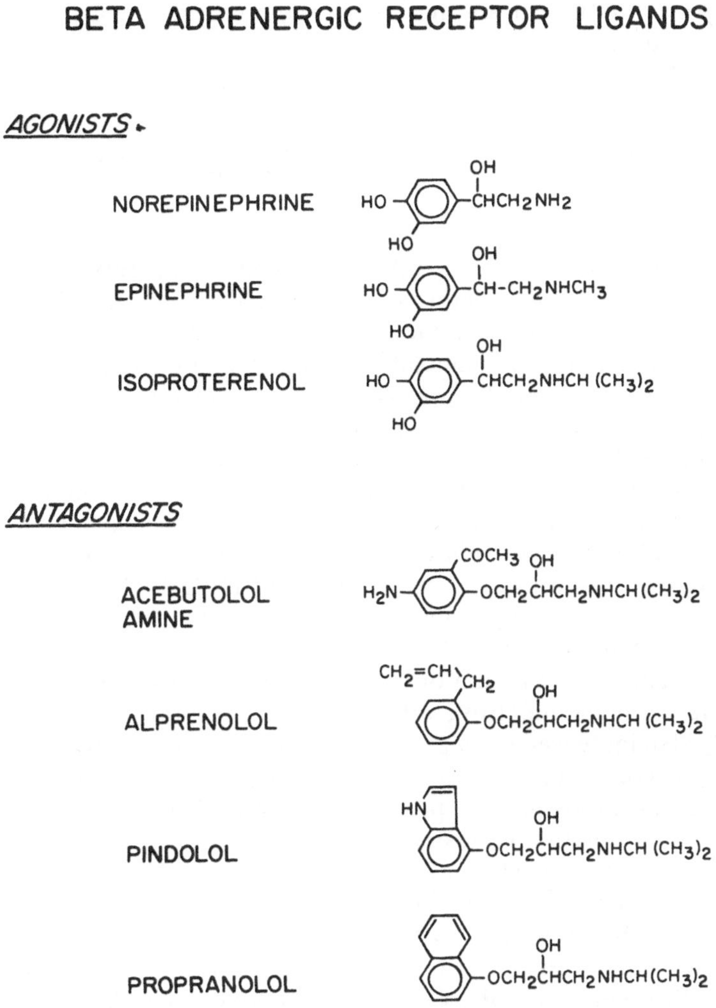

BETA ADRENERGIC RECEPTOR LIGANDS

AGONISTS ←

NOREPINEPHRINE

EPINEPHRINE

ISOPROTERENOL

ANTAGONISTS

ACEBUTOLOL
AMINE

ALPRENOLOL

PINDOLOL

PROPRANOLOL

FIGURE 12.2. Chemical structures of several common β-adrenergic receptor agonists and antagonists. From Ref. 46 with permission.

high affinity (the affinity of 1B7 for isoproterenol is 100 nM), high-affinity binding of the more hydrophobic antagonists may be expected to occur. Antibodies 5B7 and 5D9 also fit this scheme because they recognize agonists with low affinity (approximately 10^{-3} M) and have 10 to 100 times lower affinity for the antagonists than does 1B7. Similarly, antibody 2G9, which does not bind agonists, has the lowest affinity for antagonists. Thus, the rank order of potency and binding specificity typical of the β-adrenergic receptor may be a property possessed by an antibody binding site that stereoselectively recognizes agonistlike molecules. Agonist binding may prove to be a sufficient and necessary requirement for high-affinity antagonist binding.

Anti-Idiotypic Antibodies That Recognize the β Receptor

There are now several reports on the production of anti-idiotypic antibodies that recognize the β-adrenergic receptor. Schreiber et al. (47) reported that anti-alprenolol anti-idiotypic antibodies bound to the β receptor and could influence hormone stimulation of adenylate cyclase. A rabbit antibody specific for alprenolol was used to induce an anti-idiotypic response in allotype-matched rabbits. The isolated IgG fraction of the antisera was found to inhibit the binding of [³H]dihydroalprenolol ([³H]-DHA) to the anti-alprenolol idiotype in a dose-dependent manner. In addition, anti-idiotypic antibody binding to turkey erythrocytes was also demonstrated, and this binding was correlated with a decrease in specific [³H]-DHA binding to the erythrocytes. The decrease in [³H]-DHA binding sites was 65% when an anti-idiotypic antibody concentration of 5 mg/ml was employed in the binding assays. Since such a high concentration of antibody was necessary for the inhibition of ligand binding, it is possible that this may reflect a noncompetitive interaction. In addition, the anti-idiotypic antibody also increased epinephrine-stimulated adenylate cyclase activity in turkey erythrocyte membranes, suggesting that the anti-idiotypic antibody possessed agonistlike properties as well. This may, however, be more consistent with an allosteric interaction between the antibody and the receptor-cyclase system.

The production of polyclonal anti-idiotypic antibodies that recognize the β receptor has also been reported by Homcy et al. (48). Anti-alprenolol rabbit antibodies were fractionated on an acebutolol affinity resin and eluted with (−)-propranolol as described above. The class of antibody that emulated the ligand binding characteristics of the β receptor was then used to immunize allotype-matched rabbits. One anti-idiotypic antiserum was found to completely inhibit [³H]-DHA binding to the original idiotype and to an anti-acebutolol antibody with similar ligand binding properties. The anti-receptor binding properties of the anti-idiotype were determined from [³H]-DHA saturation binding curves using turkey erythrocyte membranes. The DEAE-purified IgG fraction of the anti-idiotype antiserum caused a fourfold decrease, in a competitive manner, in the binding affinity of [³H]-DHA for the receptor, with no change in the number of [³H]-DHA binding sites. In a similar manner, isoproterenol-stimulated adenylate cyclase activity in turkey erythrocyte membranes was decreased in a competitive manner by the anti-idiotypic antibody. Finally, the anti-idiotypic antibody was fractionated on an anti-alprenolol antibody-Sepharose column and a hapten-specific antibody fraction was eluted with propranolol. The idiotype-purified antibody was then characterized in direct competitive inhibition assays with [^{3}H]-DHA for turkey erythrocyte β receptors. The purified IgG (20 μg/ml) inhibited [^{3}H]-DHA binding to the receptor with a dissociation binding constant (K_d) of 50 nM.

These studies with polyclonal reagents indicate the suitability of using an elicited antibody response to obtain a population of anti-ligand antibodies whose properties mimic those of a ligand's biological receptor. The anti-idiotypic response has been used to raise a population of binding site-directed antibodies. A small fraction of these recognize the three-dimensional structural homology between the antibody's alprenolol binding site and the ligand binding site of the β-adrenergic receptor. However, due to the relatively low titer of these antibodies and the fleeting character of the anti-idiotypic response, it will be necessary to develop monoclonal anti-idiotypic antibodies via somatic cell hybridization techniques if sufficient quantities are to be produced for routine laboratory use. The potential specificity that this immunological approach affords may allow the development of anti-receptor antibodies that possess not only high affinity for the binding site but also a degree of β-subtype selectivity not available with synthetic ligands.

The production of several monoclonal anti-idiotypic antibodies directed against monoclonal anti-alprenolol antibody 37A4 has been reported by Guillet et al. (49). These authors immunized BALB/c mice with 37A4 hybridoma cells previously fixed in 1% paraformaldehyde and then fused the spleen cells of the mice with NS-1 myeloma cells. Screening of the fusion products for anti-37A4 activity resulted in 23 positive wells, 6 of which contained hybridomas secreting antibody that inhibited the binding of ligand to antibody 37A4. Of these 6 antibodies, 3 were reported to possess anti-β receptor activity (50). Studies were conducted using human A431 epidermoid cells containing β receptors, as determined by the photoaffinity labeling (51) of two polypeptides having molecular weights of 65,000 and 57,000 Da. One of these antibodies, mAb2B4 (IgM), immunoprecipitated alprenolol binding activity from digitoxin-solubilized human A431 epidermoid cells. In addition, this antibody bound predominantly to a polypeptide with a molecular weight of 55,000 Da after SDS-PAGE and nitrocellulose blotting of A431 membranes.

To determine if antibody mAb2B4 had any effect on β-adrenergic receptor function, whole-cell cAMP levels were measured in the absence and presence of antibody. Antibody mAb2B4 stimulated cAMP formation in A431 cells and this increase was specifically inhibited by the β receptor antagonist propranolol. The magnitude of the cAMP effect was less than that associated with the β-adrenergic receptor agonist isoproterenol, and on a molar basis, the antibody possessed only a fraction of the activity of the agonist. The authors suggest that the polymeric nature of mAb2B4 contributes to the agonistlike activity of the anti-idiotypic antibody. It is also possible that the increase in antibody-stimulated A431 cAMP levels results from an allosteric effect on the receptor-cyclase coupling rather than from a direct agonist effect on the receptor since the concentration of antibody used in the studies was in great excess of the number of β receptors.

Although the evidence presented above clearly indicates that polyclonal and monoclonal anti-idiotypic antibodies directed against anti-alprenolol antibodies can recognize the β-adrenergic receptor, their utility in either the purification or characterization of the receptor protein has not been demonstrated. The use of such antibodies in future studies of the structure and function of the β receptor may be limited, in light of the successful isolation of cDNA clones of each receptor subtype. Anti-receptor antibodies raised to synthetic peptides may prove to be more valuable for answering questions about the function of the β receptor at the molecular level.

Anti-Idiotypic Modulation of Antigen-Antibody Binding Affinity

By virtue of the fact that they possess an internal image of the ligand, anti-ligand anti-idiotypic antibodies can induce conformational (functional) changes in a protein. In this regard, Sege and Peterson (52) first demonstrated that anti-insulin anti-idiotypic antibodies promoted the uptake of aminoisobutyric acid in rat hepatocytes to the same extent as did insulin. It has been known for some time that antibodies can enhance the activity of certain enzymes. For example, a defective mutant of the enzyme β-galactosidase from *E. coli* has conformationally dependent determinants to which specific antibodies bind, resulting in a mutant enzyme with 1000-fold greater activity (53). The mechanism of this enhanced enzyme activity was subsequently described (54). The binding of antibody to inactive 10S dimers of the enzyme produced a conformational change in the protein, leading to the formation of active 16S tetramers of the mutant enzyme, as determined by sedimentation experiments.

These studies raise the following question: can anti-ligand anti-idiotypic antibodies regulate antigen-antibody binding affinity? Jerne's network hypothesis (55) and the model for anti-idiotypic antibody interaction with anti-ligand antibodies proposed by Sege and Peterson (52) dictate that a binding site-directed, anti-idiotypic antibody should inhibit the binding of ligand to the original idiotype. In addition, antibodies not directed to a binding site might then be expected to have little effect on ligand-antibody binding.

We have recently demonstrated that an anti-idiotypic antibody can be generated that will enhance the binding of antigen to the original idiotype (56). A rabbit anti-idiotypic antibody (R9) was raised against the anti-alprenolol monoclonal antibody 5B7 (46). Antibody R9 produced a dose-dependent increase in ^{125}I-cyanopindolol (CYP, Fig. 12.2) binding to 5B7 (Fig. 12.3) at a concentration of CYP that was significantly less than the K_D of 5B7 for the ligand (0.1 nM). Under the same conditions, no binding of CYP to 5B7 was observed when R9 was replaced with rabbit anti-mouse IgG antiserum (Fig. 12.3, inset) or with an anti-idiotypic antiserum raised against a monoclonal antibody of different specificity than 5B7 (data not shown). The binding of R9 to 5B7 resulted in a 100-fold increase in the binding affinity of 5B7 for CYP as determined by Scatchard analyses of equilibrium saturation binding experiments. In addition, no binding of CYP to R9 itself could be detected.

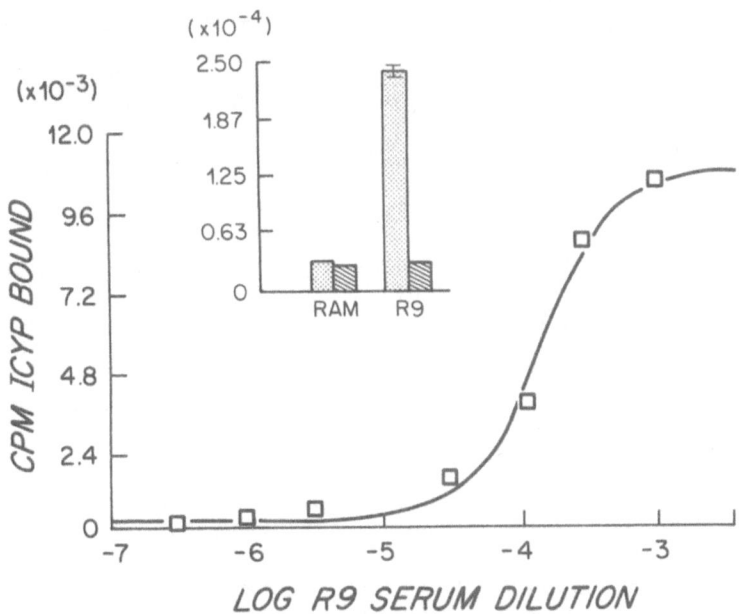

FIGURE 12.3. The effect of anti-idiotypic antibody R9 on ligand binding to antibody 5B7. Antibody 5B7 culture media (100 μl, 1:30 dilution) was incubated with ^{125}I-CYP (0.1 nM) and increasing concentrations of R9 antiserum (100 μl) for 1 h at 37°C. Alprenolol (10 μM) was used to determine nonspecific binding. Binding assays were terminated by precipitation of the antibody complex after the addition of 1 ml of ice-cold n-propanol. The precipitated protein was centrifuged for 20 min at 7,500 rpm, and the pellets were counted for radioactivity following removal of the supernatants by aspiration. Half-maximal promotion of CYP binding to 5B7 by R9 was observed at an antiserum dilution of 1/10,000. Inset: The same experiment was repeated but R9 antiserum (1/1,000 dilution) was replaced with rabbit anti-mouse IgG antiserum (RAM) at an equal serum dilution. Total CYP binding to 5B7 is depicted by the stippled bars and nonspecific binding (10 μM alprenolol) by the hatched bars.

Since the increase in ligand binding can also be demonstrated with affinity-purified F(ab) fragments of each antibody, polyvalency is not required for this promotion in ligand binding affinity. A dimer composed of
one molecule each of 5B7 and R9 appears to be responsible for the increase
in ligand binding, on the basis of size exclusion gel filtration experiments.
Kinetic analysis of saturation binding data shows that the increase in
binding affinity is directly related to a decrease in the rate of dissociation
of the ligand from the R9-5B7 complex. A 100-fold decrease in the dis-

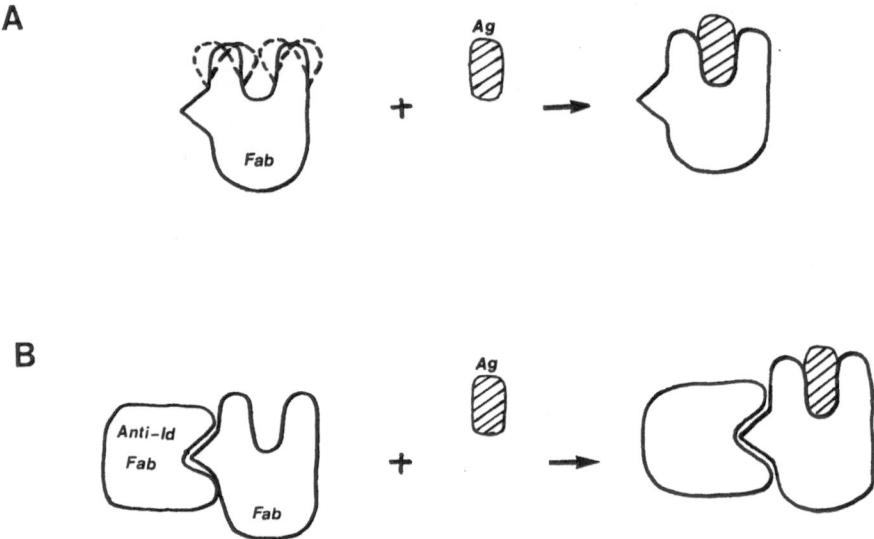

FIGURE 12.4. The proposed explanation for a decrease in the Gibbs free energy
of the rection between antigen and Fab in the presence of anti-idiotype Fab. (A)
It is proposed that the hypervariable loops forming the antibody combining site
exist in several (e.g., three) closely similar conformations when free in solution.
Gibbs energies of the three conformations, ΔG, are virtually identical and the
conformations occur in approximately equal numbers. The antigen (Ag) reacts
with only one of the possible conformations, thus effectively "freezing" the combining site structure. The process of antigen binding is therefore accompanied by
a conformational entropy loss of $-2\Delta S_{conf}$, and the value of $-T(-2\Delta S_{conf}) =$
$+2T\Delta S_{conf}$ [T = temperature K] enters into the Gibbs free energy balance of the
antigen-antibody reaction. Note that the value is positive and reduces the association constant of the reaction (K_{as}), $\Delta G = -RT \log K_{as}$. (B) If the anti-idiotypic
antibody binds to the antigen-specific Fab somewhere near the combining site,
the flexibility of the binding site is reduced and its antigen-binding conformation
becomes much more stable than the two other "nonbinding" conformations depicted above. Consequently, the reaction between the antigen and the Fab-(anti-
Fab) complex proceeds without any loss of conformational entropy and the association constant for this reaction is apparently greater than that given in (A) by
the amount of $-RT \log (2T\Delta S_{conf})$.

sociation rate constant was observed in the presence of the anti-idiotypic antibody.

It is unclear as to how the anti-idiotype produces this effect; however, at least two potential mechanisms can be envisioned. The anti-idiotypic antibody could bind to a site distant from the ligand binding domain in the framework region and effect a conformational change in the idiotype's binding site as shown in Fig. 12.4. Such a conformational change could result in a net decrease in the entropy of ligand-antibody binding and, therefore, a subsequent change in the ligand affinity binding constant. Alternatively, the anti-idiotypic antibody may bind in the near proximity of the ligand binding site itself and provide additional points of contact for that portion of the ligand that does not directly interact with residues within the binding site. This latter mechanism would preclude any major conformational changes in the idiotype with anti-idiotype causing the formation of a new hybrid ligand binding site. The formation of this "hybrid binding site" would be contingent upon the unique specificity of the anti-idiotypic antibody for the idiotype and would require that the binding affinity between the two antibodies be sufficiently high to maintain the hybrid site during the ligand binding interaction (i.e., the dissociation of the two antibodies is slower than that of the ligand from the idiotype).

It should be pointed out that these observations have been made using polyclonal anti-idiotypic antibodies. A delineation of the exact mechanism underlying the enhancement of antigen affinity will be possible only after monoclonal reagents with similar effects can be produced. Perhaps then the exact residues participating in ligand contact in the idiotypic-anti-idiotypic complex can be pinpointed. Whether or not such an interaction has any physiological implication will require further investigation.

Conclusion

Raising anti-idiotypic antibodies against anti-ligand antibodies has proved to be a useful method for developing immunological probes that can be used to characterize hormone receptors. The β-adrenergic receptor system is an excellent model system, in that it has been extensively studied and because of the large number of receptor ligands against which anti-ligand antibodies can be raised and their ligand binding properties characterized. Several groups have developed both polyclonal and monoclonal anti-idiotypic antibodies to anti-alprenolol antibodies, and these anti-idiotypes have been shown to recognize the β-adrenergic receptor and to modulate receptor function. Despite these achievements, a definitive characterization of a monoclonal anti-idiotypic antibody that clearly recognizes the receptor's ligand binding domain is still lacking. Such an antibody would serve as a probe that could be used independently of radio-labeled receptor ligands for the molecular characterization of the ligand binding site.

Acknowledgments. We thank Dr. Jiří Novotný for valuable discussions in developing models for anti-idiotypic regulation of ligand-antibody binding. This work was supported by funding from the National Institutes of Health (HL-19259, HL-26215) and by the R.J.R. Nabisco Company. D.G.S. is supported through a fellowship grant from the American Heart Association, the Massachusetts Affiliate (13-424856).

References

1. Ahlquist, R.P. 1948. Study of adrenotropic receptors. Am. J. Physiol. **153**:586.
2. Maguire, M.E., E.M. Ross, and A.G. Gilman. 1977. β-Adrenergic receptor: ligand binding properties and the interaction with adenylyl cyclase. Adv. Cyclic Nucleotide Res. **8**:1.
3. Lefkowitz, R.J., L.E. Limbird, C. Mukherjee, and M.G. Caron. 1976. The β-adrenergic receptor and adenylate cyclase. Biochim. Biophys. Acta **457**:1.
4. Murad, F., Y-M. Chi, T.W. Rall, and E.W. Sutherland. 1959. Adenyl cyclase. III. The effect of catecholamines and choline esters on the formation of adenosine 3′,5′-phosphate from preparations of cardiac muscle and liver. J. Biol. Chem. **224**:463.
5. Longabaugh, J.P., D.E. Vatner, and C.J. Homcy. 1986. The beta-adrenergic receptor/adenylate cyclase system, pp. 1097–1118. *In* H.A. Fozzard, E. Haber, R.B. Jennings, A.M. Katz, and H.E. Morgan. (eds.), The heart and cardiovascular system, Vol. 2. Raven Press, New York.
6. Homcy, C.J., and R.M. Graham. 1985. Molecular characterization of adrenergic receptors. Circ. Res. **56**:635.
7. Joseph, S.K., A.P. Thomas, R.J. Williams, R.F. Irvine, and J.R. Williamson, 1984. myo–Inositol–1, 4, 5—triphosphate. A second messenger for the hormonal mobilization of intracellular Ca^{2+} in the liver. J. Biol. Chem., **259**:3077.
8. Morgan, N.G., P.F. Blackmore, and J.H. Exton, 1983. Age-related changes in the control of hepatic cyclic-AMP levels by α_1 and β_2 adrenergic receptors in male rats. J. Biol. Chem. **258**:5103.
9. Joseph, S.K., 1983. Inositol triphosphate: an intracellular messenger produced by Ca^{2+} mobilizing hormones. Trends Biochem. Sci. **9**:420.
10. Burns, T.W., and P.E. Langley. 1975. The effect of α and β-adrenergic receptor stimulation and adenylate cyclase activity of human adipocytes. J. Cyclic Nucleotide Res. **1**:321.
11. Jacobs, K.H., W. Saur and G. Schultz. 1976. Reduction of adenylate cyclase activity in lysates of human platelets by the α-adrenergic component of epinephrine J. Cyclic Nucleotide Res. **2**:381.
12. Aktories, K., G. Schultz, and K.H. Jacobs. 1979. Inhibition of hamster fat cell adenylate cyclase by prostoglandin E₁ and epinephrine: requirement for GTP and sodium ions. FEBS Letters. **107**:100.
13. Lefkowitz, R.J., and L.T. Williams. 1977. Catecholamine binding to the β-adrenergic receptor. Proc. Natl. Acad. Sci. USA **74**:515.
14. Watanabe, A.M., L.R. Jones, A.S. Manalaw, and H.R. Busch. 1982. Cardiac autonomic receptors: recent concepts from radiolabeled ligand-binding studies. Circ. Res. **50**:161.
15. Solomon, Y., C. Londos, and M. Rodbell. 1974. A highly sensitive adenylate cyclase assay. Anal. Biochem. **58**:541.

16. Williams, L.T., R.J. Lefkowitz, A.M. Watanabe, D.R. Hathaway, and H.R. Busch. 1977. Thyroid hormone regulation of β-adrenergic receptor number. J. Biol. Chem. **252**:2787.

17. Tsai, J.S., and A. Chum. 1978. Effect of 1-tri-iodo thyronine on (−)-[³H]dihydroalprenolol binding and cyclic AMP response to (−) adrenaline in· cultured heart cells. Nature (London) **275**:138.

18. Ciaraldi, T.P., and G.R. Mainetti. 1975. Adrenergic receptors in rat heart and adipocytes and their modulation by thyroxine. Biochim. Biophys. Acta **54**:334.

19. Banerjee, S.P., and L.S. Kung. 1977. β-Adrenergic receptors in rat heart: effects of thyroidectomy. Eur. J. Pharmacol. **43**:207.

20. Harden, T.K. 1983. Agonist-induced desensitization of the β-adrenergic receptor linked adenylate cyclase. Pharmacol. Rev. **35**:5.

21. Leivtzki, A. 1985. Reconstitution of membrane receptor systems. Biochim. Biophys. Acta **822**:127.

22. Shorr, P.G.L., M.W. Strohsacher, T.N. Lavin, R.J. Lefkowitz, and M.G. Caron. 1982. The β receptor of the turkey erythrocyte: molecular heterogeneity revealed by purification and photoaffinity labeling. J. Biol. Chem. **257**:12431.

23. Homcy, C.J., and D. Sylvestre. 1982. Characterization of the mammalian β₂ receptor in 8M urea. Biochem. Biophys. Res. Commun. **108**:504.

24. Caron, M.G. and R.J. Lefkowitz. 1976. Solubilization and characterization of the β-adrenergic receptor binding site of frog erythrocytes. J. Biol. Chem. **257**:2374.

25. Vacquelin, G., P. Geynet, J. Hanoune, and A.D. Strosberg. 1977. Isolation of adenylate cyclase-free, β-adrenergic receptors from turkey erythrocyte membranes by affinity chromatography. Proc. Natl. Acad. Sci. USA **74**:3710.

26. Shorr, R.G.L., R.J. Lefkowitz, and M.G. Caron. 1981. Purification of the β-adrenergic receptor: identification of the hormone binding subunit. J. Biol. Chem. **256**:5820.

27. Homcy, C.J., J.G. Rockson, J. Countaway, and D. Egan. 1983. Purification and characterization of the mammalian β₂-adrenergic receptor. Biochemistry **22**:660.

28. Insoft, R., D.G. Sawutz, and C.J. Homcy. 1986. Purification and characterization of the β₂-adrenergic receptor from calf lung. Biochim. Biophys. Acta **861**:345.

29. Atlas, D., and A. Levitzki. 1978. Tentative identification of β-adrenoreceptor subunits. Nature (London) **272**:370.

30. Cubero, A., and C. Malbon. 1984. The fat cell β-adrenergic receptor. Purification and characterization of a mammalian β₁ adrenergic receptor. J. Biol. Chem. **259**:1344.

31. Benovic, J.L.R.G.L.Shorr, M.G. Caron, and R.J. Lefkowitz. 1984. The mammalian β₂-adrenergic receptor: purification and characterization. Biochemistry **23**:4510.

32. Graziano, M.P., C.P. Moxham, and C.C. Malbon. 1985. Purified rat hepatic β₂-adrenergic receptor: structural similarities to the rat fat cell β₁-adrenergic receptor. J. Biol. Chem. **260**:7665.

33. Dixon, R.A.F., B.K. Kobilka, D.J. Strader, J.L. Benovic, H.G. Dohlman, T. Frielle, M.A. Bolanowski, C.D. Bennett, E. Rands, R.E. Diehl, R.A. Munford, E.E. Slater, I.S. Sigail, M.G. Caron, R.J. Lefkowitz, and C.D. Strader. 1986. Cloning of the gene and cDNA for mammalian β-adrenergic receptor and homology with rhodopsin. Nature (London) **321**:75.

34. Stiles, G.L., J.L. Benovic, M.G. Caron, and R.J. Lefkowitz. 1984. Mammalian β-adrenergic receptors: distinct glycoprotein populations contain high mannose or complex type carbohydrate chains. J. Biol. Chem. **259**:8655.

35. Yarden, Y., H. Rodriquez, S.K.F. Wong, D.R. Brandt, D.C. May, J. Burner, R.N. Harkins, E.Y. Chen, J. Ramachandran, A. Ullrich, and E.M. Ross. 1986. The β-adrenergic receptor: primary structure and membrane topology. Proc. Natl. Acad. Sci. USA **83**:6795.

36. Lavin, T.N., P. Nambi, S.L. Heald, P.W. Jeffs, R.J. Lefkowitz, and M.G. Caron. 1984. ^{125}I-labeled p-azidobenzylcarozolol, a photoaffinity label for the β-adrenergic receptor: characterization of the ligand and photoaffinity labeling of $β_1$- and $β_2$-adrenergic receptors. J. Biol. Chem. 257(20):12332.

37. Wrenn, S., and E. Haber. 1979. An antibody specific for the propranolol binding site of cardiac muscle. J. Biol. Chem. **254**:6577.

38. Couraud, P.O., C. Delauier-Klutchleo, O. Durien-Trautmann, and A.D. Strosberg. 1981. Antibodies raised against β-adrenergic receptors stimulate adenylate cyclase. Biochem. Biophys. Res. Commun. **99**:1295.

39. Fraser, C.M., and J.C. Venter. 1980. Monoclonal antibodies to β-adrenergic receptors: purification and characterization of β-receptors. Proc. Natl. Acad. Sci. USA **77**:7034.

40. Strader, C.D., V.M. Pickel, T.H. Job, M.W. Strokacher, R.G.L. Shorr, R.J. Lefkowitz, and M.G. Caron. 1983. Antibodies to the β-adrenergic receptor: attenuation of catecholamine sensitive adenylate cyclase and demonstration of postsynaptic receptor localization in the brain. Proc. Natl. Acad. Sci. USA **80**:1840.

41. Temeles, G.L., J.B. Gibbs, J.S. D'Alonzo, I.S. Segal, and E.M. Scolnick. 1985. Yeast and mammalian *ras* proteins have conserved biochemical properties. Nature (London) **313**:700.

42. Haber, E. 1982. Immunological probes in cardiovascular disease. Br. Heart J. **47**:1.

43. Hoebeke, J., G. Vanguelin, and A.D. Strosberg. 1978. The production and characterization of antibodies against β-adrenergic antagonists. Biochem. Pharmacol. **27**:1527.

44. Rockson, S.G., C.J. Homcy, and E. Haber. 1980. Anti-alprenolol antibodies in the rabbit: a new probe for the study of β-adrenergic receptor interactions. Circ. Res. **46**:808.

45. Chamat, S., J. Hoebeke, and A.D. Strosberg. 1984. Monoclonal antibodies specific for β-adrenergic ligands. J. Immunol. **133**:1547.

46. Sawutz, D.G., D. Sylvestre, and C.J. Homcy. 1985. Characterization of monoclonal antibodies to the β-adrenergic antagonist alprenolol as models of the receptor binding site. J. Immunol. **135**:2713.

47. Schreiber, A.B., P.O. Couraud, C. Andre, B. Vrey, and A.D. Strosberg. 1980. Anti-alprenolol anti-idiotypic antibodies bind to β-adrenergic receptors and modulate catecholamine sensitive adenylate cyclase. Proc. Natl. Acad. Sci. USA **77**:7385.

48. Homcy, C.J., S.G. Rockson, and E. Haber. 1982. An antiidiotypic antibody that recognizes the β-adrenergic receptor. J. Clin. Invest. **69**:1147.

49. Guillet, J.G., S. Chamat, J. Hoebeke, and A.D. Strosberg. 1984. Production and detection of monoclonal anti-idiotypic antibodies directed against a monoclonal anti-β-adrenergic ligand antibody. J. Immunol. Meth. **74**:163.

50. Guillet, J.G., S.V. Kaneri, O. Darien, C. Delavier, J. Hoebeke, and A.D. Strosberg. 1985. β-Adrenergic agonist activity of a monoclonal anti-idiotypic antibody. Proc. Natl. Acad. Sci. USA 82:1781.
51. Lavin, T.N., P. Mambi, S.L. Heald, P.W. Jeffs, R.J. Lefkowitz, and M.G. Caron. 1982. ^{125}I-labeled p-azidobenzylcarazolol, a photoaffinity label for the β-adrenergic receptor. J. Biol. Chem. 257:12332.
52. Sege, K., and P.A. Peterson. 1978. Use of anti-idiotypic antibodies as cell-surface receptor probes. Proc. Natl. Acad. Sci. USA 75:2443.
53. Celada, F., J. Ellis, K. Bodlund, and M.B. Rotman. 1971. Antibody-mediated activation of a defective β-D-galactosidase: Immunological relationship between the normal and the defective enzyme. J. Exp. Med. 134:751.
54. de Macario, E.C., J. Ellis, R. Guzman, and M.B. Rotman. 1978. Antibody-mediated activation of a defective beta-D-galactosidase: dimeric form of the activatable mutant enzyme. Proc. Natl. Acad. Sci. USA 75:720.
55. Jerne, N.K. 1984. Idiotype networks and other preconceived ideas. Immunol. Rev. 79:5.
56. Sawutz, D.G., and C.J. Homcy. 1986. An anti-idiotypic antibody to an anti-alprenolol monoclonal antibody: promotion of ligand binding to the idiotype. Fed. Proc. (abstr). 45:(3):728.

13
Crime and Punishment in the Society of Lymphocytes: A Speculation on the Structure of the Putative Idiotypic Network

W. Louis Cleveland

Introduction

The clonal selection theory proposed independently by Burnet and Talmage in 1957 envisions the immune system to consist of a set of antigen-specific lymphocytes that are selected in ontogeny to avoid reactivity with self, leaving only reactivity with foreign antigens (1,2). In this theory, antibodies are viewed in the same light as any other self antigens. Given the absence of self-reactivity, interactions among antibody molecules are not predicted. Jerne, however, appreciated that immunoglobulins are different from other self antigens in that they possess an enormous sequence diversity in their v-regions. This led him to expect the occurrence, under physiological conditions, of internal interactions among v-regions, i.e., idiotype-anti-idiotype (Id-anti-Id) interactions. On the basis of this expectation, Jerne constructed a theory of immune regulation based on a network of interacting v-regions (3).

The assumptions of Jerne's original network theory can be divided into two categories: completeness assumptions and mechanistic assumptions. On the basis of the large v-region diversity, it was assumed that for every paratope in the repertoire, there was a complementary idiotope, and vice versa. It was also assumed that for every foreign epitope there was an internal cross-reactive idiotope, i.e., an "internal image." On the basis of experimental findings, it was postulated that antibodies of different specificity might share idiotopes. In addition to these completeness assumptions, there were mechanistic assumptions that concerned the activating and suppressive qualities of idiotopes and paratopes (3).

The mechanistic assumptions of Jerne's original network theory appear untenable in the light of modern findings (4). However, the completeness assumptions have not been ruled out. Indeed, the internal image concept, which arises naturally from the completeness assumptions, has recently been supported by many experimental findings (5). By themselves, the completeness assumptions do not constitute a theory of the immune system. Rather, they demarcate a large class of theories which have these assumptions in common. This is an important consideration since the completeness assumptions in themselves do not allow clear predictions. Only if appropriate mechanistic assumptions are added can definite predictions be derived and tested.

As originally appreciated by Ehrlich, the most basic feature of the immune system is self-nonself (S-NS) discrimination (6). In individuals free of autoimmune disease, there is apparently no aggressive response against S-antigens, whereas aggressive responses against foreign antigens are freely made. It is, therefore, an essential requirement of any theory of the immune system to provide a mechanism by which S-NS discrimination is established. As emphasized by Cohn, S-NS discrimination cannot be encoded in the germline and therefore must be learned by the system in ontogeny (7). As yet, none of the network theories that have been proposed have offered mechanistic assumptions that provide a clear pathway to S-NS discrimination. It is the purpose of this communication to explore a set of mechanistic assumptions that may provide such a pathway. In addition, the theoretical construction that is offered suggests a generalization of the veto phenomenon and provides an interpretation of MHC-restricted antigen recognition by T cells that is based on internal images of MHC determinants. Also incorporated into the model is the currently emerging paradigm of antigen presentation. A consideration of antigen processing has led to the proposal that the internal image concept be extended to include cross-reactions between processed v-region fragments and fragments of conventional antigens. It is further suggested that T cells specific for processed v-regions may influence immunoglobulin and T-cell receptor rearrangements.

Ontogeny of S-NS Discrimination in the B-Cell Compartment

The theory that will be presented in this discussion envisions the idiotypic network to be analogous to a human society. Human societies often can be divided into two segments: one segment that consists of an establishment composed of law-abiding citizens that do not commit crimes against each other and another segment, often referred to as a fringe element, whose members commit crimes against each other and against the establishment. Although there is continual antagonism between the establishment and the fringe element, the establishment, by definition, remains dominant. Within the establishment there is often a relatively stable stratification, with some members having higher rank than others. The mechanistic assumptions that we offer in conjunction with Jerne's completeness assumptions lead to a society of lymphocytes that has analogous properties.

In the case of the B-cell compartment, we invoke three principles of interaction which are referred to as the *principle of symmetrical stimulation,* the *principle of mutual antagonism,* and the *principle of self protection.* Consider an anti-S B cell expressing a newly emerged specificity. Upon encounter with antigen and an appropriate T helper cell, it will be stimulated to differentiate into a plasma cell and secrete antibody. On the basis of the completeness assumptions, we expect that within the expressed B-cell repertoire, there will be another B cell that is anti-Id to the anti-S B cell. The secreted anti-S antibody will bind the membrane receptors of the anti-Id B cell and set in motion a series of events that will lead to T-cell recognition, clonal proliferation, and stimulation of anti-Id secretion. The pathway which leads to the stimulation of the anti-Id B cells is, of course, a matter of critical importance and will be discussed in the section dealing with antigen presentation. What should be taken for granted at this point is that the stimulation of the anti-Id B cell can occur at very low concentrations of Id.

With time, the concentration of Id and anti-Id in body fluids will rise. When the Id and anti-Id concentrations reach sufficient levels, we assume that the clones exert antagonistic effects on each other. The principle of mutual antagonism assumes that the two clones will tend to destroy each other by antibody-dependent mechanisms. These mechanisms could include killing by complement, by cytotoxic cells of several types which possess Fc receptors that recognize antibody-coated target cells (8), and by cytotoxic T cells that recognize processed v-regions. The latter possibility will be discussed in the section dealing with antigen presentation. If no other factors were involved, one would expect that it would be a matter of chance as to whether the Id or anti-Id clone would win the conflict. However, there is another factor involved. The conflict takes place in a milieu containing the S-antigen for which the Id antibodies are specific. According to the principle of self protection, the presence of S-antigens will enable the anti-Id clone to become victorious, provided, of course, that the anti-Id clone is itself not specific for an S-antigen. This is assumed to depend on the ability of S-antigens to adsorb the anti-S antibodies, thereby reducing their concentration and biasing the outcome of the conflict in favor of the anti-Id clone. It must be acknowledged that some S-antigens are sequestered or present at very low concentrations. In these cases, the adsorption may be inadequate to provide protection and other mechanisms may be involved.

The model can be summarized in general terms as follows:

1. There is an Id-anti-Id symmetry in the expression of B-cell receptors.
2. There is a symmetry in the stimulation of Id-anti-Id pairs.
3. There is a symmetrical antagonism between Id-anti-Id pairs that is antibody dependent.
4. When the Id member of an Id-anti-Id pair is anti-S, the symmetry of antagonism is broken by the presence of S-antigen in such a way that the anti-Id clone becomes dominant.

Spontaneous Development of a B-Cell Establishment

The formation of a functional immunoglobulin gene requires DNA rearrangements in which V, D, and J gene segments are selected and made contiguous by recombination events that remove intervening sequences (9). It appears that the selection of individual segments depends on an inherently stochastic process. This implies that anti-S B cells will continually emerge throughout the life of the organism. The above model predicts that this continual emergence of anti-S clones will lead to frequent stimulation of the clones that are anti-Id to anti-S. Since these clones become dominant, frequent restimulation should therefore lead to an "establishment" of anti-Id clones whose secreted immunoglobulins can be expected to constitute a substantial portion of the immunoglobulins in body fluids. These clones can also be expected to be represented prominently in the pool of memory B cells.

It must be appreciated that there are many S-determinants and that for each of these determinants there will be many different complementary v-regions. Associated with each of these v-regions are multiple idiotopes. Clearly, the number of clones that are anti-Id to anti-S is likely to be very large. This means that there will occasionally be opportunities for conflicts among clones that aspire to become members of the establishment. In these cases, the principle of self protection does not apply. Conflicts should therefore proceed until the weaker clones are defeated, yielding a set of clones which do not interact with each other. The absence of interactions within the establishment and its quantitative dominance within the network have the important consequence of restricting the levels of immune complexes that arise as a result of network interactions.

Composition of the B-Cell Establishment

To analyze further the composition of the B-cell establishment, it is necessary to consider the types of anti-Id antibodies that might be stimulated by anti-S clones. Currently, several different kinds of anti-Id antibodies are recognized to exist. Some anti-Id antibodies recognize extremely private idiotopes (10) whereas others recognize more public idiotopes that are referred to as cross-reactive idiotopes (11). Id-anti-Id interactions may, or may not, be inhibitable by antigen (12). In addition to these types of anti-Id antibodies, which will be referred to as classical anti-Ids, there are also internal image anti-Ids (13) and epibodies (14).

Given that there are different types of anti-Ids, the representation of these types in the establishment must be considered. Membership in the establishment is based on the ability of a clone to destroy anti-S clones and the degree of protection provided by S-antigens. It is immediately apparent that internal images may be especially efficient in the destruction of anti-S clones. This follows from the fact that all clones which recognize a particular determinant, regardless of the details of their v-region sequences, should be dominated by a single clone that expresses an internal image of the determinant. This is in contrast to classical anti-Ids that recognize idiotopes that are unique to a particular clone. Many different clones recognizing private idiotopes would be needed to dominate a heterogeneous population of anti-S clones that react with a single S-determinant. The same consideration should apply, although to a lesser extent, to classical anti-Ids that recognize cross-reactive idiotopes.

As yet, there is little evidence supporting the existence of internal images of S-antigens on normal serum immunoglobulins. However, a recent report by Holmberg et al. (15) described the presence of an internal image of a self class II MHC determinant on a TNP-specific hybridoma derived from a normal neonatal spleen. The monoclonal anti-class II MHC antibody used to detect this determinant also showed reactivity with normal serum from both 12-day-old and adult mice.

In view of the fact that different types of anti-Ids may have very different efficiencies in performing the functions of establishment clones, it seems reasonable to assume that the establishment represents a stratified society in which various members have different ranks. The rank of a clone is likely to depend on the frequency with which complementary anti-S clones emerge in ontogeny, since the stimulation and proliferation of establishment clones will depend on small amounts of anti-S antibodies. Thus, it is clear that internal image clones not only are more efficient in dominating anti-S clones but are also more likely to be stimulated than those expressing classical anti-Ids.

The concept of rank can be defined further. For example, as pointed out by Jerne et al. (13), it is possible that some internal image antibodies can image two or more determinants. The more S-determinants imaged by an establishment clone the higher its rank is likely to be.

Another factor that may affect rank is specificity for foreign antigens. There seems to be no constraint preventing establishment clones from having specificity for foreign determinants. Stimulation of establishment clones by foreign antigens should add to the stimulation provided by the attempted emergence of anti-S clones. The possibility that establishment clones can have specificity for foreign antigens leads to a consideration of epibodies (14).

A basic problem facing the immune system is that foreign antigens often share determinants with S-antigens. There has to be a mechanism which restricts the response to the foreign determinants. A solution is possible within the establishment structure already described. For example, the response to shared determinants could be down-regulated by clones expressing internal images of the shared determinants. These clones could, of course, be distinct from those recognizing the foreign determinants. However, this down-regulation might also be mediated by epibodies which simultaneously are specific for the unshared determinant and bear an internal image of the shared determinant. Having both functions in a single structure would seem to offer increased reliability and efficiency in down-regulating the undesired clones. Epibody-secreting clones should have a high rank because there are two pathways of stimulation.

Self Ligand-Receptor Systems Represent a Special Case

A tacit assumption underlying the construction of our theory so far is that S-antigens do not interact with each other. As will become apparent, such an assumption is required for the absence of interactions between internal images in the establishment. However, this assumption is certainly not correct. For example, within the set of S-antigens there are numerous internal interactions, which we shall refer to as ligand-receptor systems.

To analyze the complications introduced by these interactions, it is necessary to consider the prevailing concept of antigenic determinant. Classically, antigenic determinants are structures at the surfaces of macromolecules which are intuitively regarded as "protuberances" (16). This idea is a companion to the notion that the antibody combining site is a cavity. However, a number of laboratories have raised antibodies that appear to be internal images of low-molecular ligands of physiological receptors (5). If the ligands interact with receptor combining sites that are cavities, then it follows that the combining site on the internal image antibody is a protuberance and not a cavity. It would also seem prudent to consider interactions between molecular surfaces that are neither cavities nor protuberances, but which possess complementarity as a result of interdigitating structures. As noted by Erlanger, classical concepts of antibody combining sites are likely to be oversimplified (17).

Once cavities are regarded as antigenic determinants, it becomes natural to regard anti-ligand antibodies as internal images of receptor combining sites. It should be noted that, in some cases, anti-ligand antibodies and physiological receptors have been shown to exhibit similar specificity patterns (18). In the case of a self ligand-receptor system, it seems clear that an internal image of the ligand is an anti-S antibody because of its reactivity with the receptor combining site. However, it is also anti-Id to anti-S, since it should react with an antibody to the ligand. Likewise, the anti-ligand antibody is also simultaneously anti-S and anti-Id to anti-S.

These considerations, at first sight, appear to suggest that internal images of self ligand-receptor systems should be forbidden from the establishment, since the clones will continually be in conflict and reactive with S-antigens. However, the above argument has tacitly assumed that antibody specificity is qualitative in nature. Affinity and valency effects have been neglected. Taking the latter into account, it is possible to conceive of circumstances in which the above difficulties can be avoided.

Consider a low-affinity antibody to a soluble monovalent S-ligand. This antibody, especially if it is an IgM antibody, might have quite high avidity (multivalent affinity) for the membrane immunoglobulins on a B cell bearing internal images of the ligand. Because of the low affinity of the anti-ligand antibody for soluble monovalent ligand, the ligand image B cell would not be protected. On the other hand, if the physiological receptor is a cell surface receptor, then the ligand image antibodies can bind in a multivalent fashion, leading to high avidity. Hence, anti-ligand B cells will be protected by the physiological receptor from destruction by the ligand image antibodies. This will lead to a dominance of clones producing low-affinity anti-ligand antibodies, which can be regarded as internal images of the physiological receptor combining site. If the affinity of these antibodies for physiological ligand is sufficiently low in relation to that of the physiological receptor, then no pathology should result.

The above scheme leads to an elimination of ligand images and permits the emergence of low-affinity receptor combining site images. However, it does not provide a mechanism for the elimination of clones that produce receptor combining site images having high affinity for the physiological ligand. Such clones, which might cause autoimmune disease, could, however, be eliminated by anti-Id antibodies of the classical type, since clones producing these antibodies would be protected by the ligand. Thus, even though internal images of receptor combining sites may sometimes belong to the establishment, it appears that the establishment must also contain anti-Ids of the classical type.

Maintenance of Immunological Memory in the B-Cell Compartment

The linkage within the establishment of specificity for foreign epitopes and anti-idiotypic specificity for anti-S clones creates an interesting possibility for maintaining immunological memory. The important point is that establishment clones specific for foreign epitopes can be stimulated by anti-S Id in addition to foreign antigens. This means that the anti-foreign establishment clones will continue to be periodically stimulated after the foreign antigen has been cleared. These considerations lead to the expectation that the set of B-cell establishment clones is an important repository of long-term memory in the B-cell compartment. That immunological memory could be stored in the idiotypic network is a possibility that was noted by Jerne in his original paper (3).

Interactions within the Fringe Element

Within the fringe element, clones may react with each other, with the establishment, and with both self and foreign antigens. Unlike the establishment, the fringe element is characterized by internal conflict. An important consequence of this is that secreted fringe element antibodies should often form immune complexes with other fringe element antibodies or with establishment antibodies and be cleared from circulation. This is expected to make temporary any usefulness in providing protection against pathogens and to reduce reaction with S-antigens. This also means that the persistent titer to a foreign antigen must be produced by establishment clones. As noted previously, long-lived clones of memory cells should also belong to the establishment.

Stability of the Establishment

As indicated above, establishment antibodies can form immune complexes with fringe element antibodies or with foreign antigens. This means that the stability of the establishment cannot reside in individual molecules. Similarly, individual cells in establishment clones might also be destroyed in conflicts with fringe element clones. The stability of the establishment would therefore have to reside in the longevity of its clones and ultimately in the pattern of specificities that is responsible for the periodic restimulation of these clones. The absence of stability within the fringe element would be a consequence of a continual alteration of the pattern of specificities expressed as a result of internal conflicts and conflicts with the establishment. It should be noted that Jerne has concluded from an analysis of lifetime measurements that it may be possible to divide network components into long-lived and short-lived sets (16).

Autoimmune Disease and the Establishment

A network consisting of a fringe element and an establishment provides a context in which to recast the traditional perspective of autoimmune disease. First, there is a basic distinction between an autoimmune response and an autoimmune disease. This follows from the fact that, in our model, autoimmune antibodies are normally and usefully produced by fringe element clones. Because of the transient nature and low levels of such antibody production, we assume that clinically significant autoimmune disease does not result. Clinical autoimmune disease is assumed to be a consequence of anti-S clones finding their way into the establishment as a result of aberrant regulation. Membership in the establishment should allow long-term production of relatively higher levels of autoimmune antibody. Given the stability of the pattern of specificities expressed in the establishment, one would expect disease-producing antibodies to be part of this stability. This perspective is consistent with the findings for myasthenia gravis. Tzartos and coworkers have found that the specificity pattern of antibodies to the acetylcholine receptor is remarkably constant over periods of many years in spite of therapeutic maneuvers to control the disease (19).

A Liberal or a Conservative Society?

So far it has been assumed that anti-S clones are destroyed by the establishment. This may not always happen as genomic alterations in v-regions may allow rehabilitation of anti-S clones. There is a growing body of data supporting the occurrence of frequent intraclonal somatic mutations following antigen stimulation (20–22). It also appears that multiple rearrangements can occur (9). Given these processes, it is possible that clones initially expressing anti-S reactivity can undergo genomic alterations that lead to a loss of reactivity with S-antigens and also a loss of incompatibility with the B-cell establishment. This type of mechanism would not only rehabilitate fringe element clones, but would also generate diversity as was originally foreseen by Jerne (23).

Network Interactions Involving the T-Cell Compartment

It is now known that for major classes of T cells, antigen recognition is MHC-restricted. For such T cells, antigen must be seen on membranes in association with either Class I or Class II S-MHC molecules [(24) and references therein]. This type of antigen recognition seems fundamentally different from B-cell antigen recognition, which is not restricted to membrane antigens and which does not require association with MHC molecules. It is a matter of major importance to consider whether MHC-restricted T-T interactions are a part of the putative idiotypic network and whether they might obey the interaction principles we have described for the B-cell compartment. However, current knowledge seems inadequate for the construction of a detailed mechanistic model at this time. Nonetheless, it is interesting to consider the possibility that the repertoires of Class I-restricted and Class II-restricted T cells may also contain establishments in which the clones express v-regions that bear internal images of S-determinants or specificities that are anti-idiotypic to anti-S T-cell receptors. It is also interesting to consider the interactions that may occur between the T-cell and B-cell repertoires.

Interactions between the B-Cell and T-Cell Compartments

Let us consider T cells arising in ontogeny after the B-cell establishment has already formed. If the T-cell receptor v-regions react with establishment immunoglobulins that are in body fluids, then the cells bearing them may be destroyed by the cytotoxic mechanisms outlined previously. This suggests that T-cell receptors must avoid reactivity with establishment immunoglobulins. Avoidance of reactivity can be achieved if T-cell receptors imitate establishment immunoglobulin v-regions, since the latter set of v-regions is itself free of internal interactions. It is therefore clear that the assumption that T-cell receptors bear internal images of S-antigens is consistent with the absence of interactions between soluble establishment immunoglobulins and T-cell receptors. The imitation of immunoglobulin v-regions by T-cell receptors that is sometimes seen in normal mice is absent in mice artificially devoid of B cells (25). In view of these considerations, it seems likely that the B-cell compartment has considerable influence on the idiotypic structure of the T-cell repertoire.

Are there interaction pathways that allow the T-cell repertoire to influence the B-cell repertoire? In the foregoing example, it is assumed that secreted immunoglobulins are already present in body fluids before the T cell expresses its receptor. If soluble immunoglobulin were not present, the T cell could interact directly with the B-cell surface immunoglobulins, and in the case of a cytotoxic T cell, destroy the B cell. Here a native immunoglobulin is recognized (or recognizes) in a non-MHC-restricted fashion. Recent in vitro evidence raises the possibility that such interactions may occur in vivo (26,27). It is also possible that native B-cell idiotopes can be recognized by T cells in an MHC-restricted fashion (28). This type of interaction might occur even in the presence of soluble B-cell Id, since MHC-restricted recognition is generally thought to be uninhibitable by soluble antigen. However, these examples, which involve the recognition of native immunoglobulin idiotopes, are as yet only formal possibilities. The bulk of available data indicates that T cells recognize most soluble protein antigens only after they have been processed and therefore denatured. It is possible that in vivo T cells recognize immunoglobulin v-regions only after processing. This possibility will be discussed in detail in the section dealing with antigen presentation. Although detailed predictions cannot be made with available data, it nonetheless appears that bidirectional interactions between the T-cell and B-cell repertoires are likely.

T-Cell Receptor Internal Images and the Veto Phenomenon

Among the more interesting recent discoveries in cellular immunology is the veto phenomenon, which was discovered by Miller and Derry in 1979 (29). It was first observed in studies of the mixed lymphocyte reaction (MLR). It was found that a subpopulation of B-strain cells could suppress an A anti-B MLR. Further studies suggested that these cells, which were referred to as veto cells, have the capacity to suppress Class I-restricted cytotoxic T lymphocytes (CTL) that recognize any antigens expressed on the veto cells (30). In other words, veto cells have the ability to suppress CTL that attack them. The suppression mechanism appears to be different from the classical T-cell cytolytic mechanism. Efforts to find a soluble suppressor factor have been unsuccessful. Suppression appears to be permanent (30). Veto cells have been found in bone marrow, thymus, and fetal liver of normal mice. In athymic nude mice, bone marrow and spleen contain veto cells. In spleens and lymph nodes of normal mice, veto cells can be found after a period of in vitro incubation (30). Of special interest is the finding that mature cytotoxic T cells can have veto activity (31,32).

It is clear that the veto phenomenon could play an important role in S-NS discrimination. However, according to current thinking, the veto effect is considered to be limited to antigens expressed on the veto cells (30,33). Moreover, the mechanism is regarded to be independent of the antigen-MHC specificity of the cell in the case of CTL-type veto cells (33).

The main point of this discussion is to emphasize that the presence of internal images of S-antigens on Class I-restricted T-cell receptors of mature CTL would provide a basis for extending the veto phenomenon to the entire set of S-antigens represented as internal images. In order for this scheme to work, it is necessary that veto suppression be activated when an anti-Id CTL attempts to attack the veto cell through recognition of the T-cell receptor on the veto cell. We are unaware of any evidence against this possibility. On the contrary, Fink and co-workers have found that a clonotypic antibody to the T-cell receptor on a cloned T-cell line augments the veto activity by 300-fold (34). This augmentation, which may be due, at least in part, to the induction of cell proliferation, suggests that cross-linking of the clonotypic receptor does not interfere with veto activity. Conceivably, cross-linking the clonotypic receptors by receptors on an attacking T cell might activate veto suppression as it does for other antigens expressed on the veto cell.

As noted by Miller (30), effector T-suppressor cells have been reported to bear Ly2 and Class II MHC antigens (35), raising the possibility that Class II-restricted T cells are also subject to veto control.

According to the scheme presented here, mature CTL belonging to the establishment have both dual specificity and dual effector functions. On the one hand, they are specific for targets bearing foreign antigens and can lyse these cells. On the other hand, they bear internal images of S-antigens and can suppress CTL having specificity for these S-antigens. These features are directly analogous to the properties postulated for establishment antibodies that have specificity for foreign antigen and anti-Id specificity for anti-S B-cell receptors.

T-Cell Receptor Internal Images of MHC-Determinants and MHC-Restriction

The recent extraordinary progress in characterizing the molecular genetics, amino acid sequence, and peptide structure of the T-cell receptor (36–39) has so far failed to clarify the basis of T-cell antigen recognition, which exhibits the phenomenon of MHC-restriction. On the contrary, the new data have intensified the enigma of MHC-restriction. It has been revealed that the T-cell receptor is an antibodylike molecule (40), which is puzzling since antibodies do not exhibit MHC-restriction. In the case of MHC-restricted antigen recognition, antigen must be associated with membranes bearing MHC molecules in order to be recognized. MHC-restriction appears to be a binding event phenomenon in which high-affinity complexes occur only when a ternary complex consisting of the T-cell receptor, antigen, and MHC is formed (41,42). Binary complexes consisting of antigen and T-cell receptor or MHC and T-cell receptor do not functionally activate the T-cell and do not lead to adhesion between the T-cell and a cell bearing antigen alone or antigen + incorrect MHC. The problem, therefore, is to understand why both antigen and MHC are required for a functional binding event and why neither alone will give functional binding. Essentially two types of models have been proposed to account for MHC restriction: dual recognition models which postulate the existence of two sites that recognize independent antigen and MHC determinants and single recognition models which envision a single-site receptor that recognizes a neoepitope which arises upon the formation of a complex between conventional antigen and MHC [(24) and references therein]. As yet, it is not possible to rule out either type of model. Previous formulations of both types of models are essentially "local" interpretations which consider the T-cell receptor or the antigen-MHC complex as an isolated entity. In this discussion, we offer a "non-local" interpretation of MHC-restriction that is inspired by a network perspective. This interpretation represents an extension of an allosteric dual recognition model which was previously proposed by this writer in collaboration with Erlanger (24). The unique feature of this model is that the sites for antigen and MHC are not independent but are coupled by an allosteric mechanism. The basic idea is that the binding of MHC to the T-cell receptor causes a conformational change at a second site, conferring on this site the ability to bind conventional antigen. This leads to the initial formation of a ternary complex, which may, if positive cooperativity occurs, undergo further internal conformational reorganization, to produce a tenary complex of sufficient affinity to trigger the T cell. This model is, in a sense, a mirror image of a neoepitope model, since it envisions the existence of neoparatopes that arise following interaction with conventional antigen and MHC. It encounters at least two basic difficulties, which we refer to as the *dual recognition problem* and the *neoparatope problem*.

The Dual Recognition Problem

This problem is one that arises with attempts to rationalize MHC-restriction in the context of a two-site T-cell receptor. Its essence is revealed by a consideration of an argument for dual recognition that was recently given by Parham (43). Parham starts his argument by noting that some monoclonal anti-MHC antibodies bind to cells only if they are bivalent. Apparently, the monovalent binding energy is too low to give detectable binding, whereas in the bivalent case the partial additivity of the binding energies of the two sites leads to binding. He then generalizes by imagining an antibody of mixed specificity, one site specific for MHC, the other for conventional antigen. In this way, he accounts for MHC-restriction, i.e., binding is seen only if both species are present. An important point here is that additivity of binding energy occurs only because the two species of antigen molecules are anchored in the same membrane. That is, if antigen and MHC molecules were in free solution, no additivity of binding energy would occur. This point highlights the fact that Parham's argument applies only to receptor molecules that are in free solution. Let us now suppose that the heterospecific antibodies are anchored in the membrane of a cell and that the number on a single cell may be greater than 10,000. This is the case that is relevant to the T-cell receptor, since the latter is a membrane receptor. Suppose the cell with heterospecific antibodies collides with another cell bearing self MHC and no conventional antigen. Clearly, there will be an opportunity for extensive multivalent interactions between the matrix of anti-MHC sites and the matrix of MHC antigens on the target cell. Thus, the heterospecific cells should adhere to targets when only the anti-MHC sites are occupied. This is in conflict with the basic findings of MHC-restriction. This problem, which has also been alluded to by Blanden and Ashman (44) and by Mitchison (45), applies not only to Parham's model but to dual recognition models in general.

The Neoparatope Problem

Before attempting to provide a solution to the dual recognition problem, it is worthwhile to consider another difficulty that is presented by the approach of Parham (43). In his argument he *chooses* an antibody that will not bind unless it is bivalent. There, of course, exists within the B-cell repertoire a population of antibodies that have sufficient affinity to bind MHC antigens in a monovalent fashion. This point brings up the difficulty of selecting a repertoire of T-cell receptors that have only low affinity for both antigen and self MHC at the monovalent level. Since self MHC is present in the organism, it is possible to imagine that some selection mechanism could delete receptors having a high affinity for self MHC. However, it is difficult to imagine such a mechanism for conventional antigen when such antigen is not usually present in the organism and is, moreover, unpredictable. This problem also arises in the context of the neoparatope model. In particular, it is basic to our model that the putative site for conventional antigen not react with high affinity until this site is altered by simultaneous interaction with MHC. Thus, while neoparatopes may have high affinity for conventional antigen, "old" paratopes must have only low apparent affinity. Moreover, MHC-restriction requires that the universe of neoparatopes not overlap with the universe of "old" paratopes. How can the repertoire be selected to have these properties? This is referred to as the *neoparatope problem*.

A Possible Solution to the Dual Recognition and Neoparatope Problems

As a solution to these problems, we propose that there is an internal interaction in the T-cell receptor in which the site for conventional antigen is sequestered as a result of an interaction between this site and the anti-MHC site. In short, we propose that one site is specific for the other. We further propose that the anti-A site bears an internal image of the MHC determinant that is recognized by the anti-MHC site. This scheme allows the anti-MHC site to have substantial initial affinity for MHC without leading to adherence to other cells that bear only MHC, since the internal image determinant, which is an integral component of the T-cell receptor, should be an especially effective competitive inhibitor, thereby solving the dual recognition problem. If the antigen-contacting residues associated with the anti-A site are also sequestered as a result of the postulated internal interaction, then a low apparent affinity for conventional antigen alone is expected. Moreover, the set of "old" paratopes, which are essentially nonfunctional, does not overlap with the set of neoparatopes. Thus, the neoparatope problem would also be solved. The postulated internal interaction appears compatible with the allosteric mechanism initially proposed. The internal bond between the two sites would be continually forming and breaking, allowing occasional bonds to be formed with MHC antigens on other cells. Interaction with conventional antigen, if present, at the anti-A site would interfere with the re-formation of the internal bond and allow the occurrence of the conformational changes that are assumed to be required for the formation of a high-affinity ternary complex.

Evidence for an Internal Inhibitor

A prediction of the model described above is that an incomplete T-cell receptor lacking the anti-A site, which bears the internal image of MHC, should show binding to self MHC on other cells. Data that are consistent with this possibility have been obtained by Thomas and Hoffman (46). These investigators found that antigen-specific T cells educated in vitro would not adhere to target cells in the absence of specific antigen, as expected. However, if the T cells were lightly trypsinized, adherence to target cells in the absence of antigen was observed. Moreover, the adherence phenomenon showed preference for self MHC. It seems possible that the trypsinization selectively removed the MHC internal image, leaving an incomplete T-cell receptor with an exposed anti-MHC site. With the powerful methods developed recently, it should be possible to carry out a rigorous exploration of these findings.

Evidence for Internal Images of MHC Determinants on T-Cell Receptors

Sim and co-workers, on the basis of a different theoretical argument, have been led to expect the existence of MHC internal images on T-cell receptors (47). They have provided evidence that monoclonal anti-Class II MHC antibodies as well as cytotoxic T cells specific for Class II MHC can react with Class II-restricted T cells that appear to bear no conventional Class II MHC antigens (47).

It should be emphasized that the concept of a neoparatope implies the existence of neoidiotopes and neointernal images. Therefore, if conformational changes occur upon the formation of a ternary complex, the ability to detect various idiotopes and internal images on T-cell receptors may be dependent on the occupation, or nonoccupation, of the sites for MHC and conventional antigen. This may complicate the detection of internal images on T-cell receptors.

The interpretation of MHC-restriction offered here depends on the existence of internal images of MHC determinants, which may be elements of an idiotypic network involving T-T interactions, as Müllbacher (48) and Sim and co-workers (47) have suggested. The receptor itself can be regarded as a composite structure which contains both Id (anti-MHC) and anti-Id (MHC internal image). It may seem unusual to have these structures in the same receptor (or receptor complex), since they are traditionally regarded as separate molecules in the B-cell compartment. However, the more recent epibody concept is quite reminiscent of what is proposed here for the T-cell receptor. Indeed, the T-cell receptor we envision can be regarded as an "introverted" epibody.

Implications of Antigen Processing for Network Interactions

According to a currently emerging paradigm (49), antigen recognition by Class II-restricted T cells is dependent upon the processing of conventional antigen by "antigen-presenting" cells. Antigen is internalized by the antigen-presenting cell and subjected to limited proteolytic degradation. Proteolytic cleavage leads to an unfolding of the resulting peptides and the possible exposure of sites preferentially recognized by T cells (50). Presenting cells pulsed with antigen for a few hours and thoroughly washed continue to present antigen for several days (49). This has recently led to the hypothesis that processed antigen is modified in some way that leads to a firm anchoring in the cell membrane (49). This anchoring is thought to be independent of any interaction between processed antigen and MHC, which is much too weak to account for the stable association with the cell membrane (51).

Studies by Kakiuchi et al. (52), Rock et al. (53), and Lanzavecchia (54,55) indicate that in the case of B cells, antigen presentation is greatly enhanced when the presenting B cell possesses a membrane antibody that is specific for the antigen. Using a B-cell line transformed by Epstein-Barr virus, Lanzavecchia has offered evidence that tetanus toxoid at a concentration of 10^{-13} M could be presented to cloned T cells when the B-cell antigen receptor possessed a binding constant of 10^8. B-cell lines not specific for tetanus toxoid could also present this antigen, but required antigen concentrations four orders of magnitude greater. The ability to present antigen that is present in native form at such low concentration would seem to depend on an antigen-specific "pumping" mechanism for internalization as well as firm anchoring of processed antigen in the membrane (56).

This model of antigen presentation represents an evolution of the previous antigen-bridging model of T-B cooperation which envisioned the recognition of native antigen by the T-helper cell (57). As will become evident, this new model of antigen presentation has far-reaching implications for network theories of immune regulation.

Chesnut and Grey demonstrated that B cells can present rabbit anti-mouse Ig antibodies to T cells specific for the processed Fc regions of these rabbit antibodies (58). If B cells can present the Fc region, it is reasonable to expect that they can also present the v-regions. If the v-regions of the exogenous antibody can be presented, it is likewise reasonable to expect that the endogenously synthesized v-regions can be presented, since both sets of v-regions will be internalized as an immune complex.

Mechanistic Basis of the Principle of Symmetrical Stimulation

The principle of symmetrical stimulation was previously stated as a formal principle. A mechanistic rationalization is required. Consider the immune response to a thymus-dependent protein antigen. The bulk of available data suggests that the antigen is first taken up by an antigen-presenting cell which has no intrinsic antigen specificity, such as a macrophage (59). The efficiency of uptake may be greatly enhanced by the presence of natural antibody (60). After processing, fragments of the antigen are presented on the cell membrane in association with Class II MHC. This leads to activation of T-helper cells. Antigen is also presented by antigen-specific B cells (58). However, the low precursor frequency of antigen-specific B cells suggests that they do not play a role in the initial activation of T cells. But once T cells are activated, they can deliver signals to B cells presenting the appropriate antigen, leading to clonal expansion and high-rate antibody secretion. The principle of symmetrical stimulation states that when Id B cells are stimulated, auto-anti-Id B cells are also stimulated. Experimental evidence for auto-anti-Ids is sparse, but most studies have depended on serological methods (61). Very recent studies using monoclonal antibody technology suggest that auto-anti-Ids are readily detectable (62–66). What is the pathway of stimulation? One possibility is that the internalization of antigen by the Id B cells also leads to the presentation of Id v-regions. This could, in turn, lead to the activation of T-helper cells specific for processed Id. These T-helper cells would also encounter processed Id on anti-Id B cells that have interacted with secreted Id. This interaction should stimulate the anti-Id B cells and lead to the secretion of auto-anti-Ids. In addition, interaction of soluble Id with an anti-Id B cell should induce the presentation of the endogenous anti-Id, leading to activation of a population of T cells specific for processed anti-Id. This pathway of stimulation could, in principle, continue indefinitely. However, the establishment may play a role in truncating idiotypic cascades. The important point here is that when soluble Id interacts with an anti-Id B cell, or when soluble anti-Id interacts with an Id B cell, a T cell specific for either processed Id or anti-Id will activate both specificities of B cells. Thus, there is a symmetry in the stimulation of Id-anti-Id pairs of B-cell clones by v-region specific T-helper cells, as Leserman has also appreciated (67). Moreover, this stimulation is antibody dependent. This mechanism is offered as the basis of the principle of symmetrical stimulation.

There is another aspect of the principle of symmetrical stimulation which can be rationalized in the context of the new model of antigen presentation. An essential aspect of this principle is that stimulation of B cells by soluble Id or anti-Id can occur at concentrations below the levels at which antibody-mediated cytotoxicity occurs. As noted above, antigen presentation can occur when the antigen concentration is very low. For example, a B-cell receptor having a binding constant of 10^8 could facilitate antigen internalization when the external concentration was 10^{-13} M (55). Presumably, if the binding constant were 10^{10}, presentation could occur at 10^{-15} M. Given the bivalency of IgG and the higher valency of secreted IgM, and the high concentration of mobile bivalent membrane receptors on B cells, it is likely that Id-anti-Id reactions at B-cell surfaces are characterized by high avidity constants. It should be noted that 10^{-15} M IgG is equivalent to 1.6×10^{-13} g ml^{-1}. These considerations are offered in support of our initial assumption that low concentrations of Id and anti-Id will lead to stimulation by T-helper cells without generation of significant antibody-mediated cytotoxic effects. At much higher concentrations, we expect the occurrence of cytotoxic effects mediated by Fc-receptor-bearing cells (8) and possibly by Class I- or Class II-restricted [(68) and references therein] T cells activated to mediate cytolytic effector functions. The differential effects for high and low concentrations of Id or anti-Id that we have postulated are in line with experimental observations showing that low doses of passively transferred anti-Id tend to activate whereas high doses tend to suppress Id expression, following challenge by antigen (69).

Since the mechanism of symmetrical stimulation is antibody dependent, it is necessary to consider the possible breaking of symmetry by S-antigens in cases where Id or anti-Id has an anti-S specificity. It will be recalled that the breaking of symmetry by S-antigens is a crucial factor in the antagonism of Id-anti-Id pairs. The fact that antigen presentation can take place at extremely low concentrations suggests that S-antigens may not break symmetry, since it is unlikely that adsorption would be efficient enough to reduce levels of Id or anti-Id below the threshold for presentation. Moreover, the threshold for presentation is likely to be several orders of magnitude below the threshold for cytotoxicity effects. Within this window, antigen presentation may not be a sensitive function of concentration, making partial adsorption of Id or anti-Id by S-antigens unimportant. In addition, Id-anti-Id, or S-antigen-Id, immune complexes may be taken up by macrophages and presented. The ability of these complexes to lead to presentation is in contrast with their inability to mediate cytotoxic effects relevant to Id-anti-Id conflicts. Hence, the effects of S-antigen adsorption on stimulation are basically different from the effects on antagonism.

Studies by Kohler and co-workers with the phosphorylcholine (PC) system have produced findings that are consistent with processing and presentation of v-regions. It was found that immunization with PC-hemocyanin induces Id-specific T-helper cells that can be detected by their ability to help TNP-specific B cells when challenged with TNP-T15, where T15 is a PC-specific myeloma protein that bears the idiotype that dominates the response to PC in BALB/c mice (70). This finding is consistent with our prediction that interaction of antigen with specific B cells leads to the presentation of B-cell receptor v-regions and the induction of v-region-specific T-helper cells.

The population of T-helper cells recognizing the T15 myeloma protein also recognizes the M167 myeloma protein as highly cross-reactive. M167 is also specific for PC (71). Inhibition experiments suggested that the recognized determinants are heavy-chain associated and are preserved on isolated chains (72). This cross-reactivity in T-cell recognition is in contrast to what is seen with anti-idiotypic antibodies, which generally do not detect cross-reactions with T15 and M167 (71). This suggests that at the level of native immunogenic idiotopes, these myeloma proteins are not similar. However, the cross-reactivity seen by T cells is consistent with the possibility that these cells recognize processed heavy-chain peptides. The heavy chains from these proteins are identical in the first hypervariable region and differ by only 14 of the first 125 amino acids (21). Depending on cleavage points, peptides of identical sequence could be obtained from both heavy chains, and complete identity may not be needed for cross-reactivity.

The results of Jorgensen and Hannestad (73) are also consistent with v-region processing. For example, these investigators demonstrated that reduced and fully alkylated v-domains primed T-helper cells as efficiently as native domains when the response to the native domain was monitored.

Rubinstein et al. have observed that injection of 10 μg of A48Id into BALB/c neonates leads to activation and dominance of the A48Id component of the response to bacterial levan (74,75). In unmanipulated mice this Id is normally absent. Presentation of A48Id v-regions may be involved in the activation process. Indeed, it was found that A48Id dominance was associated with the activation of T-helper cells specific for A48Id (74,75).

Residue Epitopes, Residue Idiotopes, and Residue Internal Images

The possibility that immunoglobulin v-regions may be processed and presented suggests the existence of an additional dimension of network regulation. Let us consider classical idiotopes on native immunoglobulin v-regions. Using the language of Kunkel (10), these antigenic determinants are "individual antigenic specificities" that are associated with unique aspects of the v-region sequences. This same concept can be enlarged to include unique epitopes, i.e., idiotopes, that will likely be present on the processed v-region peptides. Since these peptides represent the residue that remains after partial proteolytic degradation, we refer to these idiotopes as "residue idiotopes." It must, of course, be kept in mind that they are recognized in association with MHC antigens. In this context, an interesting possibility presents itself. Since the set of v-region peptides that can be presented by B cells is a diverse set, it is possible that peptides in this set will be cross-reactive with peptides associated with processed conventional antigens. Thus, residue epitopes on processed conventional antigens may cross-react with residue idiotopes. In particular, if a polyclonal population of T cells specific for a residue epitope also react with a residue idiotope, then the residue idiotope can be regarded as a residue internal image.

The possible existence of these residue elements suggests the existence of a new category of network interactions. It is now necessary to consider that some antibodies may bear both native and residue internal images. Consider a hapten-carrier conjugate, such as adenosine-bovine albumin. There may exist antibodies that in their native state bear internal images of adenosine (62) and in their processed state bear residue internal images of "carrier determinants," i.e., residue images of albumin residue epitopes.

These two-level internal images may be of considerable importance for the development of idiotype vaccines, as they would facilitate the induction of both T-cell and B-cell immunity to a viral protein. Noseworthy et al. may have isolated a monoclonal antibody of this type (76). Their antibody appears to bear an internal image of the reovirus determinant which interacts with the virus receptor on infectable cells (76). Interestingly, immunization with this antibody not only induces antibodies that recognize native virus, but also induces delayed-type hypersensitivity to the virus (77). The latter reactivity is generally attributed to Class II-restricted T cells that recognize processed antigen (78).

Two-level internal images may arise when there is sequence homology between the viral protein and one of the v-regions of the internal image antibody. In such a case, the homologous sequence might lead to native sequence determinants that are cross-reactive. The homologous sequence might also lead to residue determinants that are cross-reactive. It should be emphasized that the shape of the residue determinant might be quite different from the native determinant involving the same amino acids. It is interesting to note that the internal image antibody isolated by Bruck et al. (79) has a v-region stretch of amino acids that is homologous to a stretch of the viral sequence. Although two-level internal images may arise as a result of sequence homology, such homology may not always be required, given the mimicry of nonproteinaceous substances such as haptens (5) and carbohydrates (80).

The existence of two-level internal images would be of considerable theoretical importance, since it has heretofore not been obvious how the idiotypic network can store the information that a particular hapten is attached to a carrier molecule (7). Such information may also be stored using two-level epibodies. For example, an antibody specific for a native determinant on a protein could also bear a residue idiotope or residue internal image that cross-reacts with a residue epitope of the protein. Residue epibodies might play a role in the generation of auto-anti-Ids, since a carrier-specific T-helper cell specific for the residue epitope could stimulate an anti-Id B cell presenting the v-region of the residue epibody.

Residue epibodies may be relevant to the interesting data of Kim et al. (81), which reveal that some cloned T-helper lines specific for ovalbumin facilitate a phosphorylcholine response in which the T15 idiotype dominates and others do not. The T15-inducing clone, which is specific for an ovalbumin residue epitope, could also be specific for a T15 residue idiotope. In this case, the T-helper cell would see a greater density of processed antigen on B cells expressing the T15 Id, which may lead to a quantitatively greater response. This interpretation should be readily tested using synthetic peptides based on the T15 sequence and the existing T-cell clones.

A consideration of network interactions involving residue elements immediately leads to the following question: Is the set of residue idiotopes complete? Another question also arises: Is the set of T-cell receptors that recognize residue elements complete? While these questions cannot, of course, be answered on the basis of available data, some considerations can be made.

The concept of completeness is intimately related to the concept of specificity. On the basis of considerable data, it is clear that a single antibody molecule can exhibit multispecificity in antigen binding (82). This means that antibodies do not have absolute powers of discrimination. Clearly, the greater the degree of multispecificity, the greater the likelihood of interactions within the B-cell repertoire.

These considerations can be transferred to the residue compartment. The first point to be noted is that the number of native antibody v-regions is likely to be greater than the number of v-region peptides that can be presented as processed fragments. On the other hand, there is evidence that T cells exhibit a higher degree of multispecificity than B cells [(24) and references therein]. This could compensate for the smaller universe of processed v-region peptides and allow for a functional network involving residue elements. However, it should be noted that the findings of Jorgensen and Hannestad suggest that Ir gene defects may place some restrictions on completeness (73).

IgD May Play a Special Role in the Presentation of Endogenous v-Regions

An especially puzzling feature of B cells is the existence of two isotypes of membrane receptors. In addition to monomeric IgM, there is also an IgD receptor (9). The function of the IgD isotype, which occurs almost entirely as a membrane receptor (83), has continued to resist clarification. A unique feature of IgD is the proteolytic lability of its hinge and Fab regions (84). Also intriguing is the fact that immature B cells have only IgM (9). From the perspective generated by this discussion, an interesting possibility for IgD function presents itself. An essential step in antigen presentation is the proteolytic cleavage of the antigen into fragments. The greater lability of IgD hinge and Fab regions suggests that they may be processed more efficiently, leading to a greater concentration of processed fragments on the membrane and a greater potential for the stimulation of residue-idiotope-specific T-helper cells. That this would occur on mature B cells and not immature cells is interesting. Conceivably, immature B cells expressing inappropriate v-regions might first be subjected to negative selection before the expression of IgD occurs.

In this discussion, antigen presentation has been viewed as a process in which molecules in the extracellular fluid or on the cell membrane are internalized. Very recent evidence suggests that internally synthesized proteins can also be presented (68,85). This possibility makes it important to emphasize that an assumption of the model presented here is that v-regions of B cells unstimulated by antigen or anti-Id are not presented. It is assumed that antigen processing takes place in specialized compartments and that access to these compartments is a controlled process, as some experimental evidence is already beginning to indicate (86,87).

The model of v-region processing offered here differs from prior suggestions in several important features. A recent model, proposed by Leserman (67), appears heavily influenced by the determinant selection hypothesis (88). This hypothesis assumes that processed antigen adheres to the membrane of an antigen-presenting cell because of its affinity for Class II MHC. Leserman notes the difficulty of understanding how the myriad of conventional antigens can all have affinity for the several MHC molecules present in a single individual. He then assumes that presentation of conventional antigen is an incidental, occasional feature and that a uniform and centrally important consequence of the interaction of antigen with B-cell receptors is to induce the presentation of the B-cell receptors. He envisions that presentation involves antigen-induced endocytosis and limited proteolysis to expose an invariant hydrophobic site, which may be located in CH1, and which has affinity for Class II MHC molecules. Return of the processed Ig to the cell surface leads to Ig-MHC complex formation and subsequent T-cell recognition. The important point here is that the invariant site must remain connected to the hypervariable region where idiotypic determinants are located. This means that proteolysis must be relatively limited.

The model offered in this discussion assumes that membrane anchorage of processed Ig is independent of any affinity for MHC, although weak affinity may exist (51) and may play a role in the formation of the ternary complex with the T-cell receptor. The mechanism for membrane anchorage is assumed to be similar for both conventional antigen and Ig. Hence, presentation of conventional antigen and of Ig are of equal likelihood and importance. Moreover, the anchoring mechanism, which may involve the formation of covalent bonds, imposes no currently apparent restriction on the degree of proteolysis. On the basis of experiments with synthetic peptides (89), processed peptides may be as short as about 15 amino acids.

McMamara et al. (72), on the basis of their data and those of Jorgensen and Hannestad (73), have suggested that T cells "see" individual Ig chains and that antigen-presenting cells may separate Ig molecules into free polypeptides. This suggestion also envisions very limited modifications.

If we are correct in suggesting that processing for v-regions is similar to that for conventional antigens, then an experimental exploration of v-region processing is immediately accessible, given the availability of known v-region sequences for some antigen systems, the availability of T-cell clones and hybridomas, and the ease with which short peptides can be synthesized.

A Possible Mechanism for Network Control of B-Cell and T-Cell Rearrangements

Complete immunoglobulin and T-cell receptor genes are assembled by rearrangement mechanisms in which germline gene segments are selected and joined to form complete genes. To assemble, for example, an immunoglobulin heavy-chain variable region, recombination between three different germline elements (V[H], D, J) must occur. In the mouse, the v[H] segment must be selected from a set that contains 200 to 1,000 elements. The D and J sets contain 12 and 4 elements, respectively (9). As yet, a detailed understanding of the forces that regulate these recombination events has not been achieved. However, Yancopoulos and Alt have proposed a model which suggests that a single recombinase mediates the assembly of all Ig and T-cell receptor variable gene segments (90). An essential aspect of this model is that for the recombinase to be active on a particular gene segment, the gene segment must be exposed. They have observed that germline V[H] segments are transcriptionally active before rearrangement. They assume that this transcriptional activity ensures the exposure necessary for recombinase activity.

It is proposed here to extend the accessibility model of Yancopoulos and Alt to include a mechanism by which T cells can influence the selection of V[H] segments by the recombinase. Essentially three additional assumptions are needed for this extension.

Antigen presentation is usually visualized as a process in which proteins in the external fluid are internalized and then processed. Previously in this discussion, it was also assumed that endogenously synthesized membrane immunoglobulins can be internalized and processed when they interact with antigens or antibodies. In the context of the Jancopoulos-Alt model, we further assume that V[H] transcripts produce peptides which find their way into the compartments where the processing required for antigen presentation takes place. Although this could occur by an exogenous pathway in which peptides having leader sequences are brought to the cell surface and then reinternalized, it is of special interest to consider the possible existence of an endogenous pathway for antigen presentation. Evidence for this kind of pathway has recently been obtained with internal viral proteins that have no leader sequences. Recent evidence suggests that cytolytic T cells can kill infected cells through recognition of processed fragments of internal viral proteins (68,85).

The second assumption is that the transcription of unrearranged V[H] segments is a stochastic process that results in different V[H] segments being transcribed in different unrearranged cells at a particular time. Thus, while population studies may suggest transcription of large fraction of V[H] segments, we assume that in individual cells, transcription is restricted to a stochastically selected and possibly time-variable subset of V[H] segments. This means that different cells will, at a particular point in time, present the processed peptides of different sets of V[H] segments. As a result, it becomes possible for Class II-restricted T cells to "know" which V[H] segments are transcriptionally active in a particular cell and, therefore, which V[H] segments are likely targets of recombinase selection.

If it is further assumed that T cells specific for residue epitopes of V[H] translation products can deliver signals that can activate the recombinase, then a mechanism is established by which T cells can control or select rearrangements. The specificity of the control will clearly depend on the fraction of V[H] segments that are transcribed within the time interval required for T-cell recognition and activation of the recombinase. Conceivably, the effect of this control may be to focus the recombinase on a particular V[H] subfamily, rather than to select an individual V[H] segment. The sequence homology that would exist within a V[H] family or subfamily (91) should yield frequent cross-reactions among processed peptides from V[H] segments belonging to that subfamily.

This model can be extended to include DJ[H] segments which are transcriptionally active and which lead to D[μ] chains that contain 20 to 40 amino acids and also appear to contain a hydrophobic leaderlike sequence (9). Control of class switching by T cells specific for residue epitopes on processed translation products of unrearranged constant-region gene segments might also be possible, since these gene segments remain transcriptionally active after the formation of a μ-chain gene (9,92).

In this model, it has been assumed that the T cell activates the recombinase. The T cell in question might be a Class II-restricted T-helper that secretes the lymphokines appropriate for recombinase activation. It should be noted that there is already in vitro evidence that T cells together with dendritic cells can drive the maturation of B-lineage cells (92). However, it seems prudent to be open to the possibility of negative selection by Class I- or Class II-restricted CTL. Pre-B cells are reported to express Class II MHC antigens (93) and presumably express Class I MHC antigens.

As yet there is little evidence that transcripts of unrearranged V[H] segments are translated. However, Ucker and co-workers have recently discovered a cytotoxic T-cell line in which an unrearranged V[H] gene is transcribed and translated (94). Interestingly, the product appears to be on the cell membrane. From a conventional point of view, this is puzzling, since the peptide possesses no membrane-anchoring sequence. The authors suggest a noncovalent association with the heterodimeric T-cell receptor. Alt and co-workers have offered the similar suggestion that the sequence homology between V[H] segments and β2-microglobulin may allow binding to Class I MHC chains (9). Our suggestion that translation products of V[H] segments can be processed represents an alternative basis for the membrane association of the V[H] peptide described by Ucker and co-workers. However, it should be noted that processing of V[H] peptides is assumed to be controlled so that it normally occurs before rearrangement and not after. The V[H] expression observed by Ucker et al. may therefore be aberrant.

If it is granted that the T-cell repertoire is shaped by network interactions, then this model provides for network control of immunoglobulin and possibly T-cell receptor rearrangements. It is relevant to consider if this model makes sense in the context of network regulation. One consequence of T-cell selection of gene segments would be to ensure that the v-region peptides generated after antigen recognition can be recognized by the available T-cell repertoire. Except for differences due to junctional diversity and somatic mutation (9), peptides from complete v-regions should be similar to peptides from unrearranged segments. The findings of Jorgensen and Hannestad (73) demonstrate Ir gene defects in the recognition of syngeneic v-regions. Such defects may be minimized by the rearrangement control mechanism offered here.

Conclusion

Although incomplete in many respects, the model offered in this discussion is, nonetheless, sufficiently well elaborated to allow many testable predictions, only a few of which have been described because of space limitations. These predictions are of importance not only to basic questions of immune regulation but also to the practical use of monoclonal antibodies in the treatment of cancer and autoimmune diseases and as idiotype vaccines. It is hoped that the perspective generated by this model will inspire the design of experiments that will lead to the direct evidence that is needed to prove or disprove the existence of an idiotypic network.

Acknowledgments. Bernard F. Erlanger is thanked for his support and encouragement and for many stimulating discussions concerning subjects discussed in this paper. R. G. Miller, M. J. Bevan, and T. J. Braciale are also thanked for helpful discussions. This work was supported by the Evans Foundation.

References

1. Burnet, F.M. 1957. A modification of Jerne's theory of antibody using the concept of clonal selection. Aust. Sci. **20**:67.
2. Talmage, D.W. 1957. Allergy and immunology. Annu. Rev. Med. **8**:239.
3. Jerne, N.K. 1974. Towards a network theory of the immune response. Ann. Immunol. (Inst. Pasteur) **125C**:373.
4. Paul, W.E., and C. Bona. 1982. Regulatory idiotopes and immune networks: a hypothesis. Immunol. Today **3**:230.
5. Farid, N.R., and T. C.Y. Low. 1985. Anti-idiotypic antibodies as probes for receptor structure and function. Endocr. Rev. **6**:1.
6. Ehrlich, P. 1900. On immunity with special reference to cell life. Proc. Roy. Soc., Ser. B. **66**:424.
7. Cohn, M. 1981. Conversations with Niels Kaj Jerne on immune regulation: associative versus network recognition. Cell. Immunol. **61**:425.
8. Lanier, L.L., and J.H. Phillips. 1986. Evidence for three types of human cytotoxic lymphocyte. Immunol. Today **7**:132.
9. Alt, F.W., T.K. Blackwell, R.A. DePinho, M.G. Reth, and G.D. Yancopoulos. 1986. Regulation of genome rearrangement events during lymphocyte differentiation. Immunol. Rev. **89**:5.
10. Kunkel, H.G. 1970. Experimental approaches to homogeneous antibody populations. Individual antigenic specificity, cross-specificity and diversity of human antibodies. Fed. Proc. **29**:55.
11. Nisonoff, A., S-T. Ju, and F. Owen. 1977. Studies of structure and immunosuppression of a cross-reactive idiotype in strain A mice. Immunol. Rev. **34**:89.
12. Brient, B.W., and A. Nisonoff. 1970. Quantitative investigations of idiotypic antibodies. IV. Inhibition by specific haptens of the reaction of anti-hapten antibody with its anti-idiotypic antibody. J. Exp. Med. **132**:951.
13. Jerne, N.K., J. Roland, and P.-A. Casenave. 1982. Recurrent idiotypes and internal images. EMBO J. **1**:243.
14. Bona, C., S. Finley, S. Waters, and H.G. Kunkel. Anti-immunoglobulin antibodies. III. Properties of sequential anti-idiotypic antibodies to heterologous anti-gamma globulins. Detection of reactivity of anti-idiotype antibodies with

epitopes of Fc fragments (homobodies) and with epitopes and idiotypes (epibodies). J. Exp. Med. **156**:986.

15. Holmberg, D., S. Forsgren, L. Forni, F. Ivars, and A. Coutinho. 1984. Idiotypic determinants of natural IgM antibodies that resemble self Ia antigens. Proc. Natl. Acad. Sci. USA. **81**:3175.

16. Jerne, N.K. 1984. Idiotypic networks and other preconceived ideas. Immunol. Rev. **79**:5.

17. Erlanger, B.F. 1985. Anti-idiotypic antibodies: what do they recognize? Immunol. Today **6**:10.

18. Wassermann, N.H., A.S. Penn, P.I. Freimuth, N. Treptow, S. Wentzel, W.L. Cleveland, and B.F. Erlanger. 1982. Anti-idiotypic route to anti-acetylcholine receptor antibodies and experimental myasthenia gravis. Proc. Natl. Acad. Sci. USA **79**:4810.

19. Tzartos, S.J., M.E. Seybold, and J.M. Lindstrom. 1982. Specificities of antibodies to acetycholine receptors in sera from myasthenia gravis patients measured by monoclonal antibodies. Proc. Natl. Acad. Sci. USA **79**:188.

20. Cunningham, A.J. 1974. The generation of antibody diversity: its dependence on antigenic stimulation. Contemp. Top. Mol. Immunol. **3**:1.

21. Gearhart, P.J., N.D. Johnson, R. Douglas, and L. Hood. 1981. IgG antibodies to phosphorylcholine exhibit more diversity than their IgM counterparts. Nature (London) **291**:29.

22. Manser, T., L.J. Wysocki, T. Gridley, R.I. Near, and M.L. Gefter. 1985. The molecular evolution of the immune response. Immunol. Today **6**:94.

23. Jerne, N.K. 1971. The somatic generation of immune recognition. Eur. J. Immunol. **1**:1.

24. Cleveland, W.L., and B.F. Erlanger. 1984. Hypothesis: the MHC-restricted T-cell receptor as a structure with two multistate allosteric combining sites. Mol. Immunol. **21**:1037.

25. Martinez-A, C., R.R. Bernabé, A. de la Hera, P. Pereira, P-A. Casenave, and A. Coutinho. 1985. Establishment of idiotypic helper T-cell repertoires early in life. Nature (London) **317**:721.

26. Staerz, U.D., and M. Bevan. 1986. Use of anti-receptor antibodies to focus T-cell activity. Immunol. Today **7**:241.

27. Tite, J.P., J. Kaye, K.M. Saizawa, J. Ming, M.E. Katz, L.A. Smith, and C.A. Janeway. 1986. Direct interactions between B and T lymphocytes bearing complementary receptors. J. Exp. Med. **163**:189.

28. Walden, P., Z.A. Nagy, and J. Klein. 1985. Induction of regulatory T-lymphocyte responses by liposomes carrying major histocompatibility complex molecules and foreign antigen. Nature (London) **315**:327.

29. Miller, R.G., and H. Derry. 1979. A population in nu/nu spleen can prevent generation of cytotoxic lymphocytes by normal spleen cells against self antigens of the nu/nu spleen. J. Immunol. **122**:1502.

30. Miller, R.G. 1986. The veto phenomenon and T-cell regulation. Immunol. Today **7**:112.

31. Claesson, M.H., and R.G. Miller. 1984. Functional heterogeneity in allospecific cytotoxic T lymphocyte clones. I. CTL clones express strong anti-self suppressive activity. J. Exp. Med. **160**:1702.

32. Fink, P.J., H.-G. Rammensee, J. D. Benedetto, U.D. Staerz, L. Lefrancois, and M.J. Bevan. 1984. Studies on the mechanism of suppression of primary cytotoxic responses by cloned cytotoxic T-lymphocytes. J. Immunol. **133**:1769.

33. Rammensee, H-G., M.J. Bevan, and P. Fink. 1985. Antigen specific suppression of T-cell responses—the veto concept. Immunol. Today. 6:41.
34. Fink, P.J., H-G., Rammensee, J.D. Benedetto, U.D. Staerz, L. Lefrancois, and M.J. Bevan. 1984. Studies on the mechanism of primary cytotoxic responses by cloned cytotoxic T lymphocytes. J. Exp. Med. 133:1769.
35. Araneo, B.A., and R.L. Yowell. 1985. MHC-linked immune response suppression mediated by T-cells bearing I-A-encoded determinants. J. Immunol. 135:73.
36. Goverman, J., T. Hunkapiller, and L. Hood. 1986. A speculative view of the multicomponent nature of T cell antigen recognition. Cell 45:475.
37. Bushkin, Y., D.N. Posnett, B. Pernis, and C.Y. Wang. 1986. A new HLA-linked T cell membrane molecule related to the B chain of the clonotypic receptor is associated with T3. J. Exp. Med. 164:458.
38. Bank, I., R.A. DePinho, M.B. Brenner, J. Cassimeris, F.W. Alt, and L. Chess. 1986. A functional T3 molecule associated with a novel heterodimer on the surface of immature human thymocytes. Nature (London) 322:179.
39. Brenner, M.B., J. McLean, D.P. Dialynas, J.L. Strominger, J.A. Smith, F.L. Owen, J.G. Seidman, S. Ip, F. Rosen, and M.S. Krangel. 1986. Identification of a putative second T-cell receptor. Nature (London) 322:145.
40. Hood, L., M. Kronenberg, and T. Hunkapiller. 1985. T-cell antigen receptors and the immunoglobulin supergene family. Cell 40:225.
41. Watts, T.H., H.E. Graub, and H.M. McConnell. 1986. T-cell-mediated association of peptide antigen and major histocompatibility complex protein detected by energy transfer in an evanescent wave-field. Nature (London) 320:176.
42. Ashwell, J.O., and R.H. Schwartz. 1986. T-cell recognition of antigen and the Ia molecule as a ternary complex. Nature (London) 320:176.
43. Parham, P. 1984. A repulsive view of MHC restriction. Immunol. Today. 5:89.
44. Blanden, R.V., and R.B. Ashman. 1985. Speculation: Selection of pre-T cells in the thymus by unique combinations of major and minor histocompatibility antigens. Mol. Immunol. 22:827.
45. Mitchison, N.A. 1986. Antigen binding and T-cells. Nature (London) 320:106.
46. Thomas, D.W., and M.D. Hoffman. 1982. Evidence for covert cellular interaction sites expressed by activated T lymphocytes. J. Immunol. 129:1416.
47. Sim, G.K., I.A. MacNeil, and A.A. Augustin. 1986. T helper cell receptors: idiotypes and repertoire. Immunol. Rev. 90:49.
48. Müllbacher, A. 1981. Natural tolerance: a model for Ir gene effects in the cytotoxic T cell response to H-Y. Transplantation 32:58.
49. Benacerraf, B., L.D. Fallow, Jr., and K.L. Rock. 1986. Processing of native antigen by accessory cells and presentation of membrane bound MHC-associated antigen to specific T cells. In B. Pernis, S.C. Silverstein, and H.J. Vogel (eds.), Processing and presentation of antigens. Academic Press, Orlando, Florida. In press.
50. DeLisi, C., and J.A. Berzofsky. 1985. T-cell antigenic sites tend to be amphipathic structures. Proc. Natl. Acad. Sci. USA 82:7048.
51. Babbitt, B.P., P.M. Allen, G. Matsueda, E. Haber, and E. R. Unanue. 1985. Binding of immunogenic peptides to Ia histocompatibility molecules. Nature (London) 317:359.

52. Kakiuchi, T., R.W. Chesnut, and H. Grey. 1983. B-cells as antigen-presenting cells: the requirement for B-cell activation. J. Immunol. **131**:109.
53. Rock, K.I., B. Benacerraf, and A.K. Abbas. 1984. Antigen presentation by hapten-specific B lymphocytes. I. Role of surface immunoglobulin receptors. J. Exp. Med. **160**:1102.
54. Lanzavecchia, A. 1985. Antigen-specific interaction between T and B cells. Nature (London) **314**:537.
55. Lanzavecchia, A., S. Siervo, and D. Scheidegger. 1986. On the role of B-cell surface Ig in antigen presentation to T-cells. *In* B. Pernis, S.C. Silverstein, and H.J. Vogel, (eds.), Processing and presentation of antigens. Academic Press, Orlando, Florida. In press.
56. Howard, J.C. 1985. Immunological help at last. Nature (London) **314**:495.
57. Mitchison, N.A. 1971. The carrier effect in the secondary response to hapten-protein conjugates. II. Cellular cooperation. Eur. J. Immunol. **1**:18.
58. Chesnut, R., and H. Grey. 1981. Studies on the capacity of B cells to serve as antigen-presenting cells. J. Immunol. **126**:1075.
59. Unanue, E.R. 1984. Antigen-presenting function of the macrophage. Annu. Rev. Immunol. **2**:395.
60. Celis, E., and T.W. Chang. 1984. Antibodies to hepatitis B surface antigen potentiate the response of human T lymphocyte clones to the same antigen. Science **224**:297.
61. Rodkey, L.S. 1980. Autoregulation of immune responses via idiotype network interactions. Microbiol. Rev. **44**:631.
62. Shechter, Y., D. Elias, R. Bruck, R. Maron, and I. Cohen. Mice immunized to insulin develop anti-idiotypic antibody to the insulin receptor. This volume, Ch. 8.
63. Cleveland, W.L., N.H. Wasserman, R. Sarangarajan, A.S. Penn, and B.F. Erlanger. 1983. Monoclonal antibodies to the acetylcholine receptor (AChR) by a normally functioning auto-anti-idiotypic mechanism. Nature (London) **305**:56.
64. Cayanis, E., R. Rajagopalan, W.L. Cleveland, I.S. Edelman, and B.F. Erlanger. 1986. Generation of an auto-anti-idiotypic antibody that binds to glucocorticoid receptor. J. Biol. Chem. **261**:5094.
65. Ku, H-h., W.L. Cleveland, and B.F. Erlanger. Adenosine receptors: an auto-anti-idiotypic approach. Submitted for publication.
66. Cleveland, W.L., and B.F. Erlanger. 1986. The auto-anti-idiotypic strategy for preparing monoclonal antibodies to receptors. Methods Enzymol. **121**:95.
67. Leserman, L. 1986. The introversion of the immune response: a hypothesis for T-B interaction. Immunol. Today **6**:352.
68. Morrison, L.A., A.E. Lukacher, V.L. Braciale, D.P. Fan, and T.J. Braciale. 1986. Differences in antigen presentation to MHC Class I- and Class II-restricted influenza virus-specific cytolytic clones. J. Exp. Med. **163**:903.
69. Reth, M., G. Kelsoe, and K. Rajewsky. 1981. Idiotypic regulation by isologous monoclonal anti-idiotype antibodies. Nature (London) **290**:257.
70. Gleason, K., S. Pierce, and H. Kohler. 1981. Generation of idiotype-specific T cell help through network perturbation. J. Exp. Med. **153**:924.
71. McNamara, M., and H. Kohler. 1983. Idiotype-recognizing T helper cells that are not idiotype specific. J. Exp. Med. **158**:811.

72. McNamara, M., K. Gleason, and H. Kohler. 1984. T-cell helper circuits. Immunol. Rev. **79**:87.
73. Jorgensen, T., and K. Hannestad. 1982. Helper T cell recognition of the variable domains of a mouse myeloma protein (315): effect of the major histocompatibility complex and domain conformation. J. Exp. Med. **155**:1587.
74. Rubinstein, L.J., and C. Bona. 1983. Idiotype-anti-idiotype network. III. Genetic control of activation of A48Id silent clones subsequent to manipulation of the immune network. Ann. N.Y. Acad. Sci. **418**:97.
75. Bona, C.A., C. Victor-Kobrin, A.J. Manheimer, B. Bellon, and L.J. Rubinstein. 1984. Regulatory arms of the immune network. Immunol. Rev. **79**:25.
76. Noseworthy, J.H., B.N. Fields, M.S. Dichter, C. Sobotka, E. Pizer, L.I. Perry, J.T. Nepom, and M.I. Greene. 1983. Cell receptors for the mammalian reovirus. I. Syngeneic monoclonal anti-idiotypic antibody identifies a cell surface receptor for reovirus. J. Immunol. **131**:2533.
77. Sharpe, A.H., G.N. Gaulton, K.K. McDade, B.N. Fields, and M.I. Greene. 1984. Syngeneic monoclonal antiidiotype can induce cellular immunity to reovirus. J. Exp. Med. **160**:1195.
78. Miller, J.F.A.P., M.A. Vadas, A. Whitelaw, and J. Gamble. 1975. H-2 gene complex restricts transfer of delayed-type hypersensitivity in mice. Proc. Natl. Acad. Sci. USA **72**:5095.
79. Bruck, C., M.S. Co., M. Slaoui, G.N. Gaulton, T. Smith, B.N. Fields, J.I. Mullins, and M.I. Greene. 1986. Nucleic acid sequence of an internal image-bearing monoclonal anti-idiotype and its comparison to the sequence of the external antigen. Proc. Natl. Acad. Sci USA. **83**:6578.
80. Rubinstein, L.J., B. Goldberg, J. Hiernaux, K.E. Stein, and C. Bona. 1983. Idiotype-antiidiotype regulation. V. The requirement for immunization with antigen or monoclonal antiidiotypic antibodies for the activation of β2→6 and β2→1 polyfructosan-reactive clones in BALB/c mice treated at birth with minute amounts of anti-A48 idiotype antibodies. J. Exp. Med. **158**:1129.
81. Kim, J., A. Woods, E. Becker-Dunn, and K. Bottomly. 1985. Distinct functional phenotypes of cloned Ia-restricted helper T cells. J. Exp. Med. **162**:188.
82. Ghosh, S., and A.M. Campbell. 1986. Multispecific monoclonal antibodies. Immunol. Today **7**:217.
83. Rowe, D.S., K. Hug, L. Forni, and B. Pernis. 1973. Immunoglobulin D as a lymphocyte receptor. J. Exp. Med. **138**:965.
84. Spiegelberg, H.L. 1977. The structure and biology of human IgD. Immunol. Rev. **37**:3.
85. Townsend, A.R.M., J. Rothbard, F.M. Gotch, G. Bahadur, D. Wraith, and A.J. McMichael. 1986. The epitopes of influenza nucleoprotein recognized by cytotoxic T lymphocytes can be defined with short synthetic peptides. Cell **44**:959.
86. Pernis, B. 1985. Internalization of lymphocyte membrane components. Immunol. Today **6**:45.
87. Ruud, E., H.K. Blomhoff, S. Funderud, and T. Godal. 1986. Internalization and processing of antibodies to surface antigens on human B cells. Monoclonal anti-IgM antibodies are processed differently than monoclonal antibodies towards non-Ig surface receptors. Eur. J. Immunol. **16**:286.
88. Schwartz, R.H., E. Heber-Katz, and D. Hansburg. 1983. The Ia molecule contributes to the specificity of T cell activation. *In* J.W. Parker and R.L.

O'Brien (eds.), Intercellular communication in leucocyte function, pp. 117–125. John Wiley & Sons, Ltd., New York.
89. Hansburg, D., E. Heber-Katz, T. Fairwell, and E. Appella. 1983. Major histocompatibility complex-controlled antigen-presenting cell-expressed specificity of T cell antigen recognition. J. Exp. Med. **158**:25.
90. Yancopoulos, G.D., and F.W. Alt. 1985. Developmentally controlled and tissue specific expression of unrearranged V[H] gene segments. Cell **40**:271.
91. Brodeur, P. H., and R. Riblet. 1984. The immunoglobulin heavy chain variable region (Igh-V) locus in the mouse. I. One hundred Igh-V genes comprise seven families of homologous genes. Eur. J. Immunol. **14**:922.
92. Spalding, D.M., and J.A. Griffin. 1986. Different pathways of differentiation of pre-B cell lines are induced by dendritic cells and T-cells from different lymphoid tissues. Cell **44**:508.
93. Hokland, P., J. Ritz, S.F. Schlossment, and L. Nadler. 1985. Orderly expression of B cell antigens during the in vitro differentiation of nonmalignant human pre-B cells. J. Immunol. **135**:1746.
94. Ucker, D.S., Y. Kurasawa, and S. Tonegawa. 1985. An unusual clonotypic determinant on a cytotoxic T lymphocyte line is encoded by an immunoglobulin heavy chain variable region gene. J. Immunol. **135**:4204.

14
Identification of Glucose Transporter Using Anti-Idiotypic Antibodies, Anti-Transporter Antibodies, and Peptide Maps

MARGUERITE M.B. KAY

Introduction

A "neo-antigen" appears on the surface of senescent cells, leading to IgG binding and cellular removal (1–10). This "neo-antigen" is recognized by the antigen binding, Fab, region (3,10,11) of a specific immunoglobulin G (IgG) autoantibody in serum which attaches to it and initiates the removal of cells by macrophages (1–4). A number of studies performed by us (1–11) and by others (12–21) have demonstrated the presence of IgG on senescent, damaged, and stored red cells. In addition, several laboratories have recently presented evidence that IgG binding is also involved in the removal of red cells in diseases such as thalassemia (22) and sickle cell anemia (23,24). IgG has also been implicated in the removal of aging platelets (25).

We have named the "neo-antigen" on senescent red cells "senescent cell antigen" (5). Senescent cell antigen is a glycosylated 4.5-region polypeptide that appears to be derived from band 3 (5,8,10,26–28). Senescent cell antigen is located on an extracellular portion of band 3 that includes most of the 38,000-Da carboxyl terminal segment and 30% of the 17,000-Da anion transport region (28). The ~40,000-Da cytoplasmic segment and ~70% of the anion transport segment of band 3 appear to be missing from senescent cell antigen (28). The finding that anion transport is impaired in old erythrocytes (increased K_E, decreased V_{max}) is consistent with cleavage of band 3 in the anion transport region (29).

It was postulated that senescent cell antigen and glucose transporter were breakdown products of band 3 and that they were physiologically related (8). Both proteins migrate in the broad, uncharacterized 4.5 region of sodium dodecyl sulfate (SDS) polyacrylamide gels (5,30). At least four polypeptides in this region stain with antibodies to band 3 (10,27,28). One of these polypeptides, the senescent cell antigen, appears to be derived from band 3 (8,10,27,28). Band 3 is the major anion transport polypeptide of the erythrocyte membrane, and, in addition, appears to be the binding site for the glycolytic enzymes glyceraldehyde-3-phosphate dehydrogenase (31), aldolase (32), and phosphofructokinase (33). In the present study, the relationship between the as yet unidentified glucose transporter that supplies energy to cells and senescent cell antigen that signals the termination of a cell's life (1–5,7–10,26,34–37) is examined.

Early studies on the identification of the glucose transporter using affinity labeling or cytochalasin B binding suggested that glucose transport activity was attributable to band 3 (38,39). This view was later supplanted by evidence indicating that an integral membrane glycoprotein of M_r ~45,000–55,000 was probably the glucose transporter (30,40). In the latter case, the glucose transporter was identified by [^{3}H]cytochalasin B binding to Triton X-100 extracts of red cell proteins that were retarded on ion-exchange columns. Elution of the glucose transporter in these experiments did not require a pH or ion shift and yielded a heterogeneous mixture with broad peaks throughout the gels from the band 3 region to the tracking dye (40). Antibodies prepared against glucose transporter obtained in this manner reacted with polypeptides of M_r ~40,000–~80,000 (40). A review of the polyacrylamide gel electrophoresis data presented by Sogin and Hinkle (40) reveals that the trailing edge of antibody binding to glucose transporter both in isolated preparations and in membranes is in the band 3 (M_r ~100,000) region while the leading edge is in the band 6 region (M_r ~35,000). Zoccoli et al. (41) found, using gel filtration and sucrose gradient centrifugation, that the molecular mass of the glucose transporter was 225,000 Da. They suggested that the solubilized cytochalasin B binding complex contains one or two polypeptide chains (41) which could have been dimers or aggregates of band 3.

Anti-idiotypic antibodies were selected as an approach to identifying the glucose transporter because they bypass the requirement for purified transporter, thus avoiding the problems encountered by previous investigators.

Materials and Methods

Isolation of IgG from Senescent RBC and Senescent Cell Antigen

IgG was isolated as previously described (3,5). Senescent cell antigen was isolated by affinity chromatography with senescent cell IgG conjugated to Sepharose 4B as previously described (5,8).

Isolation of the Glucose Transporter

Since cytochalasin B is a specific and potent competitive inhibitor of D-glucose transport in human erythrocytes (42,43), the glucose transport carrier was isolated by affinity chromatography on cytochalasin B-Sepharose 4B columns and eluted with D-glucose. This purification procedure is both a method of isolation and a functional assay for the glucose transporter (30,40). Cytochalasin B (CalBiochem, La Jolla, Calif.) was dissolved in dimethyl sulfoxide or absolute ethanol and conjugated to Sepharose 4B. Erythrocyte ghosts were prepared as described previously (27,37,44). NaOH stripped membranes were solubilized with 0.1 M Tris hydrochloride, pH 8.0, containing 0.1 NaCl, 1 mM EDTA, 1 mM EGTA, 1 mM DFP, and 0.1% SDS (27,37,44). After centrifugation, the supernate was incubated overnight at 4°C with the cytochalasin B-Sepharose. The Sepharose was removed by centrifugation and washed five times with 5 mM phosphate buffer, pH 7.4, containing 1 mM EDTA, 1 mM EGTA, and 5 mM DFP. It was poured into a column and washed with 10 volumes of the same buffer in which the membranes were solubilized except that it contained 0.02% rather than 0.1% SDS. After washing with 2 volumes of 1.0 M NaCl and 400 volumes of 0.15 M NaCl, bound material was eluted with 0.5 M D-glucose.

Preparation of Antibodies to the Glucose Transporter

Glucose transporter was isolated by affinity chromatography with cytochalasin B. While still bound to the column, the transporter was delipidated using water/ethanol/ether (1:1:4 vol/vol) and chloroform/methanol (2:1 vol/vol). The delipidated glucose transporter was specifically eluted with D-glucose from the affinity column and injected into rabbits using a biweekly injection schedule. Rabbits were bled three out of four weeks starting one week after the third injection. The IgG fraction was isolated by affinity chromatography using protein A-Sepharose.

Anti-idiotypic antibodies to the glucose transporter were prepared according to the method of Wassermann et al. (45). Briefly, antibodies were prepared against glucose by conjugating glucosamine to Sepharose 4B and injecting rabbits biweekly with the antigen. Glucosamine rather than glucose was used for coupling procedures because an amino group was necessary to carry out the reaction. After the third injection, glucosamine conjugated to bovine serum albumin (BSA) as a carrier was injected on a biweekly schedule. The antisera were pooled, and antibodies to glucose were isolated by affinity chromatography on glucosamine-Sepharose 4B columns. Antibodies to glucose were eluted with glucose, pH 7.4. These antibodies were then injected into other rabbits to produce anti-idiotypic antibodies. The anti-idiotypic antibody rabbits were bled and the sera absorbed with BSA and red cell peripheral membrane proteins conjugated to Sepharose. The IgG fraction was isolated by affinity chromatography on protein A-Sepharose.

Antibodies to Band 3

Antibodies to purified band 3 were prepared in rabbits as previously described (10,27,37,44).

Enzymatic Treatment of Erythrocytes

Fragments are referred to by the nomenclature of Steck et al. (46,47). Washed erythrocytes were incubated with PBS containing 1 mM ATP and 200 μg/ml chymotrypsin overnight at 24°C. Chymotrypsin specifically digests band 3, yielding fragments of ~55,000 and ~38,000 Da (46,47), designated CH-55 and CH-38, respectively (46). The CH-55 fragment appears to have a molecular weight closer to 60,000 in our experiments. Digestion was terminated by the addition of 5 mM DFP. Cells were washed four times with PBS and processed in the same manner as described for intact erythrocytes.

The ~41,000-Da (TR-41) cytoplasmic segment of band 3 was produced by mild trypsin digestion of spectrin-depleted, NaOH stripped inverted vesicles (48).

The M_r~19,000 intramembranous fragment (CH-TR-19) which includes CH-17 was produced by chymotrypsin treatment of intact red cells followed by treatment of spectrin-depleted inverted vesicles with trypsin (49). Proteolysis was terminated by the addition of 5 mM DFP. Membranes were washed twice with 5 mM sodium phosphate, pH 8.0, containing 1 mM EDTA, 1 mM EGTA, and 1 mM DFP.

Sodium Dodecyl Sulfate-Polyacrylamide Gel Electrophoresis (SDS-PAGE)

Proteins were analyzed on three different gel systems: 7% SDS/polyacrylamide gels and 6–25% and 12–25% linear SDS/polyacrylamide gradient gels using the discontinuous buffer system of Laemmli (50). M_rs given in this paper are approximate and can vary by 5–10% depending on the PAGE system employed.

Immunostaining of the Membrane Proteins

Immunoautoradiography was performed by the immunoblotting technique of Towbin et al. (51) with the modifications described previously (9,10,14–16) or by the gel overlay method (27).

Two-dimensional Peptide Mapping

Two-dimensional peptide maps were obtained using the method of Elder et al. (52) with the modifications described previously (28). Chromatograms were dried and exposed to Kodak X-Omat film in Dupont cassettes with Cronex Lighting plus intensifying screens for one to three days at −80°C.

Results

Identification of the Glucose Transporter Using Antibodies Prepared Against It

Glucose transporter purified by affinity chromatography has an M_r of ~60,000 as determined by PAGE. Rabbit antibodies prepared against purified glucose transporter gave a very strong reaction with a polypeptide migrating at M_r ~78,000 (between bands 4.1 and 4.2) in SDS/polyacrylamide gels of red cell membranes. In addition, antibodies to purified glucose transporter reacted with band 3, minor bands and one major band in the 4.5 region, and polypeptides at M_r ~ 40,000, 38,000, and 18,000 in immunoblots of erythrocyte membranes (Fig. 14.1, lane B). Antibodies to band 3 reacted with these same polypeptides, suggesting that they are immunologically related to band 3 (Fig. 14.1, lane A).

Anti-idiotypic antibodies reacted with band 3, one polypeptide migrating at M_r ~78,000, and one polypeptide migrating at M_r ~45,000 in the 4.5 region in SDS-PAGE (Fig. 14.1, lane C). In addition, anti-idiotypic antibodies reacted with two faint bands in the 4.5 region (M_r ~67,000 and 58,000). Antibodies to band 3 reacted with these same polypeptides (Fig. 14.1, lane A). IgG eluted from senescent cells reacted with band 3, a 4.5-region polypeptide with a molecular weight lower than that of the glucose transporter (M_r ~67,000), a polypeptide at M_r ~45,000, and a polypeptide at M_r ~18,000 (Fig. 14.1, lane D). Specific antibodies to band 3 also reacted with these polypeptides (Fig. 14.1, lane A).

Peptide Mapping Analysis of Band 3, Glucose Transporter, and Senescent Cell Antigen

Immunological data suggest that band 3 and glucose transporter are related. Therefore, comparative two-dimensional peptide mapping was performed to determine the extent of homology between the two polypeptides. Peptide homology between glucose transporter and senescent cell antigen was also evaluated because senescent cell antigen is not only derived from band 3 but has also been localized on the band 3 molecule. Comparative analysis of peptide maps revealed that glucose transporter shares peptide homology with band 3, the CH-38 carboxyl terminal, and the CH-TR-19 anion transport segments. The proportion of peptides in each map that appears in glucose transporter is as follows: band 3, 39–43% (27 peptides appear in glucose transporter/63–69 peptides in the map of band 3); CH-38, 63% (26/41); CH-TR-19, 67% (22/33); CH-55, 20% (10/50); TR-41, 3–5% (1 or possibly 2/40). The peptides present in the glucose transporter are present in the peptide map of band 3. Glucose transporter has 7–10 peptides that do not appear in CH-38 and 6 peptides that do not appear in CH-TR-19. However, all except two of the peptides in glucose trans-

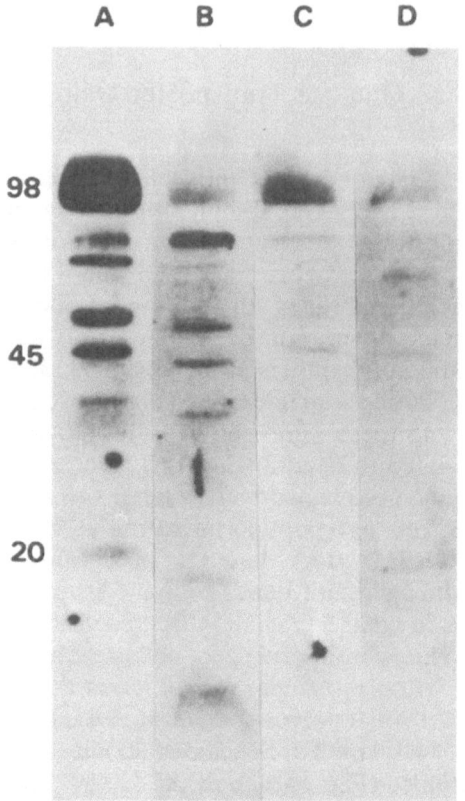

FIGURE 14.1. Immunoblots of RBC membrane proteins incubated with antibodies to band 3 (lane A), antibodies against purified glucose transporter (lane B), anti-idiotypic antibodies directed against the glucose receptor (lane C), and IgG eluted from senescent cells (lane D). M_rs on left of figure X 10^3. RBC membrane proteins were separated on 12–25% SDS-polyacrylamide gels. Reproduced from M.M.B. Kay, *Proc. Natl. Acad. Sci. USA* **82**:1731 (1985).

porter appear in the combined maps of CH-38 and CH-TR-19, as determined by superimposing the autoradiographs. These two glucose transporter peptides that are not present in CH-38 combined with CH-TR-19 may be contributed by TR-41. Alternatively, they could result from technical factors discussed previously (28). For example, degradation of a polypeptide can produce an additional peptide at each end of a fragment where it was cleaved from band 3. Senescent cell antigen contains fewer peptides than glucose transporter, but all of the peptides present in the map of senescent cell antigen are present in the map of glucose transporter. Thus, glucose transporter (Fig. 14.2, panel A) shares substantial peptide

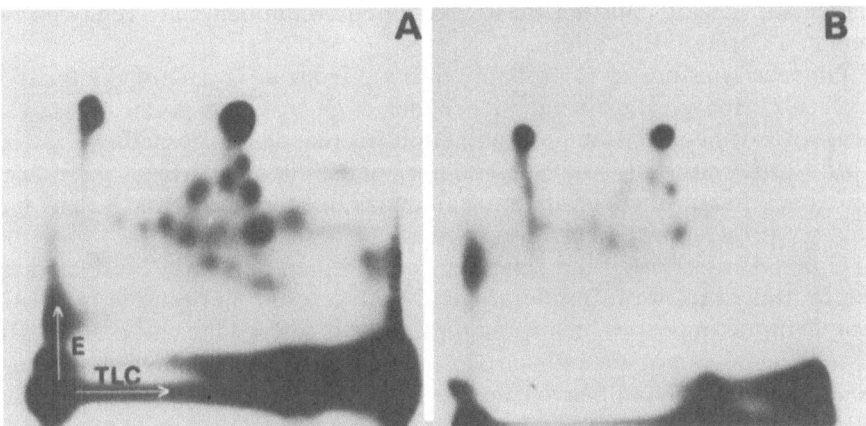

FIGURE 14.2. Two-dimensional peptide maps of glucose transporter (panel A) and senescent cell antigen (panel B). Reproduced from M.M.B. Kay, *Proc. Natl. Acad. Sci. USA* **82:**1731 (1985).

homology with senescent cell antigen (Fig. 14.2, panel B), although the two maps are not identical.

Discussion

Both the immunological and peptide mapping data indicate that glucose transporter is related to band 3. Glucose transporter appears to contain band 3 segments CH-38 and CH-17 and possibly several additional peptides. Thus, the calculated molecular weight of glucose transporter based on peptide mapping data is >55,000. Purified glucose transporter migrates at M_r ~60,000 in PAGE, which is consistent with the peptide mapping data. However, both antibodies prepared against purified glucose transporter and anti-idiotypic antibodies to the glucose transporter react with an M_r ~78,000 polypeptide in red cell membranes. The most probable explanation for the difference in apparent molecular weights between purified glucose transporter (M_r ~60,000) and the polypeptide labeled with antibodies in red cell membranes is that degradation occurred during removal from membranes or during isolation procedures. It should be noted that both antibodies to glucose transporter and anti-idiotypic antibodies reacted with lower-molecular-weight polypeptides as well as the M_r ~78,000 polypeptide in red cell membranes. It is also possible that carbohydrates, glycolipids, or other membrane components could be associated with glucose transporter in membranes, thus increasing its apparent molecular weight, but that these associated components are removed or reduced during purification.

Glucose transporter is probably derived from or is part of the band 3 molecule. However, mapping the peptides of glucose transporter to specific segments of band 3 is an approximation for reasons discussed elsewhere (28). Furthermore, if band 3 degradation occurs in situ, there is no reason to expect that in situ degradation would produce the same band 3 cleavage that we produce in vitro with enzymes.

Glucose transporter and senescent cell antigen appear to be related to each other, based on immunological and peptide mapping data. Both polypeptides appear to be located on segments toward the carboxyl rather than the amino terminus of band 3. Although senescent cell antigen appears to contain CH-38 and part of the CH-17 anion transport segment of band 3 (28), glucose transporter appears to contain CH-38 and most of CH-17. Thus, glucose transporter appears to be larger than senescent cell antigen, based on peptide mapping analysis as well as antibody staining.

Neither purified senescent cell antigen nor glucose transporter appears to contain the cytoplasmic segment of band 3. It is possible that the cytoplasmic segment of band 3 (TR-41) behaves like a cytoskeletal protein with predominantly structural functions. It extends into the cytoplasm and appears to be the binding site for glyceraldehyde-3-phosphate dehydrogenase (53), aldolase (54), and band 2.1 (55). The cytoplasmic segment may also be more sensitive to proteolysis in vivo and/or in vitro. If this is the case, TR-41 could be degraded even though protease inhibitors are present throughout all procedures. Since TR-41 is not anchored in the membrane, it may be lost during the processing of membranes. The observation that breakdown products of band 3 increase with cellular aging suggests that degradation occurs in vivo (28).

The glucose transporter was isolated by affinity chromatography with cytochalasin B, a specific and potent competitive inhibitor of D-glucose transport (42,43), and eluted with D-glucose. Thus, the method of isolation and purification, namely, cytochalasin B binding, is a functional assay that has been used to determine the presence of glucose transporter following other methods of isolation and/or enrichment (30,40,56). Antibodies prepared against the affinity-purified glucose transporter give a strong reaction with an M_r ~78,000 polypeptide, show a weaker reaction with band 3, and react with faint bands in the 4.5 region and polypeptides at M_r ~38,000 and ~18,000. In contrast, anti-idiotypic antibodies, which bypass the requirement for a purified receptor (57), bind to band 3 and its M_r ~78,000 and M_r ~45,000 breakdown products. Anti-idiotypic antibody staining of band 3 is at least as strong as staining of the M_r ~78,000 polypeptide. The differences in staining patterns between the antibodies to purified glucose transporter and the anti-idiotypic antibodies to glucose transporter are probably due to several factors. First, even though glucose transporter is treated to remove lipids, binding of associated lipid is reduced but not abolished by the procedures employed. Second, anti-idiotypic antibodies recognize the active site of the glucose transporter. Therefore, if the active site for glucose transport is a protein, then anti-idiotypes would "recognize" the protein and not associated carbohydrates or lipids. Polyclonal antibodies to purified glucose transporter, on the other hand, would recognize any associated components and bind to them regardless of the molecule on which they appear. Since a number of glycoproteins may, for example, have similar sugar groups attached, the antibodies to purified glucose transporter may react with nontransport proteins on the basis of carbohydrate content rather than active-site configuration.

The results presented here, which indicate that purified glucose transporter is structurally related to band 3 and that antibodies to affinity-purified glucose transporter react with band 3, a M_r ~78,000 polypeptide, faint bands in the 4.5 region, and lower-M_r polypeptides, explain and reconcile the apparently disparate results obtained by other groups.

Mueckler et al. (58) have sequenced a molecule that they feel is the human glucose transport protein, using cDNA cloning techniques. The sequence they have obtained is very different from their sequence of murine band 3 (58). However, the validity of the data rests on the characteristics of the antibodies used to isolate the clone that was used for DNA sequencing. The antisera were not characterized particularly as regards specificity. Another potential problem is the use of monoclonal antibodies to select a clone. If one is trying to isolate a "parent" molecule or a molecule from which a smaller molecule is derived, then polyclonal antibodies are the preferred "tool" or "probe." Polyclonal antibodies recognize many determinants, whereas monoclonals recognize a single determinant. The restricted specificity of monoclonals may allow selection of a clone that produces only a small fragment of the parent molecule. Therefore, there is some question as to whether the molecule that has been DNA sequenced is actually the glucose transport protein. We are in the process of purifying human glucose transport protein for protein sequencing using cytochalasin B affinity columns and anti-glucose transporter affinity columns. We hope this will resolve this issue. Our antibodies have been sent to a number of laboratories throughout the world. These laboratories have confirmed binding to the glucose transporter.

We would expect species differences in band 3, although we anticipate that "crucial" regions, such as the anion transport region, are conserved during evolution. Our finding that none of the glyceraldehyde-3-phosphate dehydrogenase is membrane bound in rat red cells, in contrast to human red cells where ~60–80% is membrane bound, would indicate a difference between the cytoplasmic segment of murine and human band 3 (29). There were limited sequence similarities between the cDNA sequence of murine band 3 (59) and protein sequences of human band 3 (60). For example, in only 24 out of 201 N-terminal residues were the same amino acids in the same position in both sequences of band 3 when the sequences were lined up side-by-side (e.g., 2 amino acids of residues 1–11 were the same; 3 of residues 89–119 were the same, and 2 or 3 of the residues 125–138 were the same). Kopito and Lodish (59), however, calculated 57% homology of amino acids 125–138 and considered this region highly conserved. A hydropathy plot indicates that this protein has 12 membrane spanning domains (60).

Apparently, Kopito and Lodish (59) used a method for alignment that assumes that multiple deletions occur. By inserting gaps attributed to "deletions" or DNA loops or sequences that are not transcribed and incorporated into the protein, one shifts amino acids to get a "best fit." This is valid if there are many more amino acids than are used in the protein being sequenced. Obviously, if there is an excess of amino acids, some of the generated DNA sequences must not code for amino acids in the protein but probably serve another function. The number of amino acids from the DNA sequence of band 3 (59) was approximately the same as the number of amino acids in the protein sequence of band 3 (60). In contrast, the DNA sequence of spectrin matches the protein sequence of spectrin without assuming deletions (61).

The data presented here are consistent with the possibility that band 3 itself is a set of closely related polypeptides that have several transport functions and suggest that band 3 may be responsible for glucose as well as anion transport. However, there are other possibilities. Rather than being derived from band 3, glucose transporter could be a closely associated protein. For example, band 3, or part of it, could form membrane channels for transport of many molecules. A specific transporter, such as the glucose transporter, would then attach to band 3 temporarily to transport glucose. It appears that glucose transporter and band 3 are, at least, closely associated membrane molecules and that glucose transporter may be derived from or be part of band 3. In addition, the multifunctional band 3 molecule appears to be the origin for the senescent cell antigen (10,27,28,34), a molecule whose appearance initiates a sequence of immunological events resulting in removal of "marked" cells (1–5).

Acknowledgment. This work was supported by the Veterans Administration's Research Service and the Research Corporation.

References

1. Kay, M.M.B. 1974. Mechanism of removal of senescent red cells. Gerontologist **14**:33.
2. Kay, M.M.B. 1975. Mechanism of removal of senescent cells by human macrophages *in situ*. Proc. Natl. Acad. Sci. USA **72**:3521.

3. Kay, M.M.B. 1978. Role of physiologic autoantibody in the removal of senescent human red cells. J. Supramol. Struct. **9**:555.
4. Bennett, G.D., and M.M.B. Kay. 1981. Homeostatic removal of senescent
   murine erythrocytes by splenic macrophages. Exp. Hematol. **9**:297.
5. Kay, M.M.B. 1981. Isolation of the phagocytosis inducing IgG-binding antigen
   on senescent somatic cells. Nature (London) **289**:491.
6. Kay, M.M.B. 1981. The IgG autoantibody binding determinant appearing on
   senescent cells resides on a 62,000 MW peptide. Acta Biol. Med. Ger. **40**:385.
   (Presented at the IXth International Symposium on Structure and Function
   of Erythroid Cells, Berlin, German Democratic Republic. August 27, 1980.)
7. Kay, M.M.B. 1981. The senescent cell antigen is not a desialylated glycoprotein. Blood **58**:90.
8. Kay, M.M.B. 1982. Molecular aging: a termination antigen appears on senescent cells, p. 325. *In* Protides of the biological fluids, Vol. 29. Peeters,
   Oxford. (Presented at the May 7, 1981 XXIXth Annual Colloquium on Protides
   of the Biological Fluids, Belgium.)
9. Kay, M.M.B., and G.D. Bennett. 1982. Letter to the editor. Blood **59**:1111.
10. Kay, M.M.B., P. Wong, and P. Bolton. 1982. Antigenicity, storage & aging:
    physiologic autoantibodies to cell membrane and serum proteins. Mol. Cell.
    Biochem. **49**:65.
11. Lutz, H.U., and M.M.B. Kay. 1981. An age-specific cell antigen is present
    on senescent human red blood cell membranes. Mech. Ageing Dev. **15**:65.
12. Glass, G.A., H. Gershon, and D. Gershon. 1983. The effect of donor and cell
    age on several characteristics of rat erythrocytes. Exp. Hematol. **11**:987.
13. Bartosz, G., M. Sosynski, and J. Kedziona. 1982. Aging of the erythrocyte.
    VI. Accelerated red cell membrane aging in Down's syndrome? Cell Biol. Int.
    Rep. **6**:73.
14. Bartosz, G., M. Sosynski, and A. Wasilewski. 1982. Aging of the erythrocyte
    XVII. Binding of autologous immunoglobin. Mech. Ageing Dev. **20**:223.
15. Khansari, N., and H.H. Fudenberg. 1984. Phagocytosis of senescent eryth-
    rocytes by autologous monocytes: requirement of membrane-specific autolo-
    gous IgG for immune elimination of aging red blood cells. Cell Immunol. **78**:114.
16. Alderman, E.M., H.H. Fudenburg, and R.E. Lovins. 1980. Binding of im-
    munoglobulin classes to subpopulations of human red blood cells separated
    by density-gradient centrifugation. Blood **55**:817.
17. Tannert, C.H. 1978. Untersuchungen zum altern roter blutzellen. Ph.D. dis-
    sertation, Humbolt University, Berlin, German Democratic Republic.
18. Wegner, G., C.H. Tannert, D. Maretzki, W. Schossler, and D. Strauss. 1980.
    p. 57. *In* Abstracts, IXth International Symposium on Structure and Function
    of Erythroid Cells, IgG binding to glucose depleted and preserved erythrocytes.
    Berlin, German Democratic Republic, August 27, 1980.
19. Halhuber, K.T., D. Stibenz, H. Feuerstein, W. Linss, H.W. Meyer, R. Frobes,
    E. Rumpel, and G. Geyer. 1981. Defined rearrangement of the membrane of
    banked erythrocytes. Acta Biol. Med. Ger. **40**:419.
20. Smalley, C.E., and E.M. Tucker. 1983. Blood group A antigen site distribution
    and immunoglobulin binding in relation to red cell age. Br. J. Haematol. **54**:209.
21. Khansari, N., G.F. Springer, E. Merler, and H.H. Fudenberg. 1983. Mech-
    anisms for the removal of senescent human erythrocytes from circulation:
    specificity of the membrane-bound immunoglobulin. Mech. Ageing Dev. **21**:49.

22. Galili, U., A. Korkesh, I. Kahane, and E.A. Rachmilewitz. 1983. Demonstration of a natural antigalactosyl IgG antibody on thalassemic red blood cells. Blood 61:1258.
23. Hebbel, R.P., and W.J. Miller. 1984. Phagocytosis of sickle erythrocytes: immunologic and oxidative determinants of hemolytic anemia. Blood 64:733.
24. Petz, L.D., P. Yam, L. Wilkinson, G. Garratty, B. Lubin, and W. Mentzer. 1984. Increased IgG molecules bound to the surface of red blood cells of patients with sickle cell anemia. Blood 64:301.
25. Khansari, N., and H.H. Fudenberg. 1983. Immune elimination of aging platelets by autologous monocytes: Role of membrane-specific autoantibody. Eur. J. Immunol. 13:990.
26. Kay, M.M.B., S. Goodman, K. Sorsensen, C. Whitfield, P. Wong, L. Zaki, and V. Rudloff. 1982. The senescent cell antigen is immunologically related to band 3. J. Cell Biol. 95:244a. (Presented at the XXIInd Annual Meeting of the American Society of Cell Biology, November 30–December 4, 1982, Baltimore, Md.
27. Kay, M.M.B., S. Goodman, C. Whitfield, P. Wong, L. Zaki, and V. Rudoloff. 1983. The senescent cell antigen is immunologically related to band 3. Proc. Natl. Acad. Sci. USA 80:1631.
28. Kay, M.M.B. 1984. Localization of senescent cell antigen on band 3. Proc. Natl. Acad. Sci. USA 81:5753.
29. Kay, M.M.B., G.J.C.G.M. Bosman, S.S. Shapiro, A. Bendich, and P.S. Bassel. 1986. Oxidation as a possible mechanism of cellular aging: Vitamin E deficiency causes premature aging and IgG binding to erythrocytes. Proc. Natl. Acad. Sci. USA 83:2463.
30. Kasahara, M., and P.C. Hinkle. 1977. Reconstitution and purification of the D-glucose transporter from human erythrocytes. J. Biol. Chem. 252:7384.
31. Kliman, H.J., and T.L. Steck. 1975. Association of glyceraldehydephosphate dehydrogenase with the human red cell membrane: a kinetic analysis. J. Biol. Chem. 255:6314.
32. Strapazon, E., and T.L. Steck. 1977. Interaction of the aldolase and the membrane of human erythrocytes. Biochemistry 16:2966.
33. Karadsheh, N.S., K. Uyeda, and R.M. Oliver. 1972. Changes in alosteric properties of phosphofructokinase bound to erythrocyte membranes. J. Biol. Chem. 52:3515.
34. Kay, M.M.B. 1982. Accumulation of band 3 breakdown products is a function of cell age. Blood 60:21a. (Presented at the XXIVth Annual Meeting of the American Society of Hematology, Washington, D.C., Dec. 4–6, 1982).
35. Kay, M.M.B., and J.R. Goodman. 1983. IgG antibodies do not bind to band 3 in intact erythrocytes; enzymatic treatment of cells is required for IgG binding. Biomed. Biochim. Acta 43:841. (Presented at a plenary session of the Xth International Symposium on Erythroid Cells, Berlin, August, 1983.)
36. Kay, M.M.B. 1985. Senescent cell antigen: a terminal differentiation antigen. Surv. Synth. Path. Res. 4:227.
37. Kay, M.M.B. (in press) Immunologic techniques for analysing red cell membrane proteins. In S. Shohet and N. Mohandas (eds.), Methods in hematology: red cell membranes. Churchill Livingston, Inc., New York.
38. Taverna, R.D., and R.G. Langdon. 1973. D-Glucosyl isothiocyanate, an affinity

label for the glucose transport proteins of the human erythrocyte membrane. Biochem. Biophys. Res. Commun. **54**:593.

39. Lin, S., and J.A. Spudich. 1974. Binding of cytochalasin B to red cell membrane protein. Biochem. Biophys. Res. Commun. **61**:1471.

40. Sogin, D.C., and P.C. Hinkle. 1980. Immunological identification of the human erythrocyte glucose transporter. Proc. Natl. Acad. Sci. USA **77**:5725.

41. Zoccoli, M.A., S.A. Baldwin, and G.E. Lienhard. 1978. The monosaccharide transport system of the human erythrocyte: solubilization and characterization on the basis of cytochalasin B binding. J. Biol. Chem. **253**:6923.

42. Jung, C.Y., and A.L. Rampal. 1977. Cytochalasin B binding sites and glucose transport carrier in human erythrocyte ghosts. J. Biol. Chem. **252**:5456.

43. Griffin, J.F., A.L. Rampal, and C.Y. Jung. 1982. Inhibition of glucose transport in human erythrocytes by cytochalasins: a model based on diffraction studies. Proc. Natl. Acad. Sci. USA **79**:3759.

44. Kay, M.M.B., C. Tracey, J.R. Goodman, J.C. Cone, and P.S. Bassel. 1983. Polypeptides immunologically related to erythrocyte band 3 are present in nucleated somatic cells. Proc. Natl. Acad. Sci. USA **80**:6882.

45. Wassermann, N.H., A.S. Penn, P.I. Freimuth, N. Treptow, S. Wentzel, W.L. Cleveland, and B.F. Erlanger. 1982. Anti-idiotypic route to anti-acetylcholine receptor antibodies and experimental myasthenia gravis. Proc. Natl. Acad. Sci. USA **79**:4810.

46. Steck, T.L., J.J. Koziarz, M.K. Singh, G. Reddy, and H. Kohler. 1978. Preparation and analysis of seven major, topographically defined fragments of band 3, the predominant transmembrane polypeptide of human erythrocyte membranes. Biochemistry **17**:1216.

47. Steck, T.L., B. Ramos, and E. Strapazon. 1976. Proteolytic dissection of band 3, the predominant transmembrane polypeptide of the human erythrocyte membrane. Biochemistry **15**:1154.

48. Appell, K.C., and P.S. Low. 1981. Partial structural characterization of the cytoplasmic domain of the erythrocyte membrane protein, Band 3. J. Biol. Chem. **256**:11104.

49. Markowitz, S., and V.T. Marchesi. 1981. The carboxyl-terminal domain of human erythrocyte band 3. J. Biol. Chem. **256**:6463.

50. Laemmli, U.K. 1970. Cleavage of structural proteins during the assembly of the head of bacteriophage T4. Nature **227**:680.

51. Towbin, H., T. Staehelin, and J. Gordon. 1979. Electrophoretic transfer of proteins from polyacrylamide gels to nitrocellulose sheets: procedure and some applications. Proc. Natl. Acad. Sci. USA **76**:4350.

52. Elder, J.H., R.A. Pickett, II, J. Hampton, and R.A. Lerner. 1977. Radioiodination of proteins in single polyacrylamide gel slices. J. Biol. Chem. **252**:5456.

53. Tsai, I.H., S.N.P. Murthy, and T.L. Steck. 1982. Effect of red cell membrane on the catalytic activity of glyceraldehyde-3-phosphate dehydrogenase. J. Biol. Chem. **257**:1438.

54. Murthy, S.N., T. Liu, R.K. Kaul, H. Kohler, and T.L. Steck. 1981. The aldolase binding site of the human erythrocyte membrane is at the NH_2 terminus of band 3. J. Biol. Chem. **256**:11203.

55. Bennett, V., and P.J. Stenbuck. 1979. The membrane attachment protein for

spectrin is associated with band 3 in human erythrocyte membranes. Nature **280**:468.

56. Sogin, D.C., and P.C. Hinkle. 1978. Characterization of the glucose transporter from human erythrocytes. J. Supramol. Struct. **8**:447.
57. Cleveland, W.L., N.H. Wassermann, R. Sarangarajan, A.S. Penn, and B.F. Erlanger. 1983. Monoclonal antibodies to the acetylcholine receptor by a normally functioning auto-anti-idiotypic mechanism. Nature **305**:56.
58. Mueckler, M., C. Caruso, S.A. Baldwin, M. Panico, I. Blench, H.R. Morris, W.J. Allard, G.E. Lienhard, and H.F. Lodish. 1985. Sequence and structure of a human glucose transporter. Science **229**:941.
59. Kopito, R.R., and H.F. Lodish. 1985. Primary structure and transmembrane orientation of the murine anion exchange protein. Nature **316**:234.
60. Kaul, R.K., S.N.P. Murthy, A.G. Reddy, T.L. Steck, and H. Kohler. 1983. N-terminal 201 residues of human band 3. J. Biol. Chem. **258**:7981.
61. Birkenmeier, C.S., D.M. Brodine, E.A. Repasky, D.M. Helfman, S.H. Hughes, and J.E. Barker. 1985. Remarkable homology among the internal repeats of erythroid and nonerythroid spectrin. Proc. Natl. Acad. Sci. USA **82**:5671.

15
Autoimmunization Against Hormones: A New Strategy in Animal Production

PATRICK R. CARNEGIE

Introduction

Recently, a new branch of immunology has emerged as animal scientists attempt to manipulate animal production by immunizing against hormones. Previously, the only commercial applications of immunology were in human and animal health, in diagnosis, and in forensic science. Table 15.1 lists the studies in progress in immunomanipulation of physiological responses. Now the first of the immunomodulators of animal production is being marketed and others are under development. This paper will review progress and future potential, including possible applications of anti-idiotypic antibodies to receptors.

Immunomanipulation of Reproduction

In Australia, some varieties of clover are high in estrogens which cause infertility in ewes feeding on pastures containing these clovers. When attempts were made to immunize sheep against estrogens it was unexpectedly observed that with some immunogens, there was an increase in ovulation rate (1). After a number of steroid vaccines and adjuvants were tried, it was found that a conjugate of androstenedione-7-α-carboxyethyl-thioether with human serum albumin (Fig. 15.1), when administered with DEAE-dextran as adjuvant, induced antibodies to the steroid (2). Other adjuvants produced too high a titer of antibodies which were too persistent. A fine balance of titer and timing of mating was found to be essential to induce an increase in ovulation rate in ewes. Detailed trials were commenced by C.S.I.R.O. in 1976 (3), and a commercial vaccine, "Fecundin," was developed by 1981 and marketed in 1983 by Glaxo Australia Pty. Ltd. Two subcutaneous injections are given 3 to 4 weeks apart and rams are introduced 3 to 4 weeks after the second injection, which induces a satisfactory level of antibody to the steroid (Fig. 15.2). In subsequent years, only a booster injection is required. Provided the ewes are in good con-

TABLE 15.1. Some examples of autoimmunization in animal production.

Production parameter	Immunogen conjugated to carrier
Increased fertility in sheep	Steroid hormones
Reduced libido in bulls	Luteinizing hormone-releasing hormone
Manipulation of estrus in horses	Melatonin
Increased growth in lambs	Somatostatin
Decreased carcass fat in cattle	Adipocyte membranes
Reduction of "boar taint" in pigs	5α-Androst-16-en-3-one
Manipulation of wool follicles in sheep	Epidermal growth factor

dition at mating, an increase of 20 to 30% in the lambs born can be expected depending upon the breed of sheep (4).

Precise details of the mode of action of the response to the vaccine have still to be determined, but it is thought that the antibody binds to circulating estrogen, thus interfering with the feedback loops from the ovary to the hypothalmo-pituitary system (Fig. 15.3). Ovarian follicle development is regulated by luteinizing hormone (LH) and follicle-stimulating hormone (FSH). In immunized ewes there was an increase in the frequency of pulsatile LH release during the luteal phase of the estrous cycle and an increase in the basal level of LH during the follicular phase (5).

Although Fecundin is performing as expected, its sales have been less than hoped by the developers (P.J. Kieran, Coopers Animal Health Australia Ltd., personal communication). There are two main problems facing

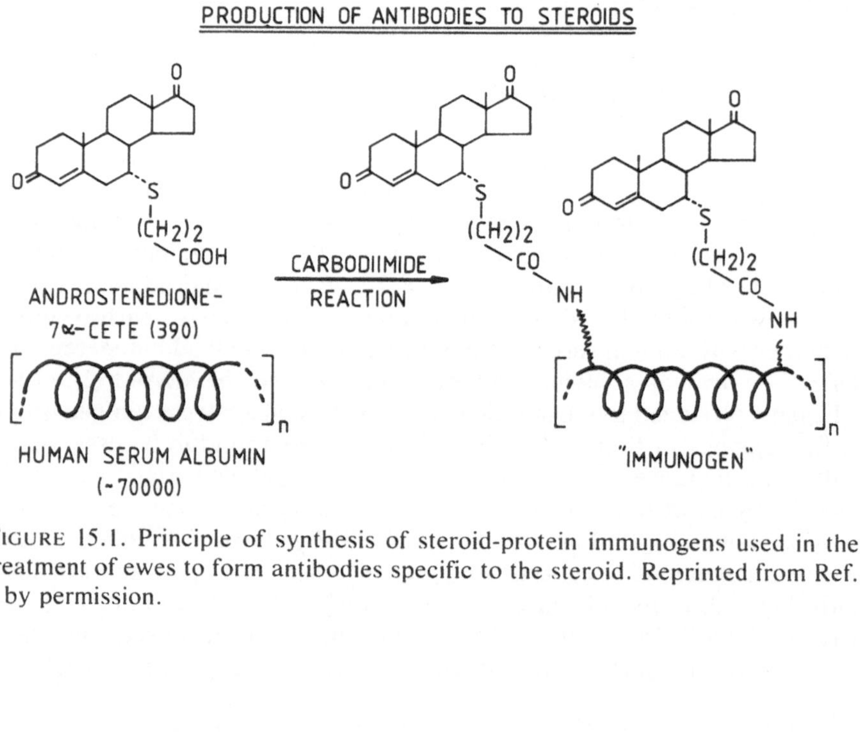

PRODUCTION OF ANTIBODIES TO STEROIDS

FIGURE 15.1. Principle of synthesis of steroid-protein immunogens used in the treatment of ewes to form antibodies specific to the steroid. Reprinted from Ref. 1 by permission.

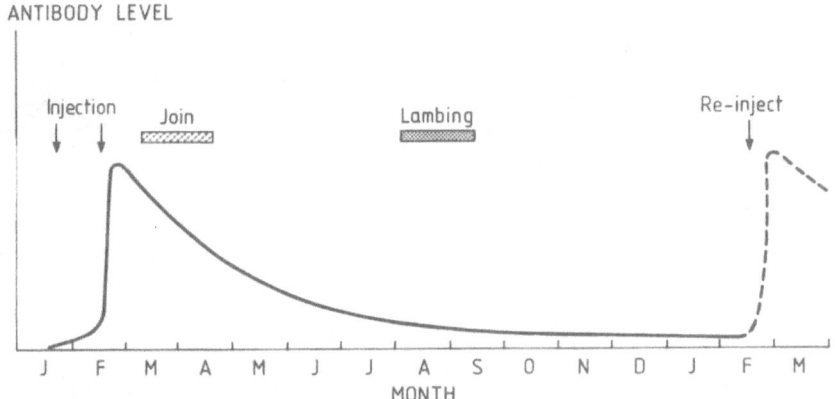

FIGURE 15.2. Schematic representation of the steroid-protein immunization sequence, antibody level changes, and appropriate time relationship with joining and lambing. Reprinted from Ref. 1 by permission.

farmers in Australia. Firstly, the ewes must be in good condition at the time of immunization; this can be difficult to achieve as drought conditions frequently produce poor-quality pasture. Secondly, the market price in Australia for fat lambs has been severely depressed in recent years, thus causing farmers to reduce, wherever possible, additional input costs.

Another application of Fecundin is in increasing the lambing performance in ewes treated with pregnant mare serum gonadotrophin to induce out-of-season mating (6).

Another possible target is inhibin, the gonadal peptide that inhibits the synthesis and release of FSH. Preliminary experiments with vaccines containing inhibin have enhanced fecundity (7,8).

Immunomanipulation of Libido

Castration of animals used for meat production is traditionally done to minimize fighting and prevent undesired breeding. However, there is a marked decrease in growth rate immediately following castration and animal welfare organizations would like to have the practice banned.

Several groups are studying the use of immunization of animals with conjugates of luteinizing hormone-releasing hormone (LHRH) to reduce libido. Schanbacher (9) reviewed work on immunization of males, which demonstrated a diminished testicular size and plasma testosterone levels. Robertson et al. (10) found that in 28-week-old bulls that developed high titers of antibody to LHRH, there was lowered serum testosterone, reduced libido and semen production, and more docile behavior. However,

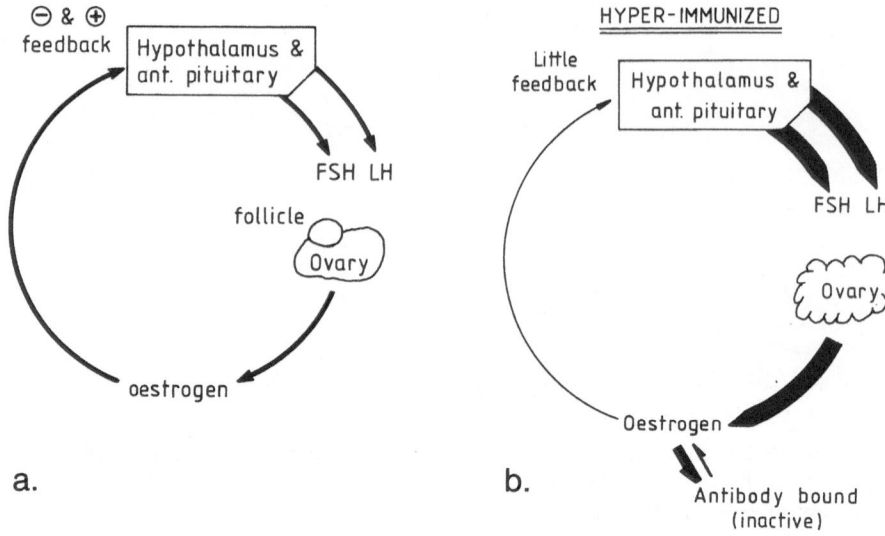

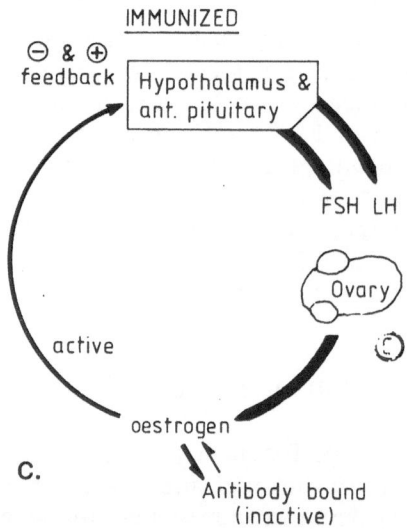

FIGURE 15.3. Model for the effects of estrogen immunization on feedback systems affecting ovarian function: (a) untreated ewe, usually with a single ovulation; (b) hyperimmune ewe with insufficient positive feedback to allow ovulation to occur; (c) immune ewe with steroid partly neutralized by antibody and tending to have multiple ovulation. Reprinted from Ref. 1 by permission.

presumably because of genetic differences in the immune response to LHRH, only 50% of the bulls developed a high titer of antibody to LHRH.

Jeffcoate and Keeling (11) reviewed studies on the influence of immunization against LHRH on estrus in the female. Estrus is induced by the rising estradiol concentrations and progesterone priming in the preovulatory period. The subsequent production of LH can be blocked by antibody to LHRH. A practical application of this type of immunomanipulation is being tried by Post (12) in heifers. In rangeland conditions in underdeveloped country, wild bulls of poor quality can disrupt programs to improve the quality of beef herds. Immunization of females with conjugates of LHRH has the potential to prevent conception in females being fattened for slaughter and in heifers that are too young for breeding. Reduction of libido by immunization could be attractive, as natural reversal of the inhibition occurs in the majority of animals after one to two years (13).

Apart from applications in the control of libido in farm animals, there are opportunities for humane control of wildlife population increase. For example, kangaroos have the potential to expand their population at a rapid rate where there is ample feed and water. Currently, controlled slaughter programs are used to prevent the buildup of overly large populations which compete with sheep and cattle for food. T. Stelmasiak (personal communication) is experimenting with an LHRH vaccine in a bullet which can be "implanted" with the aid of a rifle. This new vaccine delivery system (14) has many other applications in disease control in experimental animals.

The above studies have the potential for providing simple and humane methods for controlling libido in animals, but much more work is required to determine the optimum formulation of the vaccine and adjuvant. In addition to these applications in animal production, there is a large potential market in controlling libido in domestic pets.

Horses are seasonal breeders whose estrus is regulated by the pineal gland. Melatonin has an inhibitory effect on long-day breeders such as the horse. During spring, as day length increases, melatonin production by the pineal gland gradually decreases until a critical point is reached when reproductive activity is stimulated. As the gestation period of the horse is just over eleven months this, in combination with the occurrence of the fertile period in late spring, means that the peak of foaling is at a time in Australia when pasture quality is decreasing. There would be a distinct advantage in achieving foaling in late winter or early spring when the mare could be provided with better nutrition. Some racehorse breeders have attempted to decrease melatonin production by influencing the pineal gland with artificial light but this method is too expensive.

R.J. Fiddes and T. Stelmasiak (personal communication) prepared a conjugate of melatonin with bovine γ-globulin and have immunized horses with the conjugate in "Alhydrogel" as adjuvant. However, the production of antibody with this adjuvant was not good in contrast to the use of the immunogen with complete Freund adjuvant as adjuvant in sheep. The strong inflammatory response induced by complete Freund adjuvant is considered to render its use inadvisable in horses.

While this approach has so far not succeeded, it is theoretically sound and simple. Other adjuvants might be more effective. An alternative approach would be the use of an anti-idiotypic antibody to the antibody to melatonin to induce a temporary block of the receptors for melatonin (see below).

Goddard et al. (15) took a different approach to tackling the same problem. They immunized mares against testosterone during anestrus. Although some mares produced antibodies to testosterone, there was no evidence of increased ovarian activity. This contrasts with the results obtained by Gibb et al. (16), who found that immunization with androstenedione-6-hemisuccinate caused ewes to exhibit estrus earlier in the breeding season than controls.

Immunization with conjugates of testosterone would appear to produce a different response in pigs from that in horses and cattle in which the immunogen stimulates the development of the testes (17).

Immunomanipulation of Growth

Somatostatin from the hypothalamus inhibits the release of growth hormone from the pituitary and the release of other hormones, such as insulin, which influence growth. Ferland et al. (18) showed that passive immunization of rats with antibody to somatostatin caused an increase in plasma growth hormone and thyroid-stimulating hormone. Varner et al. (19) immunized lambs with a conjugate of somatostatin and egg albumin and found an increase in serum growth hormone levels but, unexpectedly, the control lambs gained more weight than the immunized lambs. Spencer and Garssen (20) immunized lambs with conjugates of somatostatin and human α-globulin in complete Freund adjuvant. Within five weeks an increased growth rate was observed in treated lambs, and when slaughtered at 20 weeks, there was an increase of 18% in carcass weight. It appeared that food had been utilized more efficiently in the immunized lambs (21). Because somatostatin can influence the release of motilin and other hormones having an influence on gut motility, they studied the effect of passive immunization with antibody to somatostatin on the rate of passage of feed through the gastrointestinal tract. The transit time for a marker in the food was significantly reduced in immunized lambs (22).

As the use of steroid growth promotants is coming under increasing attack in many countries, without scientific justification, the use of immunization to enhance growth has a high potential. Unfortunately, other groups (23) are having difficulty repeating Spencer and Garssen's results on the effect of active immunization against somatostatin on lamb growth.

Immunomanipulation of Carcass Quality

The production of excess fat in cattle, lambs, chickens, and pigs is a major problem. Feed is wasted, and there is increased public demand for leaner meat. While β-agonists such as clenbuterol have proved very effective in reducing the fat content of lamb carcasses (24), these drugs are unlikely to be permitted because of concern that residues in the meat might affect humans. Work on developing a commercial vaccine against adipocyte membranes is currently being carried out by Flint et al. (25) at the Hannah Research Institute in Scotland. Initial experiments have been done with sheep injected with rat adipocytes. Their approach of using whole adipocyte membranes would appear to be relatively crude and could induce a more generalized and destructive immune response. An alternative would be to immunize animals with components specific to adipocytes which are essential for the development of adipose tissue. We have recently commenced such a project with the aim of reducing the amount of fat in a pig carcass.

In intensive pig production, the use of intact males is now common in many countries. With some breeds and feeding regimes, "boar taint" in the carcass can be a problem. However, immunization against one of the main components responsible for the unpleasant flavor, 5α-androst-16-en-3-one, failed to inhibit the accumulation of the steroid in the meat (26).

Immunomanipulation of Wool Follicles

In Australia, there is a large-scale research program aimed at producing agents which will obviate the need to shear sheep. One of the potential chemical defleecing agents is epidermal growth factor (EGF) which when infused into sheep causes a temporary weakness in the wool fiber which permits the wool to be peeled off (27). Wynn et al. (28) have investigated the physiological regulation of EGF receptors on wool follicles by immunizing sheep with mouse EGF. They found a highly significant increase in the number of EGF receptors and a redistribution of the receptors. No change in wool growth was observed.

Role of Anti-Idiotypic Antibodies

Lennon and Carnegie (29) were the first to suggest that antibodies could alter the physiological function of receptors, and Carnegie and Mackay (30) emphasized the importance of anti-idiotypic antibodies to receptors.

In all the above studies on immunomanipulation of animal production, the influence of anti-idiotypic antibodies on receptors has been ignored. As immunization of animals with insulin results in the generation of anti-idiotypic antibodies to the insulin receptor (31) (see Chapter 8 of this volume), it is likely that immunization with the hormones discussed above could result in anti-idiotypic antibodies being produced to their receptors. There is a need for groups studying immunomanipulation of animal production to monitor the development of anti-receptor antibodies. Farid and Lo (32) have reviewed the interactions between anti-idiotypic antibodies to hormone receptors and the receptors. Since many of the reported anti-idiotypic antibodies cause a functional change in the receptor, then it is possible that such antibodies could be produced to manipulate a production parameter in a more reproducible way than immunization against the hormone. The animal to animal variability in response to small hormone immunogens could possibly be overcome by the use of anti-idiotypic vaccines. For example, an alternative to the use of β-agonists to decrease the fat content of meat would be to use anti-idiotypic antibodies, such as that reported by Schrieber et al. (33), which stimulate the β receptor and which should produce a leaner carcass similar to that produced by treatment with clenbuterol. In the future, the production of anti-receptor specific antibodies which produce either an inhibition or stimulation of hormone receptors could offer an alternative and, possibly, more reproducible approach to immunomanipulation of physiological responses in farm animals.

Conclusion

As the potential market for immunogens which will influence libido, growth rate, and carcass quality is very large, there is likely to be increased commercial investment in research in this area.

Manipulation of animal production parameters by drugs usually generates adverse and unscientific reaction by the community, as has happened with steroid growth promotants in Europe and even bovine growth hormone in parts of the U.S.A. In contrast, vaccination has a "good image" with the general public and permission to use new immunogens in farm animals is much more likely to be granted.

Acknowledgment. The support of Coopers Animal Health Australia Ltd. and the Australian Pig Research Council for work on studies in immunomanipulation is acknowledged.

References

1. Cox, R.I., P.A. Wilson, and M.S.F. Wong. 1985. Manipulation of endocrine systems through the stimulation of specific immune responses, p. 150. *In* Reviews in rural science 6: Biotechnology and recombinant DNA technology industries. Proceeding of symposium jointly organized by University of New England (Armidale, N.S.W.) and C.S.I.R.O.

2. Scaramuzzi, R.J., and R.M. Hoskinson. 1984. Active immunization against steroid hormones for increasing fecundity. p. 445. *In* D.B. Crighton (ed.), Immunological aspects of reproduction in mammals. Butterworths, London.
3. Scaramuzzi, R.J., W.G. Davidson, and P.F.A. Van Look. 1977. The effect of active immunization against androstenedione on oestrus and ovulation in sheep. Nature (London) **269**:817.
4. Geldard, H., C.J. Dow, and P.J. Kieran. 1984. Further developments in fecundity immunization field results. Wool Technology and Sheep Breeding **32**:69.
5. Scaramuzzi, R.J. 1984. Changes in pituitary-ovarian functions in ewes immune to steroid hormones. Proc. Aust. Soc. Anim. Prod. **15**:182.
6. Robinson, T.J., and R.J. Scaramuzzi. 1986. Immunization against androstenedione and out-of-season breeding in sheep. Proc. Aust. Soc. Anim. Prod. **16**:323.
7. O'Shea, T., L.J. Cummins, B.M. Bindon, and J.K. Findlay. 1982. Increased ovulation rate in ewes vaccinated with an inhibin-enriched fraction from bovine follicular fluid. Proc. Aust. Soc. Reprod. Biol. **14**:85.
8. Henderson, K.M., P. Franchimont, M.J. Lecomte-Yerna, N. Hudson, and K. Ball. 1984. Increase in ovulation rate after active immunization of sheep with inhibin partially purified from bovine follicular fluid. J. Endocrinol. **102**:305.
9. Schanbacher, B.D. 1984. Active immunization against LH-RH in the male, p. 345. *In* D.B. Crighton (ed.), Immunological aspects of reproduction in mammals. Butterworths, London.
10. Robertson, I.S., H.M. Fraser, G.M. Innes, and A.S. Jones. 1982. Effect of immunological castration on sexual and reproduction characteristics in male cattle. Vet. Rec. **111**:529.
11. Jeffcoate, I.A., and B.J. Keeling. 1984. Active immunization against LH-RH in the female, p. 363. *In* D.B. Crighton (ed.), Immunological aspects of reproduction in mammals. Butterworths, London.
12. Post, T.B. 1986. Immunization against LHRH as an aid in managing reproduction in grazing female beef cattle. *In* Proceedings of international symposium on use of nuclear techniques—Studies of animal reproduction and health in different environments. IAEA/FAO, Vienna.
13. Keeling, B.J., and D.B. Crighton. 1984. Reversibility of the effects of active immunization against LH-RH, p. 379. *In* D.B. Crighton (ed.), Immunological aspects of reproduction in mammals. Butterworths, London.
14. Stelmasiak, T. 1986. Vaccine and delivery system, Australian provisional patent application, pending.
15. Goddard, P.J., W.E. Allan, and J. Kilpatrick. 1985. Effect of testosterone immunization on ovarian activity in pony mares during late winter and spring. Vet. Rec. **116**:374.
16. Gibb, M., D.C. Thurley, and K.P. McNatty. 1981. Active immunization of Romney ewes with an androstenedione-protein conjugate. N.Z. J. Agric. Res. **24**:5.
17. Thompson, D.L., Jr., L.L. Southern, R.L. St. George, K.S. Jones, and F. Garza. 1985. Active immunization of prepubertal boars against testosterone: testicular and endocrine responses at 14 months of age. J. Anim. Sci. **61**:1498.
18. Ferland, L., F. Labrie, M. Jobin, A. Arimura, and A.V. Schally. 1976. Phys-

iological role of somatostatin in the control of growth hormone and thyrotropin secretion. Biochem. Biophys. Res. Commun. **68**:149.

19. Varner, M.A., S.L. Davis, and J.L. Reeves. 1980. Temporal serum concentrations of growth hormone, thyrotropin, insulin and glucagon in sheep immunized against somatostatin. Endocrinology **106**:1027.

20. Spencer, G.S.G., and G.J. Garssen. 1983. A novel approach to growth promotion using auto-immunization against somatostatin. I. Effects on growth and hormone levels in lambs. Livestock Prod. Sci. **10**:25.

21. Spencer, G.S.G., G.J. Garssen, and P.L. Bergstrom. 1983. A novel approach to growth promotion using auto-immunization against somatostatin. II. Effects on appetite, carcass composition and food utilisation in lambs. Livestock Prod. Sci. **10**:469.

22. Fadalla, A.M., G.S.G. Spencer, and D. Lister. 1985. The effect of passive immunization against somatostatin on marker retention time in lambs. J. Anim. Sci. **61**:234.

23. Hoskinson, R.M., P. Djura, R.J. Welch, and B.E. Harrison. 1986. Effect of somatostatin immunity on the growth of crossbred lambs. Proc. Aust. Soc. Anim. Prod. **16**:415.

24. Ricks, C.A., R.H. Dalrymple, P.K. Baker, and D.L. Ingle. 1984. Use of a β-agonist to alter fat and muscle deposition in steers. J. Anim. Sci. **59**:1247.

25. Flint, D.J., H. Coggrave, C.E. Futter, M.J. Gardner, and T.J. Clarke. 1985. Stimulatory and cytotoxic effects of an antiserum to adipocyte plasma membranes on adipose tissue metabolism in vitro and in vivo. Int. J. Obesity **10**:69–77.

26. Shenoy, E.V.B., M.J. Daniel and P.G. Box. 1982. The "boar taint" steroid 5 α-androst-16-en-3-one: an immunisation trial, Acta Endocrinology **100**:131.

27. Moore, G.P.M., B.A. Panaretto, and D. Robertson. 1982. Inhibition of wool growth in merino sheep following administration of mouse epidermal growth factor and a derivative. Aust. J. Biol. Sci. **35**:163.

28. Wynn, P.C., R.M. Hoskinson, and G.P.M. Moore. 1986. Active immunization of sheep against epidermal growth factor (EGF): the effect on wool growth and skin EGF receptors. Proc. Aust. Soc. Anim. Prod. **16**:436.

29. Lennon, V.A., and P.R. Carnegie. 1971. Immuno-pharmacological disease: a break in tolerance to receptor sites. Lancet **i**:630.

30. Carnegie, P.R., and I.R. Mackay. 1975. Vulnerability of cell-surface receptors to autoimmune reactions. Lancet **ii**:684.

31. Sege, K., and P.A. Petersen. 1978. Use of anti-idiotypic antibodies as cell surface receptor probes. Proc. Nat. Acad. Sci. USA **75**:2443.

32. Farid, N.R., and T.C.Y. Lo. 1985. Antiidiotypic antibodies as probes for receptor structure and function. Endocr. Rev. **6**:1.

33. Schrieber, A.B., P.O. Couraud, C. Andre, B. Vray, and A.D. Strosberg. 1980. Anti-alprenolol anti-idiotypic antibodies bind to β-adrenergic receptors and modulate catecholamine-sensitive adenylate cyclase. Proc. Nat. Acad. Sci. USA **77**:7385.

16
Anti-Idiotypes and Lymphokine Receptors: Interferon as a Model

PHYLLIS L. OSHEROFF

Introduction

Anti-idiotypic antibodies have been used as cell surface receptor probes in a number of systems (for a review, see Ref. 1). However, there have been relatively few reports which describe monoclonal anti-idiotypic antibodies that not only recognize the receptors specific for a protein or polypeptide but also trigger the known biological activity of that protein or polypeptide. The biological and potential clinical significance of various lymphokines and the generally low level of expression of cellular receptors for these biologically active proteins make the study of anti-idiotypes as probes for these lymphokine receptors especially interesting and important. The successful cloning of various human interferon cDNAs and their expression in bacteria (for a review, see Ref. 2), coupled with recent reports showing direct analysis of interferon binding to specific cellular receptors (for a review, see Ref. 3), enabled us to use the recombinant human leukocyte A interferon (rIFN-αA) as a model system to study the anti-idiotypes as interferon receptor probes. This model system can be applied to studies of receptors for other lymphokines or biologically active proteins.

Human Interferons and Their Receptors

The human interferons are a family of related proteins (4) that are classified as leukocyte, fibroblast, and immune interferons, or IFN-α, -β, and -γ, respectively. The interferons show a wide range of biological effects on target cells, notably the ability to inhibit multiplication of virus and proliferation of cells in culture and to modulate immune responses (reviewed in Ref. 5). These biological effects are thought to be mediated through binding of interferons to specific cell surface receptors (reviewed in Ref. 3). The various leukocyte interferons share 80% amino acid sequence homology while the fibroblast interferons share 30% homology with the leukocyte interferons (6,7). It has been proposed that leukocyte and fibroblast interferons interact with a common cellular binding site (8) while immune interferon appears to bind to a distinct cellular receptor (9). Although differences in specific biological activity have been reported among leukocyte interferon subtypes, suggesting the possible existence of more than one class of receptor (10), competitive binding inhibition experiments have indicated that leukocyte interferon subtypes are cross-reactive with an apparent common binding site (11; P.L. Osheroff and D.M. Manousos, unpublished results). It has been proposed that the specific activities of the different leukocyte interferons are related to differences in their receptor binding affinities (11) and the dynamics of the different interferon-receptor complexes (12). While the apparent dissociation constants of various interferons for the receptors are in the 10^{-12} to 10^{-10} M range, the estimated concentrations of receptor sites are very low (500 to 5000 sites per cell) (3). Thus, anti-idiotypic antibody appears to be an ideal probe for the study of interferon receptors.

Interferon Idiotypes

In our system, neutralizing rabbit anti-interferon antibodies are used for the induction of anti-idiotypic antibodies (anti-Id). Monoclonal anti-interferon antibodies are not used because none of the monoclonal anti-IFN-α antibodies (13) available to us is able to both bind and neutralize all of the known IFN-α subtypes (A. Palleroni and O. Bohoslawec, personal communication) which are cross-reactive with an apparent common target cell receptor. By using polyclonal idiotypic antibodies (Id) for the induction of anti-Id, there is a higher probability of obtaining the anti-Id which recognizes the interferon binding site on the receptor. The possibility that any given neutralizing monoclonal antibody may neutralize interferon activity through steric hindrance and not by direct binding to the receptor binding site on the interferon is also obviated by this approach.

Antibodies were raised in rabbits against recombinant human leukocyte A interferon (rIFN-αA), which was purified on an affinity column of monoclonal anti-IFN-α antibody LI8 (13,14). The purity of rIFN-αA was about 98%. The antisera obtained were purified by ammonium sulfate precipitation (47% saturation) and DEAE-Sepharose-CL-6B (Pharmacia) column chromatography, followed by an affinity column of rIFN-αA coupled to Affi-Gel 10 (Bio-Rad) and cross-linked with glutaraldehyde, to isolate the rIFN-αA specific antibodies. These idiotypic antibodies were assayed for interferon binding in an enzyme-linked immunosorbent assay (ELISA) and for neutralization of the antiviral activity of interferon in a cytopathic effect (CPE) inhibition assay with vesicular stomatitis virus (VSV) and using Madin-Darby bovine kidney (MDBK) cells or human amnion (WISH) cells as targets (15).

The affinity-purified Id showed apparent homogeneity on sodium do-decyl sulfate-polyacrylamide gel electrophoresis (SDS-PAGE). They showed high interferon binding titers with half-maximum binding at an antibody concentration of 60–70 ng/ml. They also showed high interferon neutralization titers with 50% neutralization of 1000–2000 units/ml of rIFN-αA at 50–100 ng/ml. These purified rabbit anti-rIFN-αA antibodies are referred to as Rb-Id and were used to induce both polyclonal and monoclonal anti-Id.

To induce polyclonal anti-Id in rabbits, the Rb-Id were cross-linked to keyhole limpet hemocyanin (KLH) (16), while non-cross-linked Rb-Id were used to induce anti-Id in mice.

Initial Screening for Internal Image Anti-Id: Interferon Neutralization Inhibition

Anti-idiotypic antisera produced in either rabbits or mice were first tested for interferonlike activity in the antiviral assay. None was detected, prob-ably due to either the transient appearance or, most likely, the very low concentration of the internal image anti-Id present in these antisera. Then it was found that these anti-idiotypic antisera had the ability to inhibit the neutralization of interferon activity by the Rb-Id. An assay was developed to test this property and was referred to as the neutralization inhibition assay. In this assay the putative anti-idiotypic antisera or hybridoma su-pernatants were incubated with various concentrations of the Rb-Id at 37°C for 1 h. The mixtures were then added rIFN-αA at 1000–2000 units/ml, incubated at 37°C for 1 h, and finally assayed for antiviral activity in the CPE inhibition assay using MDBK cells as targets. The results were compared with those from controls in which normal rabbit or mouse serum was used instead of anti-idiotypic antiserum. This is illustrated in Table 16.1. Part A shows that the antiserum of one rabbit immunized with Rb-Id-KLH is able to partially or totally inhibit the neutralization of rIFN-αA activity by various concentrations of Rb-Id, while part B shows that normal rabbit serum at the same dilution does not show such inhibition. Part C shows that the anti-idiotypic antisera of the two mice used in the hybridoma fusion also have the ability to inhibit the neutralizing effect of the Rb-Id, whereas normal mouse serum does not have such an effect (part D). In all cases, incubation of anti-idiotypic antisera with interferon in the absence of Rb-Id does not significantly change the interferon activity.

Figure 16.1 shows a schematic representation of the possible molecular interactions that may take place in the interferon neutralization inhibition. Assuming that A is the receptor binding site or "active site" of interferon which by binding to the target cell receptor R induces an antiviral state in the target cell. B is a site on one of the Id that resembles the receptor

TABLE 16.1. Inhibition of Rb-Id neutralization of rIFN-αA activity by rabbit or mouse anti-Rb-Id.

Rb-Id Concn. (µg/ml)	Interferon activity[a] (unit/ml)								
	A		B		C			D	
		rIFN-αA + Rb anti-Rb-Id serum[b]		rIFN-αA + normal Rb serum[c]		rIFN-αA + mouse anti-Rb-Id serum			rIFN-αA + normal mouse serum[f]
	rIFN-αA	rIFN-αA + Rb anti-Rb-Id serum[b]	rIFN-αA	rIFN-αA + normal Rb serum[c]	rIFN-αA	Mouse 1[d]	Mouse 2[e]	rIFN-αA	rIFN-αA + normal mouse serum[f]
1.2	<40	240	<40	<40	<40			<40	<40
0.6	<40	960	<40	<40	<40	2,560	2,560	<40	<40
0.3	160	1,920	50	80	40	3,400	3,400	<40	<40
0.15	480	1,920	420	420	160	3,400	3,400	320	480
0.075	960	2,560	850	1,280	640	2,560	2,560	960	
0	1,280	2,560	1,700	1,700	1,700	1,700	1,700	1,280	1,920

[a] Interferon activity, neutralization, and neutralization inhibition were determined by the cytopathic effect (CPE) inhibition assay with vesicular stomatitis virus and bovine kidney (MDBK) cells used as targets (15). Interferon neutralization by antibodies was determined by incubation of a pretitrated concentration of rIFN-αA (1000 units/ml) with different dilutions of anti-interferon antiserum or purified IgG for 1 h at 37°C followed by assaying for residual interferon activity in the CPE inhibition assay. Interferon neutralization was determined by comparison with controls in which no antibody was present. Interferon neutralization inhibition by anti-idiotypic antisera is described in the text. Final concentration of rIFN-αA in all assays was 6 ng/ml.
[b] Rabbit antiserum to Rb-Id cross-linked with KLH. Final dilution of antiserum in the assay, 1:20.
[c] Final dilution of normal rabbit serum in the assay, 1:20.
[d] Final dilution of Mouse 1 antiserum in the assay, 1:40.
[e] Final dilution of Mouse 2 antiserum in the assay, 1:40. Spleen cells of both Mouse 1 and 2 were used in the fusion with mouse myeloma cells.
[f] Final dilution of normal mouse serum, 1:40.

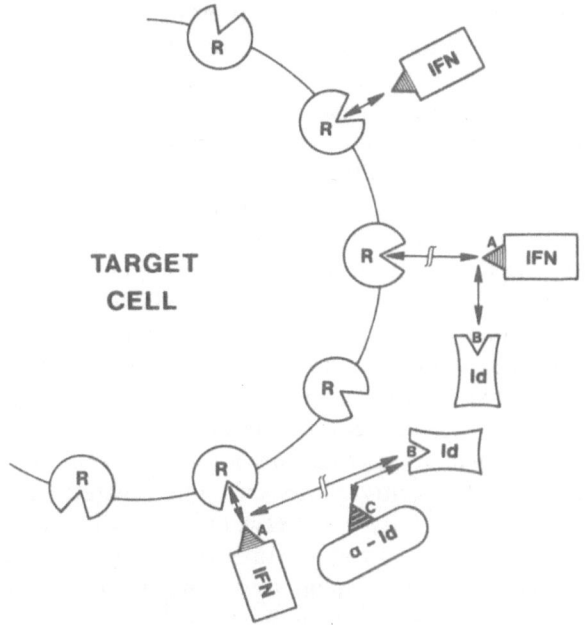

FIGURE 16.1. Schematic representation of the molecular interactions in the interferon neutralization inhibition assay. See text for details.

site and neutralizes interferon activity by binding at or near the interferon active site A, preventing it from binding to the target cell receptor. C is a site on one of the anti-Id whose variable regions resemble the interferon active site A and binds at or near the neutralizing site B on the Id and inhibits the ability of the Id to neutralize interferon activity. One advantage of this assay is that one is dealing with direct interactions between the anti-Id and the Id and between the Id and interferon. So one is selecting for an antibody which firstly is an anti-Id and secondly is relevant to the interferon activity. This assay has been successfully used in our interleukin-2 anti-idiotypic studies (manuscript in preparation) and should be applicable to other lymphokines or biologically active proteins for which a relatively simple and short-term bioassay is available.

Interferon Activity in an Anti-Idiotypic Hybridoma Antibody

Following the fusion between Rb-Id-immunized mouse splenocytes and the mouse myeloma line PAI-O (17), the hybridoma supernatants were tested in the interferon neutralization inhibition assay. Since the immunogen was polyclonal Rb-Id, some populations of which might neutralize interferon by binding to the receptor binding site of interferon while others might neutralize through steric hindrance, any monoclonal anti-Id would not be expected to completely inhibit the collective neutralizing effect of the Rb-Id. The supernatant of one of the hybridomas tested, 3-1B, consistently showed a 50% inhibition of the interferon neutralizing effect of the Rb-Id but did not show any interferonlike antiviral activity at the concentration tested. However, as part of an in vivo passage of some hybridoma lines which seemed to have become dependent on the hypoxanthine and thymidine in the media (18), 3-1B cells were injected into mice, and hybridoma cells in the ascitic fluids of two individual mice were centrifuged, resuspended, and cultured separately in vitro. One day later, samples of these two culture supernatants (containing traces of ascitic fluids) were assayed both for interferon neutralization inhibition and for antiviral activity. It was found that one of these two supernatants showed significant interferonlike antiviral activity when tested with both MDBK and WISH cells. The activity was between 2.2 x 10^5 and 2.5 x 10^5 units/ml supernatant. Thus, in vivo passage of 3-1B cells in that particular mouse might have served as a subcloning step. The in vivo conditions in this mouse selectively favored the growth of a subpopulation of cells which secreted immunoglobulins that exhibited interferonlike activity. This subpopulation of hybridoma cells was presumably those of a subclone of 3-1B and is designated "3B1." The antiviral activity in the 3B1 supernatant was not due to murine interferon that might be present in the traces of mouse ascitic fluid: firstly, this supernatant did not show any antiviral activity when tested with mouse L cells which possess murine interferon receptors; and secondly, natural murine interferon at this concentration showed no detectable antiviral activity when tested with both MDBK and WISH cells. The antiviral activity in the 3B1 supernatant which was kept at 4°C remained stable when tested after over one and a half years. The culture supernatant of the other in vivo passaged 3-1B cells did not show any antiviral activity.

Evidence that the antiviral activity in the culture supernatant of 3B1 was due to immunoglobulins is shown in Table 16.2. Up to 75% of the antiviral activity exhibited by the 3B1 supernatant could be absorbed by amylose-bound goat anti-mouse immunoglobulin serum (Roche Diagnostics) while interferon was not absorbed. Quantitative absorption was not achieved when the same amount of immunosorbent was incubated with 50% of the supernatant. This could be due to the fact that whole goat serum and not purified IgG was bound to the amylose. Table 16.2 also shows that the antiviral activity in the 3B1 supernatant could be quantitatively absorbed by Rb-Id immobilized on Affi-Gel 10 and could be partially eluted at pH 2.5. The acid eluate maintained its antiviral activity if brought to neutral pH immediately but lost its activity if left at acid pH. Since IFN-α and -β are stable at low pH (e.g., 0.2 M acetic acid, pH 2.3)

TABLE 16.2. Absorption of interferon activity in 3B1 hybridoma supernatant by amylose-bound goat-anti-mouse immunoglobulin serum (G-A-M amylose) and Rb-Id-Affi-Gel 10 (Rb-Id-Ag10).[a]

| | Interferon activity (unit/ml) | |
| | 3B1 supernatant dilution [b] | |
Gel added	1:50	1:100
None	6,466	1,920
G-A-M amylose suspension, 150 μl	3,233	960
G-A-M amylose suspension, 300 μl	2,155	480
Rb-Id-AG10 suspension, 10 μl	25 (3,233)[c]	

[a]G-A-M amylose (20% suspension) or Rb-Id-AG10 (30% suspension) was washed twice with PBS and resuspended in 1% BSA/PBS to the original volumes. Varying amounts of these suspensions were then aliquoted and centrifuged. The pellets were added 50 μl of 1:50 and 1:100 dilutions (in 1% BSA/PBS) of the 3B1 hybridoma supernatant, and the mixtures were incubated at room temperature for 1 h with constant mixing. The mixtures were centrifuged and the supernatants were assayed for residual antiviral activity in the CPE inhibition assay along with controls in which no gels were added.

The Rb-Id-AG10 pellets were washed once with PBS and once with 1% BSA/PBS and were suspended in 50 μl of 0.1 M glycine, pH 2.5–0.12 M NaCl and centrifuged. The supernatants were immediately neutralized with 1 M HEPES, pH 7.6, and assayed for antiviral activity eluted from the gels.
[b]Data for the two different 3B1 supernatant dilutions were obtained from separate experiments.
[c]Value in parentheses is for eluate.

and immunoglobulins are not, these data suggest that the antiviral activity in this supernatant is due to the immunoglobulins and not interferon. Immunodiffusion tests showed that the 3B1 hybridomas secreted immunoglobulin of the IgG subtype (γ1 heavy chain and κ light chain).

Competition with Interferon Binding to Rb-Id by the Anti-Idiotypic Hybridoma Antibody

Table 16.3 shows that the antibody in the 3B1 supernatant is able to compete with interferon for binding to the Rb-Id. rIFN-αA was radiolabeled using the lactoperoxidase-glucose oxidase Enzymobead radioiodination reagent (Bio-Rad, Richmond, CA) and purified on Sephadex G25 followed by monoclonal anti-rIFN-αA antibody (LI-8) columns as previously described (19). Undiluted 3B1 supernatant or a 1:10 dilution inhibits the binding of ^{125}I-rIFN-αA to the Rb-Id by 50 to 70%, whereas normal mouse IgG at comparable or higher concentrations does not show any inhibition. The anti-idiotypic antiserum of one of the mice used in the hybridoma fusion also does not show significant inhibition, probably due to a very low concentration of the relevant antibodies in the polyclonal antiserum.

TABLE 16.3. Competition with binding of ^{125}I-rIFN-αA to Rb-Id by 3B1 hybridoma supernatant.[a]

Preincubation	^{125}I-rIFN-αA bound (cpm)	% Inhibition of binding
None	556	0
3B1 supernatant		
Undiluted	163	71
Diluted 1:10	281	50
Diluted 1:20	348	37
Anti-Rb-Id antiserum of Mouse 2		
Diluted 1:5	381	31
Diluted 1:10	428	23
Diluted 1:20	366	34
Normal mouse IgG		
0.5 mg/ml	615	0
1.0 mg/ml	592	0

[a]Wells of a flexible 96-well polyvinyl chloride plate were coated with 50 μl of 1 μg/ml of Rb-Id, or 1 μg/ml of normal rabbit IgG as control, and blocked with 5% BSA/PBS for 2 h at 37°C. Various amounts (0.5, 1, 2, 3, 5 μl) of ^{125}I-rIFN-αA (purified on Sephadex G25 and monoclonal antibody LI-8 affinity columns; see text) were added and incubated for 2 h at 37°C with shaking. The wells were washed 4 times with PBS and individual wells were cut out and counted in a Beckman r300 counter. Control plates coated with normal rabbit IgG (1 μg/ml) did not bind any ^{125}I-rIFN-αA (not shown). The amount of ^{125}I-rIFN-αA sufficient to reach saturation binding to the Rb-Id (1 μl, 3050 cpm) was used in the test for the ability of the 3B1 supernatant to compete with ^{125}I-rIFN-αA to bind to the Rb-Id. In the competition test, the microtiter plate was coated with Rb-Id and blocked with BSA as above. Ten microliters of various dilutions of the 3B1 supernatant, the anti-Rb-Id antiserum of Mouse 2, and mormal mouse IgG were added and incubated at room temperature for 1 h. ^{125}I-rIFN-αA (1 μl) was then added and incubated at room temperature for 1 h. The wells were washed 4 times with PBS and cut out individually and ^{125}I-rIFN-αA counted in a Beckman r300 counter.

Competition with Interferon Binding to MDBK Cells by the Anti-Idiotypic Hybridoma Antibody

Figure 16.2 shows the binding of ^{125}I-rIFN-αA to receptor-containing MDBK cells. Binding assay was performed as previously described (19). The binding is specific and saturable at a ligand concentration of 200–400 pM. Scatchard analysis of the binding data from six experiments shows that the apparent dissociation constant (K_d) of interferon is 5.1 (±0.5) × 10^{-10} M and the number of binding sites per cell is 1056±95, which is similar to reported values obtained from experiments performed under similar conditions (20).

Receptor binding competition experiments were performed to determine whether the IgG in the 3B1 supernatant bound to the same cell surface receptor as did interferon. In these experiments, receptor-bearing MDBK cells were preincubated at 0°C for 1 h with the 3B1 supernatant before incubation with a saturating amount of ^{125}I-rIFN-αA (2.5×10^{-10} M) at 0°C

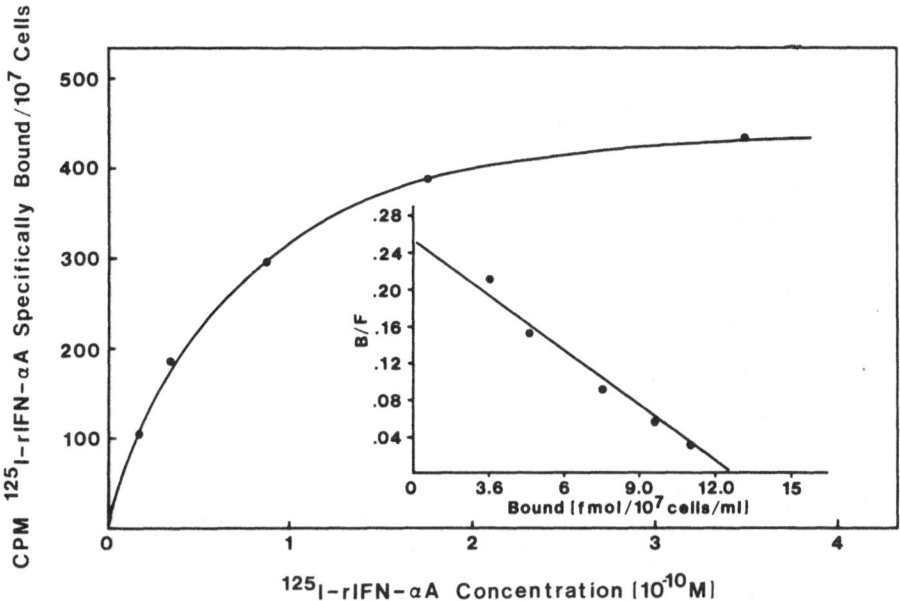

FIGURE 16.2. Specific binding of ^{125}I-rIFN-αA to MDBK cells. Inset: Scatchard analysis of the binding data (see text).

for 2 h. Table 16.4 shows that the 3B1 supernatant at 1:2 dilution causes a 75% inhibition of the binding of ^{125}I-rIFN-αA to MDBK cells, suggesting that the 3B1 antibody binds to the same cell surface receptor for interferon. Neither normal mouse IgG nor the anti-idiotypic antiserum of the mouse used in the hybridoma fusion causes any significant inhibition (10–14%) of ^{125}I-rIFN-αA binding, suggesting that the binding of 3B1 is specific. The other in vivo passaged 3-1B hybridoma supernatant, which does not show any antiviral activity, does not cause comparable level of inhibition (26%) of the interferon binding.

Summary and Conclusions

An anti-idiotypic hybridoma antibody 3-1B, directed against affinity-purified rabbit idiotypic antibodies to recombinant human leukocyte A interferon, was able to inhibit the neutralization of the antiviral activity of interferon by the Rb-Id. An in vivo passage of uncloned 3-1B cells yielded a subclone, 3B1, which exhibited interferonlike antiviral activity (2×10^5–5×10^5 units/ml supernatant) when assayed with both MDBK cells and WISH cells. Evidence suggested that this antiviral activity was not due to murine interferon and could be absorbed by polymer-bound goat anti-

TABLE 16.4. Competition with binding of ^{125}I-rIFN-αA to MDBK cells by 3B1 hybridoma supernatant.

Competing agent	^{125}I-rIFN-αA bound (cpm)	% Inhibition of binding
None	793	0
rIFN-αA (5×10^{-8} M)[b]	38	95
3B1 supernatant		
Diluted 1:2	198	75
Diluted 1:5	438	45
Diluted 1:10	490	38
Diluted 1:50	641	20
Diluted 1:200	680	14
Inactive 3-1B supernatant, undiluted	584	26
Anti-Rb-Id antiserum of Mouse 2, diluted 1:20	709	10
Normal mouse IgG, 1.5 µg	683	14

[a] 10^7 MDBK cells in 1-ml EMEM complete medium were preincubated at 0°C with various competing agents at the indicated concentrations or dilutions for 1 h. Controls received no competing agents. The mixtures were then added saturating concentration (2.5 × 10^{-10} M) of ^{125}I-rIFN-αA and incubated at 0°C for 2 h. The mixtures were washed and counted as described for the binding assay (see text).
[b] 200-fold excess of unlabeled rIFN-αA.

mouse immunoglobulin serum and by immobilized Rb-Id, and could be partially eluted from the latter at pH 2.5. This antibody was able to compete with ^{125}I-labeled interferon for binding to the Rb-Id. It could also compete with ^{125}I-labeled interferon for binding to interferon receptor-bearing MDBK cells.

Internal-image anti-idiotypic antibodies to biologically active proteins or polypeptides, e.g., the lymphokines, can be used as powerful reagents for cell surface receptor isolation and purification, receptor gene cloning, and structure-function studies of receptors. In addition, a monoclonal anti-idiotypic antibody whose variable regions resemble the receptor binding site or the "active site" of the biologically active protein is also important in the study of the active site structure of the protein. The Fab fragment of this monoclonal anti-idiotypic antibody can be crystallized for X-ray studies (21,22) as a probe for the active site structure. This internal image anti-idiotypic antibody can also be used to affinity isolate and purify the idiotypic antibodies which are specific to the receptor binding site or active site of the biologically active protein. These "active site-specific" idiotypic antibodies can then be used to test the binding of various molecules regarded as potential agonists or antagonists of the active protein.

Acknowledgments. I would like to thank T.-R. Chiang for her excellent technical assistance. I also thank D. Manousos for her participation in this study. Special thanks are due to J. Wardwell, L. Petervary, P. Familletti, and A. Gruarin for performing the antiviral assay. The secretarial assistance of L. Bowen is appreciated.

References

1. Strosberg, A.D., J.G. Guillet, S. Chamat, and J. Hoebeke. 1985. Recognition of physiological receptors by anti-idiotypic antibodies: molecular mimicry of the ligand or cross-reactivity? Curr. Top. Microbiol. Immunol. **119**:91.
2. Pestka, S. 1983. The human interferons—from protein purification and sequence to cloning and expression in bacteria: before, between, and beyond. Arch. Biochem. Biophys. **221**:1.
3. Aguet, M. and K.E. Mogensen. 1983. Interferon receptors, pp. 1–22. *In* I. Gresser (ed.), Interferon 1983, Vol. 5. Academic Press, New York.
4. Friedman, R.M. 1981. Interferons: a primer. Academic Press, New York.
5. Stewart, W.E., II. 1979. The interferon system. Springer-Verlag, New York.
6. Taniguchi, T., N. Mantei, M. Schwarzstein, S. Nagata, M. Muramatsu, and C. Weissmann. 1980. Human leukocyte and fibroblast interferons are structurally related. Nature **285**:547.
7. Goeddel, D.V., D.W. Leung, T.J. Dull, M. Gross, R.M. Lawn, R. McCandliss, P.H. Seeburg, A. Ullrich, E. Yelverton, and P. Gray. 1981. The structure of eight distinct cloned human leukocyte interferon cDNAs. Nature **294**:768.
8. Branca, A.A., and C. Baglioni. 1981. Evidence that type I and II interferons have different receptors. Nature (London) **294**:768.
9. Anderson, P., Y.K. Yip, and J. Vilcek. 1982. Specific binding of ^{125}I-human interferon-γ to high affinity receptors on human fibroblasts. J. Biol. Chem. **259**:11301.
10. Rehberg, E., B. Kelder, E. Hoal, and S. Pestka. 1982. Specific molecular activities of recombinant and hybrid leukocyte interferons. J. Biol. Chem. **257**:11497.
11. Aguet, M., M. Grobke, and P. Dreiding. 1984. Various human interferon α subclasses cross-react with common receptors: their binding affinities correlate with their specific biological activities. Virology **132**:211.
12. Uze, G., K.E. Mogensen, and M. Aguet. 1985. Receptor dynamics of closely related ligands: "fast" and "slow" interferons. EMBO J. **4**:65.
13. Staehelin, T., B. Durrer, J. Schmidt, B. Takacs, J. Stocker, V. Miggiano, C. Stahli, M. Rubinstein, W.P. Levy, R. Hershberg, and S. Pestka. 1981. Production of hybridomas secreting monoclonal antibodies to the human leukocyte interferons. Proc. Natl. Acad. Sci. USA **78**:1848.
14. Staehelin, T., D.S. Hobbs, H.-F. Kung, C.-Y. Lai, and S. Pestka. 1981. Purification and characterization of recombinant human leukocyte interferon (IFLrA) with monoclonal antibodies. J. Biol. Chem. **256**:9750.
15. Rubinstein, S., P.C. Familletti, and S. Pestka. 1981. Convenient assay for interferons. J. Virol. **37**:755.
16. Takemori, T., H. Tesch, M. Reth, and K. Rajewsky. 1982. The immune response against anti-idiotope antibodies. I. Induction of idiotope-bearing antibodies and analysis of the idiotope repertoire. Eur. J. Immunol. **12**:1040.
17. Stocker, J.W., H.K. Forster, V. Miggiano, C. Stahli, G. Staiger, B. Takacs, and T. Staehelin. 1982. Generation of 2 new mouse myeloma cell lines "PAI" and "PAI-O" for hybridoma production. Research Disclosure **1982**:155.

18. Zola, H., and D. Brooks. 1982. Techniques for the production and charac-terization of monoclonal hybridoma antibodies, pp. 1–57. *In* J.G. Hurrell (ed.), Monoclonal hybridoma antibodies: techniques and applications. CRC Press, Boca Raton, Florida.
19. Osheroff, P.L., T.-R. Chiang and D. Manousos. 1985. Interferon-like activity in an anti-interferon anti-idiotypic hybridoma antibody. J. Immunol. **135**:306.
20. Zoon, K., D.Z. Nedden, and H. Arnheiter. 1982. Specific binding of human α interferon to a high affinity cell surface binding site on bovine kidney cells. J. Biol. Chem. **257**:4695.
21. Gibson, A.L., J.N. Herron, D.W. Ballard, E.W. Voss, Jr., X.M. He, V.A. Patrick, and A.B. Edmundson. 1985. Crystallographic characterization of the Fab fragment of a monoclonal anti-ss-DNA antibody. Mol. Immunol. **22**:499.
22. Edmundson, A.B., and K.R. Ely. 1985. Binding of *N*-formylated chemotactic peptides in crystals of the Mcg light chain dimer: similarities with neutrophil receptors. Mol. Immunol. **22**:463.

17
Anti-Idiotypes, Retrovirus Receptors, and an Idiotypic Network in a Model of Retrovirus-Induced Thymic Leukemia

BLAIR ARDMAN AND SUSAN BURDETTE

Introduction

The concept that idiotypes of immunoglobulin variable regions can mimic epitopes on other kinds of molecules has been applied to the production of antibodies against several kinds of cell surface receptors (1). Starting with a ligand for a receptor, one can prepare an anti-ligand antibody and then an anti-anti-ligand antibody which mimics the ligand and binds to its cell surface receptor. This approach is particularly useful in obtaining antibodies to receptors which are poorly immunogenic, expressed in small amounts on cell surfaces, or cannot easily be copurified with their ligands. One class of receptors which meet these criteria and thus have defied identification and characterization are those for oncogenic retroviruses.

The understanding of how oncogenic retroviruses enter and thereby infect their target cells has lagged behind the rapid growth in information about the molecular biology of retrovirus-induced neoplasia. Oncogenic retroviruses can be divided into two categories: oncogene-containing and those containing no identifiable acute transforming sequences. Oncogene-containing retroviruses induce neoplasia in vivo and cellular transformation in vitro soon after infection; their transforming capability can be traced to the oncogene which they carry. For these viruses, cellular receptors may only be vehicles for virus entry and serve no other function relevant to the process of transformation.

Retroviruses which contain no known oncogenes can also induce malignancies, but the mechanism by which they do so is unknown. A well-studied example of transforming but oncogene-deficient retroviruses are the murine leukemia viruses (MuLV), some of which induce thymic or T-cell leukemias. A characteristic feature of the thymic leukemias caused by MuLV is that they occur only after a lengthy latent period from the time of infection. There is evidence to suggest that leukemia induction by MuLV is a result of integration of retroviral sequences with promoter or enhancer activity near oncogenes in the host-cell genome (2,3). However, unlike oncogene-containing retroviruses, MuLV cannot transform fibroblasts susceptible to infection nor can they infect their target cell of transformation, the murine thymocyte, in vitro. This distinction in transforming capabilities between those retroviruses which have oncogenes and those which do not, combined with the in vivo restriction of MuLV infection, suggests that other factors, in addition to oncogene activation, are required in MuLV-induced leukemogenesis.

Observations of the biology of MuLV-induced T-cell leukemia (which are discussed in this chapter) indicate that the MuLV receptor is an integral component in the pathogenesis of leukemic transformation. However, until the MuLV receptor can be identified and characterized, we can only speculate about its role in the leukemic process. In the present chapter, we will discuss two approaches we have used to obtain anti-idiotypic antibodies which bind to retrovirus receptors. The data we have obtained support the notion that anti-idiotypes to antiviral antibodies can be used to identify MuLV receptors. In addition, it appears that such anti-idiotypes are spontaneously produced in mice which are genetically programmed to develop thymic leukemia, creating an idiotypic network which connects antiviral antibodies and the target cells of leukemic transformation.

The Viral Ligand and Its Receptor

Murine leukemia viruses bind to their cellular receptor via a 70,000-dalton glycoprotein termed gp70 (4,5), the outermost structure on the retroviral envelope. MuLV which cause T-cell leukemias possess gp70s which are highly polymorphic because they arise from recombination between envelope gene sequences of endogenous viruses (6). The endogenous viruses arise from proviruses which are germline genes within the host's cellular DNA. Some of the endogenous viruses are replication competent and are generally nonleukemogenic (7). In contrast, their progeny, which contain recombinant gp70 molecules, can be leukemogenic (7,8).

MuLV which contain recombinant gp70 (rec-gp70) were first detected by their ability to infect both murine and nonmurine cells (9), a characteristic shared by none of their putative nonrecombinant parent viruses. This novel host range of the recombinant MuLV (i.e., viruses which contain rec-gp70) was believed to result from use of receptors different from those used by the nonrecombinant viruses. This has been confirmed through analysis of viral interference patterns of infection. Recombinant

MuLV are able to infect cells which are constitutively infected with non-recombinant MuLV and vice versa (10). However, cells infected with a recombinant retrovirus cannot usually be infected with a second recombinant retrovirus, and cells infected with a nonrecombinant retrovirus cannot be infected by another, similar virus (10,11). Such interference operates at the membrane receptor level.

Detailed examination of the gp70 molecule has suggested certain features which could determine the particular receptor to which a MuLV will bind. For example, one notable feature of the rec-gp70 molecules of leukemogenic MuLV is that no two are identical. Peptide mapping of different rec-gp70s (7,12) and oligoribonucleotide fingerprinting of the genes which encode them (6,7) have revealed their heterogeneity. This diversity among rec-gp70 molecules suggested not only that recombinant MuLV might bind to receptors different from those to which their nonrecombinant parents bind but also that each individual recombinant virus might bind to its own specific receptor. In the context of T-cell leukemias induced by recombinant MuLV, McGrath and Weissman (13) postulated that such specific viral receptors may be the same as those used by T cells for antigen recognition. This hypothesis formed the basis for our first approach to identify MuLV receptors on T cells, that is, to determine if anti-idiotypes to specific anti-rec-gp70 antibodies can bind to the T-cell antigen receptor.

An alternative hypothesis about the nature of rec-gp70 receptors on T cells is that such receptors, like those on fibroblasts, can bind several distinct rec-gp70 molecules by recognition of conserved regions among these envelope glycoproteins. Although each rec-gp70 molecule is unique when viewed in toto, peptides can be shared among different rec-gp70s (7,12). In addition, sequencing of rec-gp70 encoding regions has revealed two regions which are conserved among different recombinant virus isolates but are not found in related retroviruses with a different host range (J. Stoye, personal communication). Moreover, these two regions encode for amino acids which are likely to be located externally on the gp70 molecule. Thus, conserved elements among unique rec-gp70 molecules have been demonstrated at the biochemical and genetic level, and such elements, because of their predicted orientation, could participate in ligand-receptor interactions.

Attempts to characterize MuLV receptors biochemically have resulted in the description of several different proteins of various molecular weights (14–18). The usual technique for such experiments involved coprecipitation of receptors via a gp70 intermediate with an anti-gp70 antibody. As yet, none of the virus binding structures thus detected have been shown to facilitate retrovirus entry into susceptible cells. Thus, antibodies which bind directly to the receptor structures will be required to distinguish those receptors which actually mediate viral infection from cell surface molecules which merely bind virus.

Antibody$_1$ (Anti-Ligand) Selection: Criteria for Choice

Despite recent insights into the molecular organization of the rec-gp70 molecule, the receptor binding region of this large and complex ligand remains unknown. Prerequisite for the successful production of anti-receptor antibodies (e.g., those antibodies which contain idiotopes mimicking receptor binding epitopes of rec-gp70) is knowledge of whether unique or conserved regions (or both) of rec-gp70 participate in receptor binding. For example, if a region unique to a particular gp70 participates in receptor binding, then an anti-rec-gp70 antibody which binds to that unique region would be an appropriate immunogen with which to produce an anti-receptor anti-idiotype. Alternatively, if conserved regions of different rec-gp70 molecules participate in receptor binding, then an anti-rec-gp70 antibody which binds to several different rec-gp70s would be a logical starting point for anti-idiotype production.

Antibodies which neutralize virus in vitro may also prove to be good candidates from which to choose an Ab$_1$ because they necessarily bind to externally displayed epitopes on the gp70 molecule which, in turn, mediate receptor binding. This assumption has proved valid for the mammalian reovirus where receptor-binding anti-idiotypes were produced using a neutralizing, antiviral antibody as the Ab$_1$ immunogen [19,20]. Yet, anti-idiotypes to antibodies which neutralize Coxsackie virus B$_4$ were found not to bind to the Coxsackie virus receptor [21]. Thus, an antibody's ability to neutralize virus is not always sufficient to guarantee its utility as an immunogen for the production of anti-idiotypes which can bind to viral receptors.

The problem surrounding the selection of an appropriate Ab$_1$ for use in anti-receptor Ab$_2$ production is not restricted to the recombinant, leukemogenic MuLV. Polymorphic envelope glycoproteins are the rule in retroviruses which cause leukemia in birds and cats, as well as in the human immunodeficiency virus (HIV), the etiologic agent of AIDS [22]. Despite the polymorphism of HIV envelope glycoproteins, different isolates bind to the same cell surface antigen, CD4 [23,24]. This observation suggests that for HIV, conserved regions rather than unique ones are the likely participants in receptor binding.

Antigen Receptor Hypothesis: Virus Receptor is T-Cell Receptor for Antigen

When we began our experiments in 1982, we decided to approach the problem of obtaining an appropriate Ab$_1$ (an antibody which binds to a receptor-binding region of gp70) by testing a hypothesis concerning the nature of the putative retrovirus receptor on thymocytes. McGrath and Weissman [13] had proposed that the retroviral receptor on T cells is actually the T-cell receptor for antigen. In this model, each thymic leukemia would arise from a subset of target cells whose receptors bind specifically to the rec-gp70 of a particular leukemogenic retrovirus. As support for their hypothesis, McGrath and Weissman (25) showed that a retroviral-induced thymic leukemia binds preferentially to the retrovirus it produces. In addition, Zielinski et al. (26) found that different cloned, leukemogenic retroviruses will induce T-cell leukemias which express relatively predictable surface phenotypes, suggesting the presence of different subsets of target cells expressing unique receptors among the thymic T-cell compartment. Moreover, gp70 is capable of inducing T-cell proliferation (27) and lymphokine production (28). Taken together, these data suggested that the antigen receptor on T lymphocytes may also serve as a retroviral receptor. Thus, if a monoclonal antibody (Ab_1) could be obtained which bound selectively to a rec-gp70 (i.e., antigen) of one particular retrovirus, the paratope of such an antibody may resemble, to a degree, a T-cell antigen receptor. The corresponding anti-idiotype ($Ab_{2\beta}$) should then bind to that T-cell antigen receptor.

We began our experiments with a monoclonal antibody that recognized the rec-gp70 molecule of a cloned, polytropic leukemogenic virus termed P_1. This antibody (mAb 1416) was selected because it failed to react with two other retroviruses (P_2 and P_5) which have closely related but distinct rec-gp70 molecules (29). In addition, mAb 1416 did not react with non-recombinant gp70. Thus, mAb 1416 appeared to express a paratope resembling a receptor specific for the P_1 retrovirus. If the receptor for the P_1 retrovirus was also the T-cell receptor for antigen, such a receptor should be present on a tumor induced by the P_1 retrovirus and not present or detectable on an unselected population of normal thymocytes nor on T-cell leukemias induced by retroviruses different from P_1. Correspondingly, an anti-1416 antibody ($Ab_{2\beta}$) bearing an "internal image" of P_1 gp70 should bind only to cells expressing the putative clonotypic P_1 gp70 receptor.

We prepared an anti-idiotypic antiserum by immunizing a rabbit with mAb 1416. The purified anti-idiotype ($R\alpha1416$) from this serum reacts with the variable region of mAb 1416 because (i) prior incubation of 1416 with the P_1 retrovirus completely blocks the binding of the anti-idiotype to the anti-gp70 antibody (29); and, (ii) the P_1 virus can inhibit $R\alpha1416$ from binding to hybridoma cells which express mAb 1416 on their surface (unpublished data).

The $R\alpha1416$ antiserum was tested for binding to cells thought to express P_1 gp70 receptors (e.g., a P_1-induced thymic leukemia) and to cells on which no such receptors, according to the antigen receptor hypothesis, should be present or detectable (e.g., a P_2-induced thymic leukemia, normal thymocytes). We found that $R\alpha1416$ bound to both P_1- and P_2-induced thymic leukemias but not to normal thymocytes (Fig. 17.1, row A). mAb 1416 only bound to the P_1-induced thymic leukemia cells because they

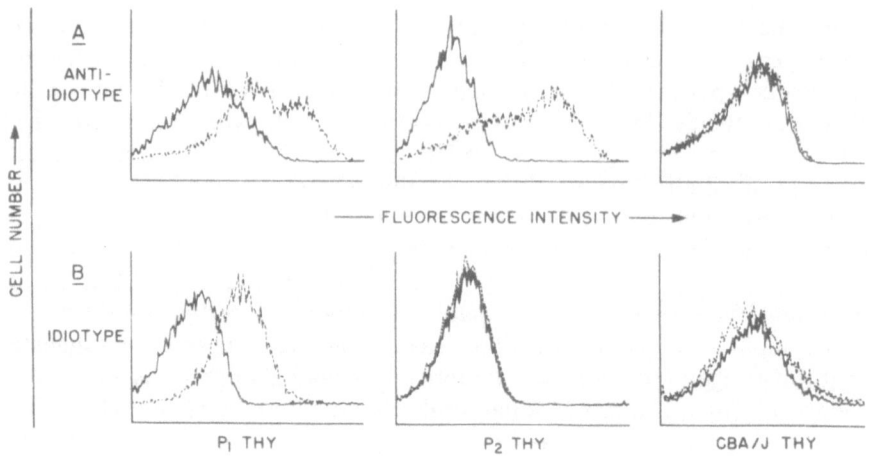

FIGURE 17.1. Fluorescence histograms of Rα1416 and mAb 1416 binding to leukemic and normal thymocytes. (A) Indirect immunofluorescence of cells stained with 25 µg/ml of affinity-purified Rα1416 (----) or affinity-purified control rabbit Ig (————). (B) Indirect immunofluorescence of cells stained with 50 µg/ml of mAb 1416 (----) or affinity-purified MOPC 195b (————). Reproduced from The Journal of Experimental Medicine, 1985, Volume 161: p. 676, by copyright permission of The Rockefeller University Press.

expressed P_1 gp70 on their surfaces (Fig. 17.1, row B). Further testing of Rα1416 revealed binding to 13 of 14 independently derived thymic leukemia cell lines, one Friend virus-induced erythroleukemia, and one Abelson virus-transformed pre-B-cell leukemia. Rα1416 did not bind to normal spleen or lymph node mononuclear cells. These results indicated that the Rα1416 serum did not identify a clonotypic determinant, (e.g., P_1-specific antigen receptor) but that it recognized a membrane structure which is expressed on murine leukemias irrespective of their cell lineage. In addition, if Rα1416 did indeed recognize the P_1 gp70 receptor, then such receptors appeared to be present on non-T cells, even though the P_1 virus does not normally induce non-T-cell leukemias.

It is probable that Rα1416 contains both internal images of P_1 gp70 ($Ab_{2\beta}$) and antibodies against framework idiotopes of 1416 ($Ab_{2\alpha}$). Because of this, the relative contribution of these Ab_2 subsets to cell binding could not be ascertained. It is also possible that the internal images of P_1 gp70 contained in the Rα1416 serum were only approximate, so that the specific features of the P_1 gp70 were imprecisely mimicked by the idiotopes contained in the polyclonal anti-idiotypic antiserum. Such imprecision would explain the binding of Rα1416 to leukemias which express receptors for genetically distinct retroviruses.

To determine if the leukemia-associated, idiotypic determinant detected by Rα1416 was a virus-binding structure, we tested the anti-idiotype for binding to cells known to express available receptors for retroviruses (uninfected, susceptible fibroblasts) and to fibroblasts whose receptors should be partially or completely blocked by a preexistent retroviral infection. For this purpose, the mouse fibroblast line, *Mus dunii*, was used because it possesses receptors for both recombinant and nonrecombinant retroviruses (30). Four cell lines were tested: (1) uninfected *M. dunii;* (2) P_1 (recombinant) virus-infected *M. dunii;* (3) P_2 (recombinant) virus-infected *M. dunii;* and (4) ecotropic (nonrecombinant) virus-infected *M. dunii*. In this assay, the uninfected and the ecotropic virus-infected cells were expected to show the greatest binding of the anti-idiotype because their receptors for rec-gp70 are unoccupied by this glycoprotein. Similarly, the recombinant virus-infected cells (P_1-*M. dunii;* P_2-*M. dunii*) were expected to show the least anti-idiotype binding, as their receptors are partially or completely occupied by rec-gp70.

The binding of Rα1416 to the fibroblasts was detected by indirect immunofluorescence and measured on a flow cytometer. Rα1416 bound less to the P_1- and P_2-infected cells than to the uninfected or ecotropic-virus-infected cells (Table 17.1). Although only a partial decrease in Rα1416 binding was observed, the decrease was specific for cells infected with recombinant virus. Several factors could account for the incomplete inhibition of anti-idiotype binding to the recombinant virus-infected cells. First, receptor excess on infected cells could make total inhibition impossible. Indeed, *M. dunii* cells have been shown not to follow the rules of viral interference. *M. dunii* cells infected with one recombinant virus can still be infected with another recombinant virus (11). Therefore, infected cells might express receptors which are not occupied by retroviral envelope glycoproteins. Second, it is also possible that the K_a of the antibody-receptor interaction is greater than that of the original ligand for its receptor. Support for such an idea has been provided by Colonno et

TABLE 17.1. Binding of Rα1416 anti-idiotype to *Mus dunii* cells.

Cell[a]	% Binding	
	Anti-idiotype[b]	Anti-gp70[c]
M dunii	100	0
Eco- *M. dunii*	100	>95
P_1- *M. dunii*	68	>95
P_2- *M. dunii*	65	>95
Eco- + P_1- *M. dunii*	52	>95
Eco- + P_2- *M. dunii*	47	>95

[a]Cells were uninfected or infected with one or two retroviruses.
[b]Results calculated with respect to anti-idiotype binding to uninfected *M. dunii* cells (designated 100%).
[c]Test performed with monoclonal antibodies to either recombinant (P_1, P_2) or nonrecombinant (Eco) gp70.

al. (31), who showed that anti-receptor antibodies can displace rhinovirus from its cellular receptor. A third and intriguing possibility for the observed partial inhibition is that recombinant virus-infected cells may shed a greater amount of rec-gp70 receptors from their surfaces than do uninfected or ecotropic-infected cells. In our laboratory, metabolic labeling experiments of receptor-bearing cells have revealed the presence of a protein in their culture supernatants which can be immunoprecipitated by Rα1416 (unpublished data). Although quantitative comparisons of receptor shedding among infected and uninfected cells have yet to be performed, other cell surface antigens are shed in retrovirus-infected cells, some of which are associated with the retrovirus envelope glycoproteins (32,33).

The pattern of Rα1416 binding suggests that it does identify a retrovirus receptor on *M. dunii* fibroblasts and on several independently derived murine leukemias. It also appears that such a receptor is unlikely to be related to the T-cell receptor for antigen. Additional support for this latter conclusion was recently provided by Owen et al. (34), who found that many of the AKR thymic leukemias to which Rα1416 bound do not produce T-cell receptor β-chain mRNA.

Conserved Region Hypothesis: Recombinant Retroviruses Share Receptor-Binding Elements

When evidence began to emerge that recombinant retrovirus receptors on murine leukemias may not be clonotypic structures, we hypothesized that T cells, like fibroblasts, express a common receptor for different rec-gp70 molecules. As mentioned above, fibroblasts which are infected with one recombinant retrovirus generally are resistant to superinfection by another recombinant retrovirus. In most cases, the resistance to superinfection operates at the cell membrane level and is believed to be the result of receptor blockade by gp70. Because different recombinant retroviruses bind to the same receptor on susceptible fibroblasts, this binding could be mediated via a region which is conserved among different rec-gp70 molecules. An Ab_1 recognizing such a conserved region and binding to several different rec-gp70s might be more likely to resemble the rec-gp70 receptor than an Ab_1 that binds to only one particular rec-gp70. An Ab_2 raised against the more broadly reactive Ab_1 could, in principle, resemble the receptor-binding region for all recombinant retroviruses.

A suitable mAb with which to test this hypothesis became available when Chesebro et al. (35) produced an anti-gp70 mAb which reacted with 27 of 28 recombinant retroviruses tested. This antibody (termed 514) did not react with ecotropic (nonrecombinant) retroviruses. We have found that mAb 514 neutralizes recombinant retroviruses in vitro in the absence of complement. These characteristics of mAb 514 indicated that it bound to a region of rec-gp70 which was shared by most recombinant retroviruses and suggested that this region, because of its accessibility to mAb 514 binding in vitro, could also participate in retrovirus binding to cells.

We hypothesized that anti-idiotypic antibodies to mAb 514 would be part of the immune response to leukemogenic retroviruses in AKR mice, a murine strain which has a high incidence of spontaneous thymic leukemia (36). When recombinant retroviruses appear in AKR mice during the pre-leukemic period (age 7–9 months), their novel viral envelope glycoproteins should stimulate the production of anti-recombinant gp70 antibodies. Jerne's network theory (37) would predict that AKR mice produce Ab_2 as part of their immune response to rec-gp70. Moreover, by mimicking rec-gp70, some species of Ab_2 should bind to cells which express receptors for rec-gp70. A search for spontaneous anti-anti-gp70 antibodies in AKR mice was initiated, using 514 as the idiotype with which to screen for the presence of anti-idiotypes.

In order to produce anti-idiotypes, hybridomas were made from spleen cells of unimmunized, preleukemic AKR mice. Hybridomas were screened for the presence of binding to mAb 514 by ELISA. Of over 500 clones tested from two fusions, six clones were chosen for further study on the basis of their specificity for mAb 514. All six antibodies were obtained from the same mouse and all were IgMs.

Idiotype specificity was tested by ELISA competition assay. Supernatants from the six anti-idiotypes were incubated with the idiotype mAb 514 or with control IgMs and were then tested for their ability to bind to 514. Of the six IgMs tested, only the idiotype itself, mAb 514, inhibited the binding of each of the six anti-idiotypes to the idiotype. The anti-idiotypes were then tested for binding to the antigen binding site, or paratope region, of mAb 514. ^{125}I-labeled mAb 514 (0.05 µg/ml) was incubated with each of the six anti-idiotypes and then tested for binding to recombinant viral antigen. Five of the six anti-idiotypes were capable of inhibiting the binding of mAb 514 to viral antigen (Table 17.2). Fifty percent inhibition

TABLE 17.2. Properties of monoclonal anti-idiotypes: Inhibition of the Id-virus interaction is related to leukemia cell binding affinity.

Anti-idiotype	Inhibitory concentration (µg/ml)[a]	Binding affinity[b] to:	
		A29 cells	A71 cells
359	<0.1	650	580
572	0.1	97	97
341	1.0	78	85
328	11.5	4	18
321	9.0	6	8
384	>200	<1	<1

[a]Amount of anti-idiotype producing 50% inhibition of the Id-virus interaction.
[b]Binding affinity is expressed as the percent fluorescent cells/concentration of antibody. Leukemia cell lines A29 and A71 were derived from two AKR mice.

was obtained at concentrations of anti-idiotype ranging from < 0.1 μg/ml (mAb 359) to 11.5 μg/ml (mAb 328). mAb 384 produced less than 10% inhibition of mAb 514 binding to viral antigen at a concentration of 200 μg/ml. Addition of excess unrelated IgM antibody (MOPC-104E, 100 μg/ml) did not alter the inhibition results, substantiating that the anti-idiotypes specifically bound to the variable region of mAb 514. Also, competition between the idiotype and anti-idiotype for binding to viral antigen was not responsible for the inhibition because preincubation of the virus-coated tubes with anti-idiotype did not prevent subsequent binding of mAb 514. Thus, five of the six anti-idiotypes appeared to bind to an antigen binding region of mAb 514 while the sixth (mAb 384) bound to a separate framework idiotope.

The six monoclonal anti-idiotypes were then assayed for their ability to bind to AKR leukemia cell lines, all of which should express retrovirus receptors. Five of the six anti-idiotypes bound to several different AKR leukemias, and their relative binding affinities were similar for each cell line tested. The binding affinities to two representative cell lines are shown in Table 17.2. The sixth anti-idiotype, mAb 384, showed no detectable binding to leukemia cells. This was the same anti-idiotype which appeared to bind to a framework region, rather than the paratope, of mAb 514. It should be noted that the anti-idiotypes with the greatest cell binding affinity were also the best inhibitors of mAb 514 binding to recombinant virus. This relationship between cell binding and the ability to inhibit the idiotype-virus interaction could denote $Ab_{2\beta}$ anti-idiotypes that most closely resemble a receptor-binding region of rec-gp70.

Because they showed the best binding to leukemia cells, mAb 359 and mAb 572 were tested by indirect fluorescence for binding to cells not transformed by recombinant MuLV. These anti-idiotypes showed no binding to a Moloney-virus-infected pre-B cell line or to normal lymphocytes obtained from thymus, spleen, lymph node, or bone marrow of a 1-month-old AKR mouse. Therefore, the binding of the monoclonal anti-idiotypes appeared to have specificity for cells transformed by a recombinant retrovirus.

Based on cell binding affinity and inhibition of the idiotype-virus interaction, mAb 359 appeared to be the best candidate to demonstrate molecular mimicry of rec-gp70. Therefore, mAb 359 was tested to determine if it could serve as a surrogate immunogen for virus and induce an anti-gp70 immune response. Six mice were immunized with 100 ng of unconjugated mAb 359 in an alum precipitate. A control group of mice was immunized with a monoclonal IgM, MOPC-104E, at the same concentration and in the same preparation. Mice were immunized twice (days 0 and 14) and the sera were tested 14 days after the second immunization (day 28). All six sera from the mice immunized with mAb 359 showed increased binding to recombinant viral antigen, whereas the control sera showed no increase in binding (Fig. 17.2). This pattern was present for three recom-

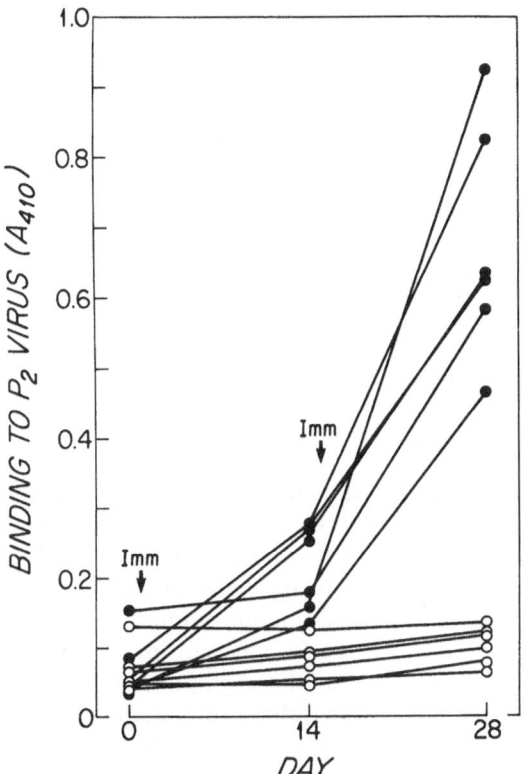

FIGURE 17.2. Induction of anti-retroviral antibodies by monoclonal anti-idiotype. C57B/6 mice were immunized with mAb 359 or MOPC-104E using the conditions stated in the text. Preimmune and immune sera from mAb 359 (•) and MOPC 104E (○) immunized mice were tested by ELISA for binding to sucrose density gradient-purified retrovirus (P_2). Results are the mean of triplicate tests.

binant viruses tested. The ability to induce an Ab_3 which can bind to viral antigen provided further evidence that mAb 359 contains a determinant that mimics rec-gp70.

Although all of the monoclonal anti-idiotypes described in this chapter are IgMs and are derived from one mouse, subsequent fusions of preleukemic AKR splenocytes have routinely yielded anti-(anti-gp70) antibodies of both IgG and IgM subclasses (unpublished data). Some of these anti-idiotypes can bind to leukemia cells. We have also been able to detect anti-anti-gp70 antibodies in sera from preleukemic AKR mice. After removal of rheumatoid factors from sera of 6-month-old (preleukemic) AKR mice, anti-idiotypic activity against mAb 514 can be demonstrated. These same sera also bind to a monoclonal anti-gp70 antibody derived from an unimmunized, preleukemic AKR mouse. Such findings indicate that the

TABLE 17.3. Preleukemic T-cell proliferation induced by anti-idiotypes.

Stimulator[a]	[^{3}H] thymidine incorporation (cpm)[b]
Concanavalin A	7368 ± 378
mAb 359	3036 ± 196
mAb 572	1960 ± 573
mAb 341	2368 ± 279
MOPC-104E	393 ± 35
Media	299 ± 24

[a]Fresh spleen and thymus cells from a preleukemic AKR mouse were incubated in microtiter wells with concanavalin A (2 μg/ml), monoclonal antibody (10 μg/ml), or media alone. MOPC-104E (IgM:λ) was used as an antibody control.
[b][^{3}H] thymidine incorporation was measured at 72 h. Results are expressed as the mean of triplicate measurements ± SD.

production of auto-anti-idiotypes to anti-gp70 antibodies is a natural phenomenon during the preleukemic period. Whether these anti-idiotypes represent part of the network response to rec-gp70 or their appearance precedes that of rec-gp70 remains to be determined. This latter possibility is intriguing because it implicates an autoimmune response to receptor as an early event in AKR leukemogenesis, setting the stage for the recombinant retrovirus to infect a stimulated population of target T cells. Preliminary experiments have shown that preleukemic AKR thymocytes and splenocytes will proliferate when incubated with selected $Ab_{2\beta}$ monoclonal anti-idiotypes (Table 17.3). Perhaps by promoting proliferation of the T cells which are the targets of retroviral infection, auto-anti-idiotypes may contribute to leukemogenesis.

Summary and Conclusions

We have utilized anti-idiotypes to anti-retroviral antibodies in order to identify cell surface receptors for recombinant MuLV and to study the role of such receptors in retrovirus-induced thymic leukemia. The results described in this chapter suggest that (i) anti-idiotypes to anti-rec-gp70 antibodies can mimic gp70 and bind to rec-gp70 receptors; (ii) rec-gp70 receptors are not clonotypic structures on T-cell leukemias and are, therefore, not T-cell receptors for antigen; (iii) rec-gp70 receptor expression by hematopoetic cells is leukemia-associated; and, (iiii) natural anti-idiotypes to anti-rec-gp70 antibodies exist in preleukemic mice and such anti-idiotypes may participate in leukemogenesis.

There are several examples of membrane structures which bind to viruses but which also have properties inherent to normal cell function. The reovirus receptor is also a β-adrenergic receptor (38); the Epstein-Barr virus receptor is also a receptor of a component of serum complement

(39); a human immunodeficiency virus (HIV) receptor is the T4 lymphocyte antigen (23,40); Semiliki Forest Virus receptors are also histocompatibility antigens (41). The conserved nature of the cell surface structure among hematopoetic cells which is bound by the Rα1416 serum and its association with the leukemic state suggests this structure performs some physiological function, perhaps in the regulation of cell growth. This speculation is strengthened by the demonstration that monoclonal anti-anti-gp70 antibodies can induce proliferation of preleukemic AKR T cells.

The development of retrovirus-induced leukemia is accompanied by the appearance of an idiotypic network which interconnects virus, antibody, and the target cells of transformation (Fig. 17.3). In mice responding to retrovirus infection, anti-idiotypes likely participate in the regulation of the anti-gp70 immune response. Since some of these anti-idiotypes are also anti-receptor antibodies, cells which express retrovirus receptors are connected to the idiotypic network. If proliferation of preleukemic T cells occurs in the mouse as it does in vitro, the anti-idiotypic antibodies could expand a T-cell population susceptible to virus infection, thereby increasing the probability of infection and subsequent malignant transformation. Once the final transformation event occurs, several roles for cell-binding anti-idiotypes can be envisioned. For example, they could continue to have a proliferative effect if the rec-gp70 receptor is actually a growth factor receptor which can be activated by antibody binding. Alternatively,

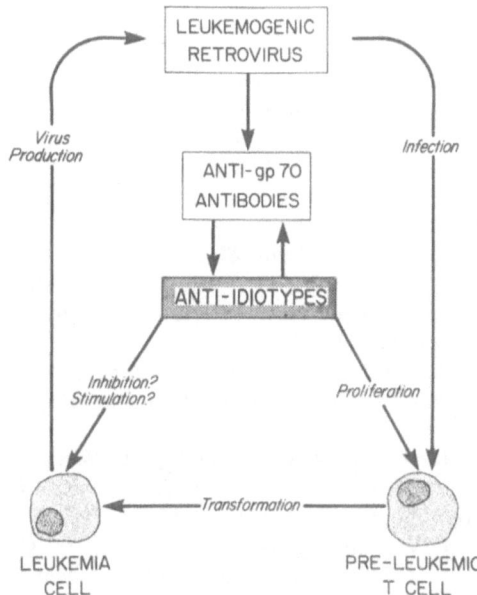

FIGURE 17.3. Model of an idiotypic network in retrovirus-induced, murine T-cell leukemia.

by regulating lymphocyte clones reactive with rec-gp70, auto-anti-idiotypes may create tolerance in the host for both recombinant retroviruses and transformed cells which express rec-gp70. This model implies a genetic predisposition for susceptibility to retrovirus-induced leukemia manifested as an imbalance in immunoregulation. Finally, by binding to leukemia cells which express available rec-gp70 receptors, auto-anti-idiotypes could inhibit, rather than promote, tumor growth. Such an effect also could result in the selection of tumor variants which express less retrovirus receptors or have receptors which are completely blocked by retrovirus, providing a growth advantage for cells which are receptor-negative or excessive virus producers.

Based on our present data, we believe that auto-anti-idiotypes are an integral component in the process of leukemic transformation induced by MuLV, and not merely epiphenoma of an immune response to virus. Current investigations into auto-anti-idiotype modulation of T-cell growth and proliferation combined with the study of their in vivo effects on leukemogenesis may shed light on the pathogenetic mechanisms triggered by leukemogenic retroviruses.

Acknowledgments. We thank Dr. Robert S. Schwartz for critical discussion and support for this work. We also thank Marie Wininger for her assistance in preparing this manuscript.

References

1. Strosberg, A.D., J.G. Guillet, S. Chamat, and J. Hoebeke. 1985. Recognition of physiological receptors by anti-idiotypic antibodies: molecular mimicry of the ligand or cross-reactivity? Curr. Top. Microbiol. Immunol. **119**:91.
2. Corcoran, L.M., J.M. Adams, A.R. Dunn, and S. Cory. 1984. Murine T lymphomas in which the cellular *myc* oncogene has been activated by retroviral insertion. Cell **37**:113.
3. Cuypers, H.T., G. Selten, W. Quint, M. Zijlstra, E.R. Maandag, W. Boelens, P. van Wezengeek, C. Melief, and A. Berns. 1984. Murine leukemia virus induced T-cell lymphomagenesis: integration of proviruses in a distinct chromosomal region. Cell **37**:141.
4. DeLarco, J., and G.J. Todaro. 1976. Membrane receptors for murine leukemia viruses: characterization using the purified viral envelope glycoprotein, gp71. Cell **8**:365.
5. Fowler, A.K., D.R. Twardzik, C.D. Reed, O.S. Weislow, and A. Hellman. 1977. Binding characteristics of Rauscher leukemia virus envelope glycoprotein gp71 to murine lymphoid cells. J. Virol. **24**:729.
6. Rommelaere, J., D.V. Faller, and N. Hopkins. 1978. Characterization and mapping of RNase T1-resistant oligonucleotides derived from the genomes of Akv and MCF murine leukemia viruses. Proc. Natl. Acad. Sci. USA **75**:495.
7. Green, N., H. Hiai, J.H. Elder, R.S. Schwartz, R.H. Khiroya, C.Y. Thomas,

P.N. Tsichlis, and J.M. Coffin. 1980. Expression of leukemogenic recombinant viruses associated with a recessive gene in HRS/J mice. J. Exp. Med. **152**:249.
8. Cloyd, M.W., J.W. Hartley, and W.P. Rowe. 1980. Lymphomagenicity of recombinant mink cell focus-inducing murine leukemia viruses. J. Exp. Med. **151**:542.
9. Hartley, J.W., N.K. Wolford, L.J. Old, and W.P. Rowe. 1977. A new class of murine leukemia virus associated with development of spontaneous lymphomas. Proc. Natl. Acad. Sci. USA **74**:789.
10. Rein, A. 1982. Interference grouping of murine leukemia viruses: a distinct receptor for the MCF-recombinant viruses in mouse cells. Virology **120**:251.
11. Chesebro, B., and K. Wehrly. 1985. Different murine cell lines manifest unique patterns of interference to superinfection by murine leukemia viruses. Virology **141**:119.
12. Elder, J.H., J.W. Gautsch, F.C. Jensen, R.A. Lerner, J.W. Hartley, and W.P. Rowe. 1977. Biochemical evidence that MCF murine leukemia viruses are envelope (env) gene recombinants. Proc. Natl. Acad. Sci. USA **74**:4676.
13. McGrath, M.S., and I.L. Weissman. 1978. A receptor-mediated model of viral leukemogenesis: hypothesis and experiments, p. 577. In Cold Spring Harbor Symposium on Normal Neoplastic Hematopoietic Cell Differentiation.
14. Twardzik, D.R., A.K. Fowler, O.S. Weislow, G.A. Hegamyer, and A. Hellman. 1979. Cell surface binding proteins for the major envelope glycoprotein of murine leukemia virus. Proc. Soc. Exp. Biol. Med. **162**:304.
15. Landen, B., and C.F. Fox. 1980. Isolation of BPgp70, a fibroblast receptor for the envelope antigen of Rauscher murine leukemia virus. Proc. Natl. Acad. Sci. USA **77**:4988.
16. Schaffar-Deshayes, L., J. Choppin, and J.-P. Levy. 1981. Lymphoid cell surface receptor for Moloney leukemia virus envelope glycoprotein gp71 II. Isolation of the receptor. J. Immunol. **126**:2352.
17. Robinson, P.J., G. Hunsmann, J. Schneider, and V. Schirrmacher. 1980. Possible cell surface receptor for Friend murine leukemia virus isolated with viral envelope glycoprotein complexes. J. Virol. **36**:291.
18. Johnson, R.A., and M.R. Rosner. 1986. Characterization of murine-specific leukemia virus receptor from L cells. J. Virol. **58**:900.
19. Nepom, J.T., H.L. Weiner, M.A. Dichter, M. Tardieu, D.R. Spriggs, C.F. Gramm, M.L. Powers, B.N. Fields, and M.I. Greene. 1982. Identification of a hemagglutinin-specific idiotype associated with reovirus recognition shared by lymphoid and neural cells. J. Exp. Med. **155**:155.
20. Noseworthy, J.H., B.N. Fields, M.A. Dichter, C. Sobotka, E. Pizer, L.L. Perry, J.T. Nepom, and M.I. Greene. 1983. Cell receptors for mammalian reovirus I. Syngeneic monoclonal anti-idiotypic antibody identifies a cell surface receptor for reovirus. J. Immunol. **131**:2533.
21. McClintock, P.R., B.S. Prabhakur, and A.L. Notkins. 1986. Anti-idiotypic antibodies to monoclonal antibodies that neutralize Coxsackievirus B₄ do not recognize viral receptors. Virology **150**:352.
22. Coffin, J.M. 1986. Genetic variation in AIDS viruses. Cell **46**:1.
23. Dalgleish, A.G., P.C.L. Beverley, P.R. Clapham, D.H. Crawford, M.F. Greaves, and R.A. Weiss. 1984. The CD4(T4) antigen is an essential component of the receptor for the AIDS retrovirus. Nature (London) **312**:763.

24. Sattentau, Q.J., A.G. Dalgleish, R.A. Weiss, and R.C.L. Beverley. 1986. Epitopes of the CD4 antigen and HIV infection. Science 234:1120.
25. McGrath, M.S., and I.L. Weissman. 1979. AKR leukemogenesis: identification and biological significance of thymic lymphoma receptors for AKR retroviruses. Cell 17:65.
26. Zielinski, C.C., S.D. Waksal, L.D. Tempelis, R.H. Khiroya, and R.S. Schwartz. 1980. Surface phenotypes in T-cell leukemia are determined by oncogenic retroviruses. Nature (London) 288:489.
27. Enjuanes, L., J.C. Lee, and J.N. Ihle. 1979. Antigenic specificities of the cellular immune response of C57BL/6 mice to the Moloney leukemia/sarcoma virus complex. J. Immunol. 122:665.
28. Ihle, J.N., J.C. Lee, and L. Rebar. 1981. T cell recognition of Moloney leukemia virus proteins. III. T cell proliferative responses against gp70 are associated with the production of a lymphokine inducing 20α-hydroxysteroid dehydrogenase in splenic lymphocytes. J. Immunol. 127:2565.
29. Ardman, B., R.H. Khiroya, and R.S. Schwartz. 1985. Recognition of a leukemia-related antigen by an anti-idiotypic antiserum to an anti-gp70 monoclonal antibody. J. Exp. Med. 161:669.
30. Chattopadhyay, S.K., M.R. Lander, S. Gupta, E. Rands, and D.R. Lowy. 1981. Origin of mink cytopathic focus-forming (MCF) viruses: comparison with ecotropic and xenotropic murine leukemia virus genomes. Virology 113:465.
31. Colonno, R.J., P.L. Callahan, and W.J. Long. 1986. Isolation of a monoclonal antibody that blocks attachment of the major group of human rhinoviruses. J. Virol. 57:7.
32. Bubbers, J.E., and F. Lilly. 1977. Selective incorporation of H-2 antigenic determinants into Friend virus particles. Nature (London) 266:458.
33. Scheinberg, D.A., and M. Strand. 1981. 55,000-Dalton, retrovirus-associated, cell membrane glycoprotein: purification and quantitative measurements of expression in viruses, cells, and tissues. Mol. Cell. Biol. 1:144.
34. Owen, F.L., W.M. Strauss, C. Murre, A.D. Duby, H. Hiai, and J.G. Seidman. 1986. AKR murine thymic leukemias are from a distinct thymic cell lineage and do not express the β-chain of the T cell antigen receptor. Proc. Natl. Acad. Sci. USA 83:7434.
35. Chesebro, B., W. Britt, L. Evans, K. Wehrly, J. Nishio, and M. Cloyd. 1983. Characterization of monoclonal antibodies of strains of Friend MCF and Friend ecotropic murine leukemia virus. Virology 127:134.
36. Furth, J., H.R. Seibold, and R.R. Rathbone. 1933. Experimental studies on lymphomatosis of mice. Am. J. Cancer 19:521.
37. Jerne, N.K. 1974. Towards a network theory of the immune system. Ann. Immunol. (Inst. Pasteur) 125C:373.
38. Co, M.S., G.N. Gaulton, A. Tominaga, C.J. Homcy, B.N. Fields, and M.I. Green. 1985. Structural similarities between the mammalian beta-adrenergic and reovirus type 3 receptors. Proc. Natl. Acad. Sci. USA 82:5315.
39. Nemerow, R.R., R. Wolfert, M.E. McNaughton, and N.R. Cooper. 1985. Identification and characterization of the Epstein-Barr virus receptor on human B lymphocytes and its relationship to the C3d complement receptor (CR2). J. Virol. 55:347.
40. Klatzmann, D., E. Champagne, S. Chamaret, J. Gruest, D. Guetard, T. Her-

cend, J.-C. Glukman, and L. Montagnier. 1984. T-lymphocyte T4 molecule behaves as the receptor for human retrovirus LAV. Nature (London) **312**:767.

41. Helenius, A., B. Morein, E. Fries, K. Simons, P. Robinson, V. Schirrmacher, C. Terhorst, and J.L. Strominger. 1978. Human (HLA-A and HLA-B) and murine (H-2K and H-2D) histocompatibility antigens are cell surface receptors for Semliki Forest virus. Proc. Natl. Acad. Sci. USA **75**:3846.

18
An Idiotype Approach for a Vaccine Against Hepatitis B Surface Antigen

YASMIN THANAVALA

Background and Some Theoretical Considerations

The current interest in the molecular mimicry of viral, bacterial, parasite, and tumor antigens using anti-idiotypic antibodies is based on the concept set forth by Niels Jerne (1) over a decade ago as the network theory of the immune system. He envisaged the immune system as a complex network of cells held together via complementary recognition mediated by structures (idiotopes) encoded by the variable-region genes of the antibody molecule. The network would thus consist of idiotype-anti-idiotype (Id-anti-Id) interactions between lymphocytes, with all lymphocytes being in a state of dynamic equilibrium as a result of these interactions. External antigen, by perturbing this equilibrium, would provoke an immune response.

One of the predictions of Jerne's theory is that for every exogenous antigen there is an "internal image" counterpart within the immune system. Work from several laboratories has verified the existence of such internal image antibodies. Such internal image anti-idiotypic antibodies can mimic the original external antigen and can therefore function as surrogate antigens. They thus have the potential of being used in vaccine programs (2,3) and may also play a role in the etiology and regulation of autoimmune diseases. It is their potential role in vaccine development that we are concerned with in this chapter.

Conceptual changes in vaccine development have been occurring for several years. While the effectiveness of currently available vaccines is not the issue here, the small but existing risk factors of certain essential vaccines are becoming unacceptable and have in the one instance of vaccination against pertussis, even resulted in a decline in the vaccination rate. There is a real need for safer vaccines in situations where large segments of the population require vaccination.

The network hypothesis offers an elegant way for developing vaccines which are not based on the conventional approach of using nominal antigenic material. The strategy for developing anti-idiotypic vaccines takes advantage of the fact that the repertoire of external or nominal antigens is mimicked by idiotypic structures on immunoglobulin and possibly also on receptors and products of T cells.

There are several circumstances in which the use of conventional vaccines may present certain problems, and in these cases internal image anti-id vaccine may prove advantageous.

1. Difficulty in obtaining adequate amounts of antigen has been one of the key factors in our present failure to produce appropriate vaccines for a wide range of diseases such as malaria, trypanosomiasis, leprosy, and leishmaniasis. In certain cases like hepatitis B, it is not even possible to culture the virus in vitro. In all these situations, monoclonal anti-id which would serve as surrogate antigen could be easily produced on a large scale.
2. Products obtained by gene cloning may not be of value for use as vaccines if they require glycosylation or the presence of lipid or a nucleic acid core to attain the configuration of the native antigen. Similarly, a synthetic peptide may not adopt the three-dimensional structure of the original antigen and may therefore prove to be less immunogenic. Internal image anti-id molecules would have the correct conformation and may thus prove to be good antigens.
3. The inherent hazards associated with the use of putatively killed vaccines and the risk of attenuated strains reverting to the virulent form constitute two of the main disadvantages of conventional vaccine programs. Anti-id vaccines would not be infectious agents, and their use would circumvent the dangers associated with both killed and attenuated microorganisms. Internal image anti-ids may also offer tremendous promise in providing immunity to toxins. The native toxin cannot be used due to its inherent toxicity, and the denatured molecule could lose the antigenicity necessary to produce protective neutralizing antibodies.
4. Sometimes determinants of a microbial antigen not needed for immune protection or complexes of the organism with body components may provoke autoantibodies that could be damaging. Thus, vaccines directed to the entire organism might also produce reactivity against normal body components. It would therefore be an advantage to immunize only against selected protective antigenic determinants. Monoclonal anti-id vaccines, by mimicking individual epitopes, would provide a means of immunizing against single antigenic determinants.
5. A major advantage of the anti-id approach to vaccination would be that it would allow the presentation of, say, carbohydrate antigen epitopes as mimicked by the structure and conformation of the protein surrogate (i.e., anti-id molecule).

It may also be possible that the presentation in a different molecular context may break tolerance that may exist to some antigens and allow the expression of otherwise silent clones.

Anti-id vaccines will not be useful against organisms like influenza virus that have epitopes that show antigenic drift.

Hepatitis B Virus Infection—Magnitude of the Problem

With over 200 million people chronically infected, hepatitis B is a public health problem of worldwide importance. In the Far East and Africa, 10% of the population are chronic carriers (4) of hepatitis B virus (HBV), and for these people two of the major causes of mortality are chronic active hepatitis and liver cirrhosis.

The immune response to HBV infection is variable, with the great majority of people clearing the hepatitis B surface antigen (HBsAg) and developing antibodies to HBsAg. However, a small percentage do not resolve the infection. Viral synthesis persists and the individual becomes a chronic carrier. Epidemiological studies have shown the significance of HBV in the incidence of hepatocellular carcinoma (5,6), thus making this one of the few viruses known to be associated with a human cancer.

When infection with HBV occurs, very large quantities of viral particles can be detected in the blood. The hepatitis B virus is 42 nm in diameter and consists of an envelope and a nucleocapsid. During chronic infection, the serum contains not only complete virus particles but also a vast excess of empty viral envelope (7,8). The viral envelope exists in the form of 22-nm-diameter spherical or filamentous particles and carries the hepatitis B surface antigen. Serologically, HBsAg has a group-specific a determinant and two sets of mutually exclusive determinants d/y and w/r giving four major serotypes (9,10), adw, ayw, adr, and ayr, which, however, have an unequal distribution worldwide (11), the y and r determinants being absent in the Far East and Africa, respectively. Recovery from natural infection with hepatitis B virus is associated with antibodies directed to the common group specific a determinant (12,13).

The worldwide problem of HBV infection and its association with chronic liver disease and hepatocellular carcinoma have necessitated the development of a safe and effective vaccine. The hepatitis B virus cannot be propagated in tissue culture or in animals other than the chimpanzee, and the absence of naturally occurring attenuated strains has prevented the development of a conventional live vaccine. Thus, alternative strategies for vaccine production have become necessary. The currently licensed vaccine consists of purified HBsAg particles obtained from the plasma of HBV chronic carriers (14). The vaccine has been shown to be both safe and effective in high-risk adult populations (15–17) and in newborn infants (18–20), but about 10% of the recipients of this vaccine do not make detectable antibodies to HBsAg (21–23). They therefore constitute a nonresponder population perhaps comparable to the nonresponder strains observed in murine studies and perhaps, even more significantly, comparable to the group of humans who do not mount a response to HBsAg upon natural infection with HBV and progress to becoming chronic carriers of the virus.

It is widely recognized that an urgent need still exists for the development of alternative second-generation vaccines. This is because of the limited availability of human plasma from suitable hepatitis B carriers, with the attendant problems of elaborate purification procedures (24) which make such vaccines beyond the financial reaches of the health care programs of many countries where there is a need for mass vaccination (6). There is also the inherent problem of the biological safety in the use of plasma from individuals who may be at high risk of exposure to other unrelated or unrecognized agents. One other problem in the use of the vaccine is in immunocompromised patients, e.g., those on maintenance hemodialysis, in whom seroconversion rates are often observed to be less than 70% (25).

The greatest challenge for a successful hepatitis B vaccine will be the reduction in the number of chronic carriers, especially the children born to hepatitis B carrier mothers living in highly endemic regions. The carrier mothers have a very high load of virus replication, and transmission to the baby occurs during birth. These babies run a greater than 80% risk of acquiring chronic hepatitis B. Administration of immunoglobulin containing high levels of anti-HBsAg merely delays the onset of the disease.

Among the viable alternative approaches to the licensed vaccine are (i) the preparation of polypeptides in water-soluble micelle form (26) from HBsAg particles, (ii) the development of recombinant DNA vaccines with expression either in bacteria or yeast (27), (iii) the production of HBsAg particles in mammalian cells like the Chinese hamster ovary (28), (iv) the construction of a live vaccine using as a vector a virus capable of replicating in humans, e.g., vaccinia virus (29), and (v) the use of synthetic peptides representing the amino acid sequence corresponding to the primary gene product of the surface antigen gene (30–32). The above all constitute variations on conventional vaccines. However, the approach in which we, as well as a number of other groups, are interested is the use of internal image anti-idiotype vaccine which can act as surrogate antigens (33–39). This strategy may be regarded as a nonconventional vaccine since the anti-id is not derived using material from the pathogen in question (e.g., killed or attenuated pathogen) nor is it based on structural information about the pathogen, as are synthetic peptide vaccines.

Experimental Approach and Results

Production of Monoclonal Anti-Idiotypes

For the production of monoclonal anti-idiotypes, we used a syngeneic monoclonal idiotype which was raised to a conserved epitope of hepatitis

B surface antigen. Eight-week-old BALB/c mice received three intraperitoneal injections, two weeks apart, of 50 μg of purified monoclonal idiotype (H3F5). The first injection was given in complete Freund adjuvant, the next in incomplete adjuvant, and the third in saline. All the animals responded by making a good serum anti-idiotypic response, as measured by a solid-phase radioimmunoassay, and spleen cells from one animal were fused with a nonsecreting myeloma cell line JK Ag8 6.5.3 in a ratio of 4:1 using 35% polyethylene glycol (PEG 4000GK, Merck). Immediately after fusion the cells were plated out in 96-well microtiter plates on a feeder layer of BALB/c macrophages. The cells were maintained in a humidified incubator in an atmosphere of 6% CO_2 in air at 37°C.

Detection of Hybridomas Producing Monoclonal Anti-Idiotypic Antibodies

Hybrid cells were observed in 160/400 wells, and when the spent supernatants were tested in a solid-phase radioimmunoassay, six hybrids binding selectively to the immunizing idiotype and not with pooled normal mouse immunoglobulin (39) were detected (Table 18.1).

Do the Anti-Idiotypic Antibodies Recognize Paratopic or Nonparatopic Idiotopes?

Four of the monoclonal anti-idiotypic antibodies were purified from ascitic fluid and studied in a solid-phase inhibition assay designed to test whether

TABLE 18.1. Solid-phase RIA for the detection of hybridomas producing monoclonal anti-idiotypes to H3F5 (anti-HBs) idiotype.[a]

Clone E_{max}	Binding (cpm) to plates coated with:	
	Idiotype	Normal mouse Ig
2E7	7,225	123
2F10	8,569	150
3G9	1,625	80
3H1	4,532	132
4D4	6,077	83
4D5	3,045	76
Controls		
Tris-BSA	146	82
Anti-Id serum 1:100	7,304	235

[a] 96-well polyvinyl plates were coated with 5 μg/ml of idiotype/normal mouse Ig at room temperature for two days. Excess sites were blocked with 0.02 M Tris containing 0.5% BSA. Test supernatants (100 μl) were incubated on the coated plates overnight at room temperature. Total counts of ^{125}I-labeled idiotype (H3F5) added to each assay well were 20,000 cpm. The anti-idiotype serum used as a positive control was from the same animal whose spleen was used for producing the hybrids.

they were directed to paratopic (antigen binding site related) or nonparatopic epitopes (non-antigen binding site). As can be seen in Fig. 18.1, all the anti-idiotypes were able to inhibit the binding of the hepatitis B surface antigen to the idiotype-coated solid phase, suggesting that they recognize idiotypic determinants very close to, if not identical with, the paratopic regions of the combining site.

Detection of Internal Image Anti-Idiotypic Antibodies (Ab2β)

If the anti-idiotype is behaving as an internal image, then, like antigen, it should bind polyclonal anti-HBs sera raised in different species, in which, it is likely that idiotypes unrelated to antigen binding would be identical. Thus, cells from the six anti-idiotypic hybridomas were tested for the

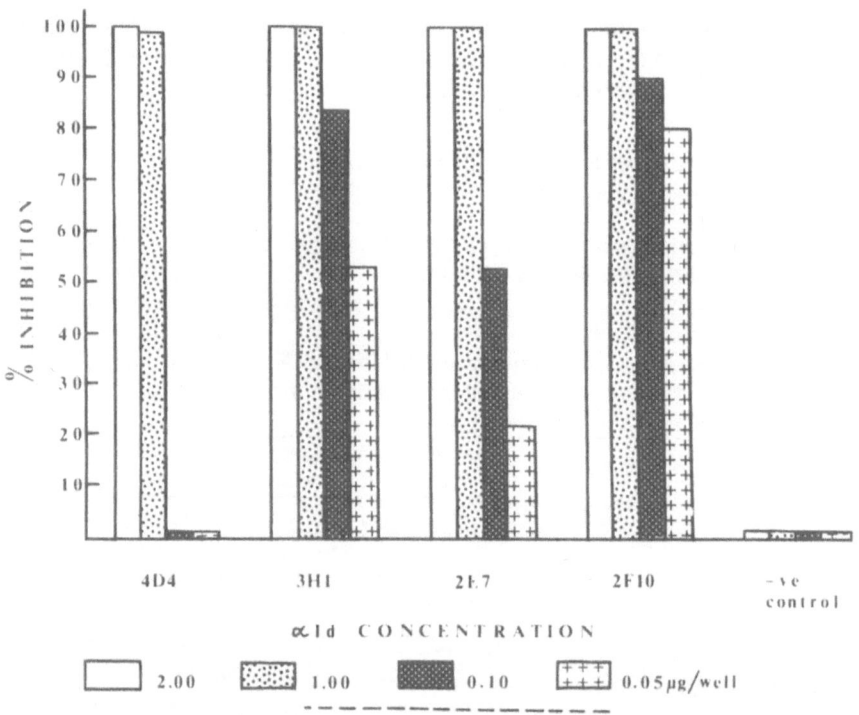

FIGURE 18.1. Inhibition of the binding of radiolabelled HBsAg to solid phase monoclonal idiotype in the presence of different concentrations of monoclonal anti-idiotypes, to determine if the anti-idiotypes are directed to paratopic or nonparatopic idiotypes. Polyvinyl plates were coated with the immunizing idiotype (H3F5) at 5 μg/ml. The negative control was an irrelevant monoclonal Ig lacking reactivity with the idiotype. The reaction was also inhibited by serum containing HBsAg: a 1:10 dilution produced 49.7 percent inhibition, 1:25, 35.8 percent and 1:50, 25.1 percent inhibition.

presence of internal image using an indirect immunofluorescence technique (40). Cytocentrifuge smears of the anti-id-producing clones were incubated with either preimmune sera or anti-HBs containing sera from one goat, two rabbits, two piglets, and five humans.

Of the six anti-idiotypic hybridomas studied, two (4D4 and 2F10) stained strikingly with all the polyclonal anti-HBs sera tested and with sera from five human subjects immunized with HBsAg vaccine (Merck Sharp & Dohme) (Table 18.2, Fig. 18.2). The technique was controlled by the lack of staining of 4 hybridomas also producing IgG1 antibodies to the same idiotype and by 10 randomly selected IgG1-secreting hybridomas not showing reactivity with the immunizing idiotype. Furthermore, no staining was seen with control sera lacking anti-HBs antibodies.

Specificity of staining was further confirmed in experiments where aliquots of the goat anti-HBs serum were passed over an immunoabsorbent of Sepharose-HBsAg or Sepharose-ovalbumin. There was good correlation between the presence of antibodies and the degree of immunofluorescent staining (Table 18.3). The staining that was seen with sera from five human subjects who had been immunized with the Merck Sharp & Dohme HBsAg vaccine, where their preimmunization samples were negative, also supported the case for specificity of the immunofluorescence technique.

We draw particular attention to the failure of the two swine antisera to stain the two internal image hybridomas after selective absorption of antibodies to the *a* but not the *d* or *y* determinants, showing the relationship of internal image reactivity to a restricted part of the hepatitis antigen, as would be predicted by the hypothesis, since the anti-idiotype should only mirror one of the epitopes on the original antigen.

We have also investigated the immunochemical characteristics of two types of surrogate HBsAg antigenic epitopes consisting of (i) synthetic peptides representing amino acid residues 139–147, a hydrophilic region corresponding to a part of the *a* determinant of the surface antigen and (ii) our panel of monoclonal anti-idiotypes raised against a monoclonal anti-HBs antibody, two of which behave as an internal image of an "a" determinant. We were able to demonstrate that these surrogate antigens show concordance in that the internal image anti-idiotypes inhibit the binding of both monoclonal and polyclonal anti-HBs antisera to the cyclical and linear peptides (41).

The four monoclonal anti-idiotypic antibodies (2F10, 4D4, 2E7, and 3H1) were used in assays to assess their ability to inhibit the binding of ^{125}I-labeled cyclical peptide 139-147 by a panel of mouse monoclonal anti-HBs antibodies and a range of polyclonal anti-HBs antisera from a variety of species (Fig. 18.3). The binding of pooled human anti-HBs Ig and a monoclonal anti-HBs to the cyclical form of a peptide representing amino acids 124–137 of the *y* subtype of HBsAg could not be inhibited by the anti-ids, thus providing a good negative control (see Fig. 18.4B).

The results expressed in Fig. 18.4 represent the inhibition of binding of cyclical peptide 139-147 obtained with 5 μg of the purified anti-ids.

TABLE 18.2. Immunofluorescent staining of anti-idiotype hybridoma cells for HBsAg internal image with anti-HBs sera from different species.[a]

		Staining with sera from:											
		Human		Goat		Rabbit		Swine anti-HBs		Absorbed swine anti-HBs	
Hybridoma clone	No. tested	Anti-HBs	Preimmune	Anti-HBs	Normal	Anti-HBs	Normal	Anti-ay	Anti-ad	Anti-y	Anti-d
Internal image anti-Id	2	+	-	+++	-	++	-	+	+	-	-
Non-internal image anti-Id	4	-	-	-	-	-	-	-	-	-	-
Unrelated hybrids	10	-	-	-	-	-	-	-	-	-	-

[a]The hybridomas were stained as described in the legend to Fig. 18.2. Anti-idiotype clones 4D4 and 2F10 behaved as internal images. The two pig anti-HBs sera raised against ay or ad failed to stain after absorption of antibodies to the HBsAg a determinant. Anti-id clones 3H1, 2E7, 4D5, and 3G9 were negative for internal image. Clones producing antibodies to unrelated anti-HBs idiotypes were unstained. Slides were coded before reading.

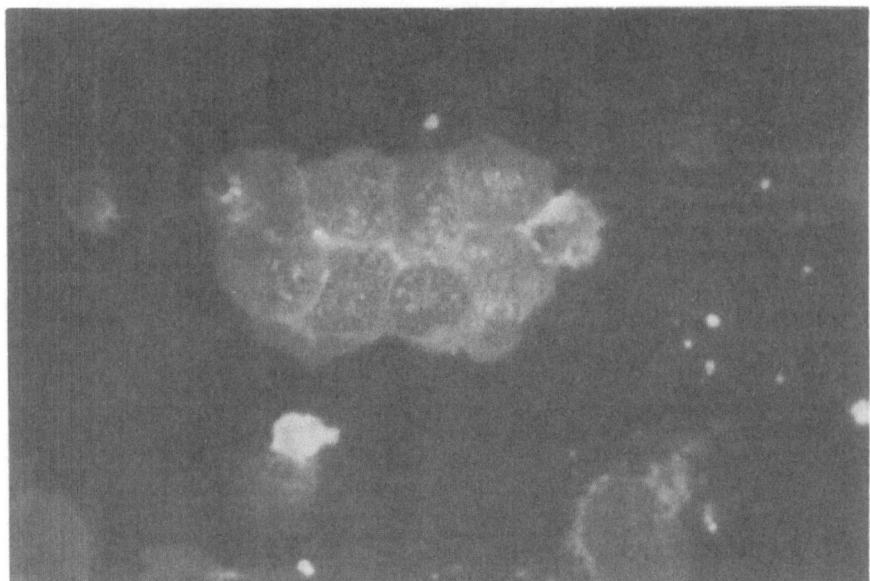

FIGURE 18.2 (Top) Lack of staining of anti-idiotype producing hybridoma cells with a hyperimmune goat anti-HBsAg serum.

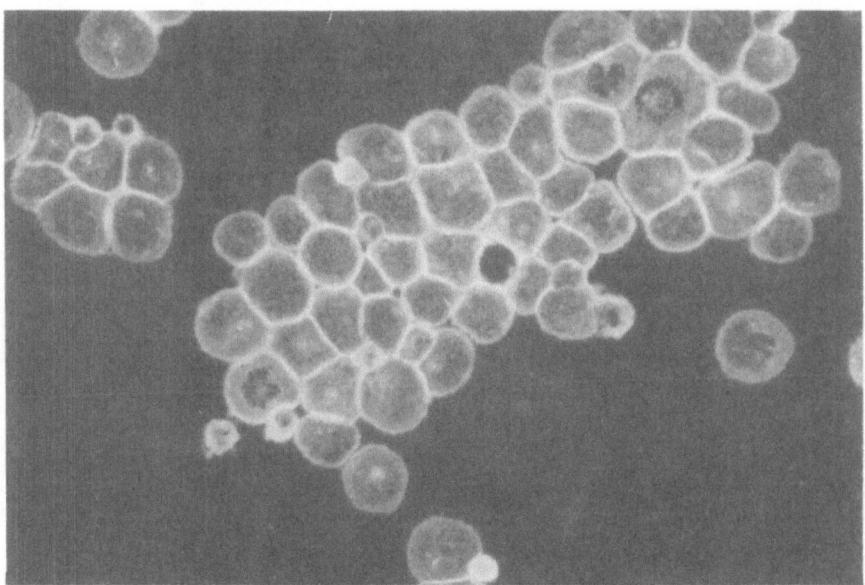

FIGURE 18.2. (Bottom) Strong positive staining of 'internal image' anti-idiotype producing hybridoma cells with a hyperimmune goat anti-HBsAg serum.

TABLE 18.3. Specificity of staining of the internal image hybridoma cells.[a]

Samples tested	Dilution relative to starting serum sample	Antibody activity (Iµ/ml)	Immunofluorescent staining
Unabsorbed goat anti-HBs	1:20	117	+ + +
	1:40	59	+ + +
Sepharose-HBsAg column			
Unbound fraction	1:20	0.10	+
	1:40	0.05	+ / −
	1:80	0.03	−
Eluted fraction	1:15	48	+ + +
	1:30	24	+ +
	1:60	12	+
Sepharose-ovalbumin column			
Unbound fraction	1:15	99	+ + +
	1:30	50	+ + +
	1:60	25	+ +
	1:120	12.5	+
Eluted fraction	1:15	0.02	−

[a]The unbound fractions from each column came through in 0.02 M Tris, pH 8.0, and the columns were then eluted with 4 M potassium thiocyanate. Antibody activity in the various fractions was tested using a two-site immunoradiometric assay which used solid-phase HBsAg and ^{125}I-labeled HBsAg as a second reagent.

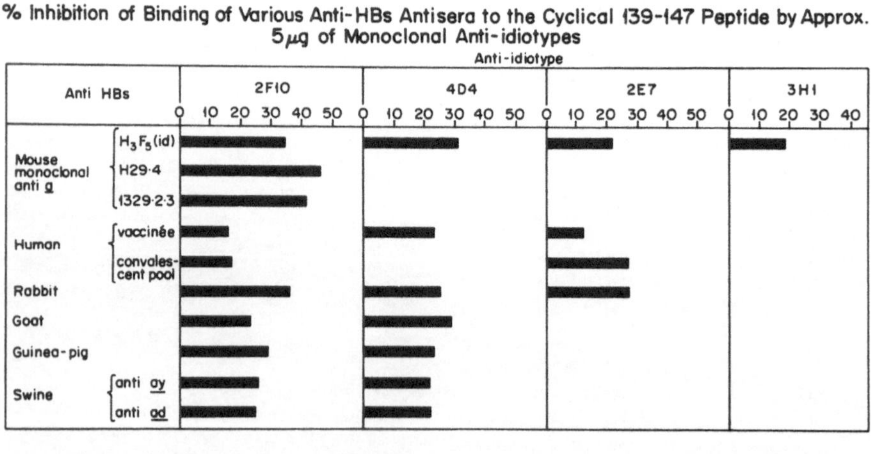

% Inhibition of Binding of Various Anti-HBs Antisera to the Cyclical 139-147 Peptide by Approx. 5 µg of Monoclonal Anti-idiotypes

FIGURE 18.3. Inhibition of binding of a panel of monoclonal and polyclonal anti-HBs antisera to the cyclical peptide 139–147 by approximately 5 µg of monoclonal antiidiotypes.

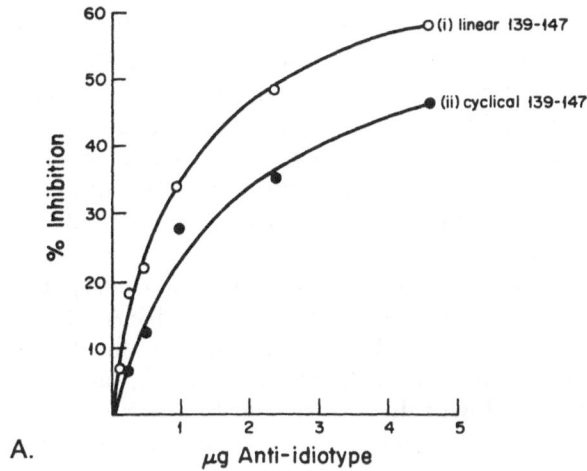

Inhibition of Binding of Monoclonal Anti-HBs
to HBs Peptides by Monoclonal Internal Image
Anti-Idiotype 2FIO

A.

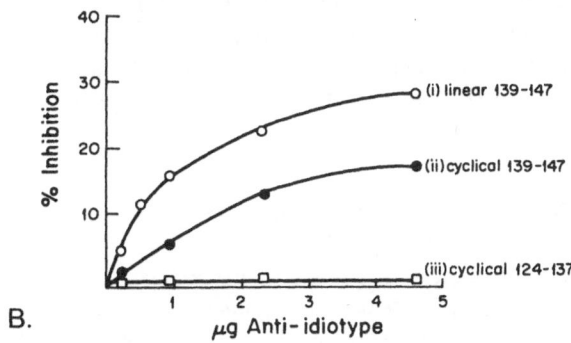

Inhibition of Binding of Pooled Human Anti-HBs
to Peptides by Monoclonal Internal Image Anti-
Idiotype 2FIO

B.

FIGURE 18.4. Inhibition of binding of (A) monoclonal anti-HBs H29.4 and (B) polyclonal human anti-HBs antibody to (i) the linear and (ii) the cyclical peptide 139–147 by the monoclonal internal image antiidiotype 2F10.

Curves for the inhibition of binding by a range of internal image anti-id (2F10) concentrations of the monoclonal anti-HBs and pooled human anti-HBs antibodies to both the linear and cyclical forms of peptide 139-147 are shown in Fig. 18.5.

It should be noted that the binding of the linear 139-147 peptide by the monoclonal and polyclonal anti-HBs was more readily inhibited by the anti-id than was the binding of the cyclical peptide, and this is consistent with a lower affinity of the anti-HBs sera for the linear form. These results may also indicate that the cyclized peptide has a reasonable similarity to the conformation of the native determinant on HBsAg and that there is a significant loss of entropy in adapting the linear peptide to fit the corresponding antibodies.

Experiments were also done with the gp30 p25 polypeptide complex which expresses all the antigenic determinants of HBsAg. Both the internal image anti-ids (2F10 and 4D4) inhibited the binding of the gp30 p25 complex by mouse, human, and swine anti-HBs (Fig. 18.5), but the levels of inhibition were slightly lower than those seen when the synthetic peptides were used as antigens. The binding of swine anti-HBs *ad* and *ay* antibodies to the gp30 p25 complex was inhibited by both 2F10 and 4D4 monoclonals. However, these anti-ids did not inhibit the binding of the polypeptide complex by swine anti-*y* and anti-*d* antibodies (see Fig. 18.5), confirming the immunofluorescence data on the mimicry by the anti-ids of only the *a* determinant of HBsAg.

We also measured the affinity of the antigen-antibody reaction, as such measurements would give us an indication of the similarity of the surrogate antigenic epitope (anti-id or peptides) to that on the original antigen HBsAg that induced the antibody response in vivo. Thus, the affinities of human, rabbit, and goat anti-HBs antisera for the surrogate antigens (anti-ids and

% Inhibition of Binding of Anti-HBs Antisera to the gp 30 p25 Complex by Four Monoclonal Anti-idiotypes

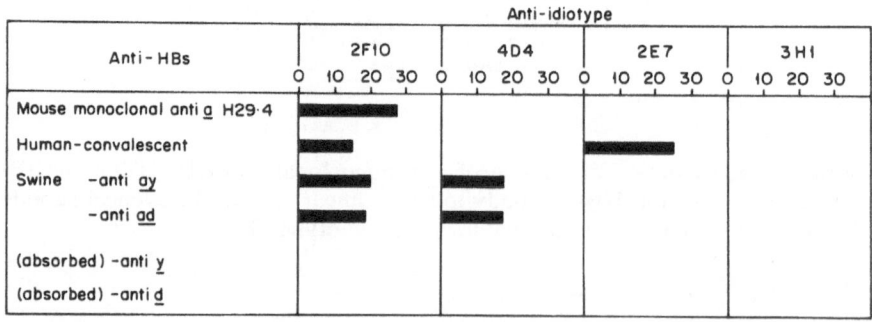

FIGURE 18.5. Inhibition of binding of anti-HBs antisera to the gp30p25 complex by four monoclonal anti-idiotypes.

peptide 139-147) were assessed and compared with their affinities for the gp30 p25 complex. Analysis of the binding curves revealed that the internal image anti-ids were bound with higher affinity by the various monoclonal antibodies and heteroantisera than were the peptides. One might also envisage that the intact divalent anti-id molecule would be bound by the B-cell surface receptors in vivo with a higher avidity than would be expected with the presumably monovalent peptide. Also of interest was the finding that the level of antibody, expressed as picomoles of antigen-combining sites, of polyclonal sera for the peptides was greater than that for the anti-id. This indicates that the peptide can combine with a wider spectrum of the antibodies in a polyclonal anti-HBs serum than can the monoclonal anti-id, which is therefore more restricted in its simulation of the original HBsAg. The implication of these observations for the use of these surrogate antigens for potential vaccines is that while anti-ids may be capable of stimulating B cells of higher affinity, they would react with a more restricted range of B-cell specificities than would the peptides. Use of several internal anti-ids could, however, override this limitation.

Summary of Anti-id Production in Other Viral, Bacterial, and Parasite Systems

Viruses

Several examples of the use of anti-id antibodies in the induction of immunity to viruses have been reported. These include Sendai virus (34), reovirus (35), poliovirus (42), tobacco mosaic virus (43), rabies virus (44), and hepatitis B virus (36). The anti-ids utilized have been in some instances monoclonal syngeneic reagents and also polyclonal antisera. The protection induced has been shown to be mediated both by cellular and humoral responses.

Bacteria

Anti-idiotypic antibodies directed at antibodies specific for the capsular polysaccharides of *E. coli* offered protection from subsequent infection with this bacteria (37). Similarly, enhanced survival of BALB/c mice injected with *Streptococcus pneumonia* was reported (38) when these animals were immunized with a monoclonal anti-idiotype against the T15 idiotype.

Parasites

Preimmunization of mice with an allogeneic polyclonal anti-idiotypic serum to a number of monoclonal antibodies directed to a glycoprotein of *Trypanosoma rhodesiense* protected the mice from a subsequent infection with the homologous strain of the parasite (33). However, the antibodies produced were clone specific and the induction of immunity was genetically restricted by genes linked to the Igh locus.

It would be fair to conclude based on the experimental evidence accumulated to date that internal image anti-idiotypes have tremendous promise as potential new vaccines.

Acknowledgments. This work was carried out in the laboratory of Prof. I.M. Roitt at The Middlesex Hospital Medical School, England. Prof. Roitt collaborated in all aspects of this study. Dr. M. Steward, Dr. C. Howard, and Dr. S. Brown collaborated on the synthetic peptide experiments. I would also like to thank Mrs. D. Ovak and Mrs. C. Zuber for excellent secretarial help.

References

1. Jerne, N.K. 1974. Towards a network theory of the immune system. Ann. Immunol. (Inst. Pasteur) 125C:373.
2. Roitt, I.M., A. Cooke, D.K. Male, F.C. Hay, G. Guarnotta, P.M. Lydyard, L. deCarvalho, Y.M. Thanavala, and I. Ivanyi. 1981. Idiotypic networks and their possible exploitation for manipulation of the immune response. Lancet i:1041.
3. Nisonoff, A., and E. Lamoyi. 1981. Implications of the presence of an internal image of the antigen in anti-idiotypic antibodies: possible application to vaccine production. Clin. Immunol. Immunopathol. 21:397.
4. Maupas, P., P. Coursaget, A. Goudeau, F. Barin, J.P. Chiron, and B. Raynaud. 1981. p. 3. In P. Maupas and P. Goesky (eds.), Hepatitis B vaccine. Elsevier/ North Holland Biomedical Press, New York.
5. Beasley, R.P. and L.Y. Hwang. 1984. The epidemiology of hepatocellular carcinoma, pp. 209–224. In G.N. Vyas (ed.), Viral hepatitis and liver disease. Grune and Stratton, New York.
6. Maupas, P., and J.L. Melnick. 1981. Hepatitis B virus and primary hepatocellular carcinoma. Prog. Med. Virol. 27:1.
7. Almeida, J.D. 1972. Individual morphological variations seen in Australia antigen positive sera. Am. J. Dis. Child. 123:303.
8. Robinson, W.S. 1977. The genome of hepatitis B virus. Annu. Rev. Microbiol. 31:357.
9. Levene, C., and B.S. Blumberg. 1969. Additional specificities of Australia antigen and the possible identification of hepatitis carriers. Nature (London) 221:195.
10. LeBouvier, G.L. 1971. The heterogeneity of Australia antigen. J. Infect. Dis. 123:671.
11. Courouce-Pauty, A.M., A. Plancon, and J.P. Soulier. 1983. Distribution of HBsAg subtypes in the world. Vox Sang. 44:197.
12. Gerety, R.J., E. Tabor, R.H. Purcell, and F.J. Tyeryar. 1979. Summary of an International Workshop on hepatitis B vaccines. J. Infect. Dis. 140:642.

13. Szmuness, W., C.E. Stevens, E.J. Harley, E.A. Zang, W.R. Oleszko, D.C. William, R. Sedovsky, J.M. Morrison, and A. Kellner. 1980. Hepatitis B vaccine: Demonstration of efficacy in a controlled clinical trial in a high risk population in the United States. N. Engl. J. Med. **303**:833.
14. Hilleman, M.R., E.B. Buynak, W.J. McAleer, and A.A. McLean. 1981. p. 120. In S. Krugman and S. Sherlock (eds.), Proceedings of the European Symposium on Hepatitis B, Merck, Sharpe & Dohme International.
15. Crosnier, J., P. Jungers, A.M. Courouce, A. Laplanche, E. Benhamou, F. Degos, B. Lacour, P. Prunet, Y. Cerisier, and P. Guesry. 1981. Randomised placebo-controlled trial of hepatitis B surface antigen vaccine in french haemodialysis units: I Medical Staff. Lancet **i**:455.
16. Laplanche, A., A.M. Courouce, E. Benhamou, P. Jungers, and J. Crosnier. 1982. Responses to hepatitis B vaccine. Lancet **i**:222.
17. Szmuness, W., C.E. Stevens, E.J. Harley, E.A. Zang, H.J. Alter, P.E. Taylor, A. DeVera, G.T.S. Chen, A. Kellner, and the Dialysis Vaccine Trial Study Group. 1982. Hepatitis B vaccine in medical staff of hemodialysis units. Efficacy and subtype cross-protection. N. Engl. J. Med. **307**:1481.
18. Maupas, P., J.-P. Chiron, F. Barin, P. Coursaget, A. Goudeau, J. Perrin, F. Denis, and I. Diap Mar. 1981. Efficacy of hepatitis B vaccine in prevention of early HBsAg carrier state in children. Controlled trial in an endemic area (Senegal). Lancet **i**:289.
19. Maupas, P., L.Y. Huang, R.P. Beasley, S.H. Chen, and T.Y. Lee. 1983. Immunogenicity of hepatitis B virus vaccine in healthy Chinese neonates. J. Infect. Dis. **148**:526.
20. Prozesky, O.W., C.E. Stevens, W. Szmuness, H. Rolka, E.J. Harley, M.C. Kew, J.E. Scholtz, and A.D. Mitchell. 1983. Immune response to hepatitis B vaccine in newborns. J. Infect. Dis. **7**(Suppl. 1):53.
21. Szmuness, W., C.E. Stevens, E.A. Zang, E.J. Harley and A. Kellner. 1981. A controlled clinical trial of the efficacy of the hepatitis vaccine (hepta vax B): a final report. Hepatology **1**:377.
22. Walker, M.E., W. Szmuness, C. Stevens, and P. Rubinstein. 1981. Proc. Am. Assoc. Blood Banks, No. 4.
23. Maupas, P., A. Goudeau, F. DuBois, P. Coursaget, and F. Barin. 1981. p. 117. In P. Maupas and P. Guesry (eds.), Hepatitis B vaccine. Elsevier/North Holland Biomedical Press, New York.
24. McAuLiffe, V.J., R.H. Purcell, and J.L. Gerin. 1980. Type B hepatitis: a review of current prospects for a safe and effective vaccine. Rev. Infect. Dis. **2**:470.
25. Stevens, C.E., W. Szmuness, A.J. Goodman, S.A. Wesley, and M. Fonno. 1980. Hepatitis B vaccine: Immune responses in haemodialysis patients. Lancet **ii**:1211.
26. Skelly, J., C.R. Howard, and A.J. Zukerman. 1981. Hepatitis B polypeptide vaccine preparation in micelle form. Nature (London) **290**:51.
27. Valenzuela, P., A. Medina, W.J. Rutter, G. Ammerer, and B.D. Hall. 1982. Synthesis and assembly of hepatitis B virus surface antigen particles in yeast. Nature (London) **298**:347.
28. Michel, M.L., P. Pontisso, E. Sobczak, Y. Malpiece, R. E. Streeck, and P. Tiollais. 1984. Synthesis in animal cells of hepatitis B surface antigen particles carrying a receptor for polymerized human serum albumin. Proc. Natl. Acad. Sci. USA **81**:7708.

29. Smith, G.L., M. Mackett, and B. Moss. 1983. Chemically synthesized peptides predicted from the nucleotide sequence of the hepatitis B virus genome elicit antibodies reactive with native envelope protein of some particles. Nature (London) **302**:490.

30. Ionescu-Matiu, I., R.C. Kennedy, J.T. Sparrow, A.R. Culwell, Y. Sanchez, J.L. Melnick, and G.R. Dreesman. 1983. Epitopes associated with a synthetic hepatitis B surface antigen peptide. J. Immunol. **130**:1947.

31. Lerner, R.A., N. Green, H. Alexander, F.T. Lui, J.G. Sutcliffe, and T.M. Shinnick. 1981. Proc. Natl. Acad. Sci. USA **78**:3403.

32. Bhatnager, P.K., E. Papas, H.E. Blum, D.R. Milich, D. Nitecki, M.J. Karels, and G.N. Vyas. 1982. Immune response to synthesis peptide analogues of hepatitis B surface antigen specific for the a determinant. Proc. Natl. Acad. Sci. USA **79**:4400.

33. Sacks, D.L., and A. Sher. 1983. Evidence that anti-idiotype induced immunity to experimental African trypanosomiasis is genetically restricted and requires recognition of containing site-related idiotopes. J. Immunol. **131**:1511.

34. Ertl, H.C.J., E. Homans, S. Tournas, and R.W. Finberg. 1984. Sendai virus-specific T cell clones. V. Induction of virus-specific response by antiidiotypic antibodies directed against a T helper cell clone. J. Exp. Med. **159**:1178.

35. Sharpe, A.H., G.N. Gaulton, K.K. McDade, B.N. Fields, and M.I. Greene. 1984. Syngeneic monoclonal antiidiotype can induce cellular immunity to reovirus. J. Exp. Med. **160**:1195.

36. Kennedy, R.C., and G.R. Dreesman. 1984. Enhancement of the immune response to hepatitis B surface antigen. J. Exp. Med. **159**:655.

37. Stein, K.E., and T. Soderstrom. 1984. Neonatal administration of idiotype or anti-idiotype primes to protection against E. coli K13 infection in mice. J. Exp. Med. **160**:1001.

38. McNamara, M.K., R.E. Ward, and H. Kohler. 1984. Monoclonal idiotype vaccine against *Streptococcus pneumoniae* infection. Science **226**:1325.

39. Thanavala, Y., A. Bond, R. Tedder, F.C. Hay, and I.M. Roitt. 1985. Monoclonal 'internal image' anti-idiotypic antibodies of hepatitis B surface antigen. Immunology **55**:197.

40. Thanavala, Y., A. Bond, F.C. Hay, and I.M. Roitt. 1985. Immunofluorescent technique for the detection of monoclonal internal image anti-idiotypic antibodies of hepatitis B surface antigen. J. Immunol. Methods **83**:227.

41. Thanavala, Y., S.E. Brown, C.R. Howard, I.M. Roitt, and M.W. Steward. 1986. A surrogate hepatitis B virus antigenic epitope represented by a synthetic peptide and an internal image antiidiotype antibody. J. Exp. Med. **164**:227.

42. Uytdehaag, F.G.C.M., and A.D.M.E. Osterhaus. 1985. Induction of neutralizing antibody in mice against poliovirus type II with monoclonal anti-idiotypic antibody. J. Immunol. **134**:1225.

43. Francotte, M., and J. Urbain. 1984. Induction of anti-tobacco mosaic virus antibodies in mice by rabbit antiidiotypic antibodies. J. Exp. Med. **160**:1485.

44. Reagan, K.J., W.H. Wunner, T.J. Wiktor, and H. Kroprowski. 1983. Antiidiotypic antibodies induce neutralizing antibodies to rabies virus glycoprotein. J. Virol. **48**:660.

19
Regulation of Cell Growth and Immunity by Reovirus Anti-Receptor Antibodies

GLEN N. GAULTON AND MARK I. GREENE

Introduction

The study of ligand-receptor interactions has been greatly aided by the application of anti-idiotypic antibody probes which mimic the specificity of ligand binding. Anti-idiotypic antibodies which represent internal images of viral attachment proteins have been utilized in our laboratories as probes of the biochemistry, cell biology, and immunology of the mammalian reovirus and its cellular receptor. The basis of this approach rests on the network theory of immune regulation originally proposed by Niels Jerne (1). Current applications of this technique have been recently reviewed by us (2,3). In brief, anti-idiotypic anti-receptor antibodies have been produced which recognize receptors for cellular growth factors (4,5), physiological regulators (4,6), and virus binding proteins (7,8). In these examples, anti-receptor antibodies were defined by immunoprecipitation and purification of receptor molecules or by their capacity to modify receptor function or ligand signal delivery. Anti-idiotypic antibodies have also been successfully employed as immunological mimics of antigen-ligand. For example, anti-idiotypic antibodies were used to stimulate T and B cell-mediated immunity to a variety of parasite, bacterial, and viral pathogens (9–12).

The mammalian reovirus anti-idiotype system demonstrates properties of each of the previously mentioned biochemical and immunological studies, and additionally offers the unique advantage of a well-defined monoclonal internal image anti-idiotope which was constructed in syngeneic mice. Studies in our laboratories have focused on using monoclonal anti-idiotopes to study: (i) the molecular biology of reovirus receptors; (ii) the role of receptor perturbation by virus and anti-idiotope in alterations of cell growth and function; and (iii) the use of anti-idiotope antibodies as viral vaccines. The results presented here demonstrate our most recent progress in each of these areas.

Construction and Characterization of Monoclonal Anti-Idiotopes

The specificity of reovirus binding to its cellular receptor is governed by the specificity of the minor outer capsid protein S1, which also serves as the viral hemagglutinin (HA) (13). In addition, the reovirus HA determines the serotype specificity of all immune reactions tested to date (14,15). Thus, an internal image anti-idiotope to the binding domain of the HA should mimic viral tropism and serotype specific immunity. The selection of monoclonal antibodies (Ab$_1$) which interact with this domain of the serotype 3 HA (HA3) was accomplished by measuring the inhibition of polyclonal anti-idiotype binding to HA3 by Ab$_1$ using a solid-phase radioimmunoassay, as previously described (16). Syngeneic monoclonal anti-idiotopes (Ab$_2$) to competing Ab$_1$ were then constructed and screened for internal image activity (17,18).

One Ab$_2$, termed 87.92.6, displayed the following characteristics which are diagnostic of an internal image anti-idiotope to the reovirus HA3: the binding of Ab$_1$ to HA3 is blocked by Ab$_2$ (18); the binding of Ab$_2$ to a panel of cells mirrors type 3 virus tropism (17); and prior incubation of cells with Ab$_2$ inhibits viral binding (7) and infectivity on neuronal targets (Table 19.1). In this instance, preincubation of neuroblastoma cells with purified 87.92.6 inhibited the production of infectious viral progeny by 82% compared to controls. As Ab$_2$ does not bind virus, these blocking effects represent specific competition for binding to the reovirus cellular receptor. Further confirmation of the internal image nature of Ab$_2$ is presented in the following sections. Henceforth we term the monoclonal 87.92.6 as anti-Id3.

TABLE 19.1. Anti-idiotope inhibition of viral infectivity.

Pretreatment	Viral protein synthesis (% inhibition)
None	24,859
Defective reo3	1,492 (94%)
Anti-Id3	4,723 (81%)
Control anti-Id	23,864 (4%)

Pretreatments were conducted by incubation for 30 min at 37°C followed by washing and the addition of intact virus. Viral protein synthesis was determined by immunoprecipitation of radiolabeled viral protein 20 h following incubation of L cells with intact reovirus type 3.

Anti-Idiotope as a Receptor Probe

The isolation and characterization of reovirus receptors was accomplished using both polyclonal and monoclonal anti-idiotypes. Immunoprecipitates from surface-radioiodinated R1.1 thymoma cells display a single M_r 67,000 band not seen in controls, which is also the major membrane band that cosediments with surface-bound virus (7,8). Immunoprecipitations have been performed on a number of different cell types and donor species, and in each instance receptors were of identical molecular weight (7). Purified receptors are sensitive to digestion by protease and neuraminidase and are monomeric with a slight charge heterogeneity of pI 5.8–6.0. Demonstration that the protein band seen in anti-Id3 immunoprecipitates serves as the reovirus type 3 attachment protein was achieved by Western blot analysis. Labeled reovirus and anti-Id3 bound to the same M_r 67,000 band on blots of either anti-Id3-purified receptor or membrane extracts run on SDS-PAGE. Prior incubation of blots with unlabeled anti-Id3 completely blocked the binding of labeled reovirus to this protein (G.N. Gaulton, unpublished).

Similarities of Reovirus and β-Adrenergic Receptors

A number of recent observations indicate that many viruses utilize integral membrane proteins which provide essential cellular functions as specific viral attachment sites. For example, the membrane receptor for lactate dehydrogenase virus has now been identified as the Ia molecule (19), and the Epstein-Barr virus has been shown to bind to the C3d receptor CR2 on B lymphocytes (20). Our initial experiments indicate that the mammalian reovirus type 3 may utilize β-adrenergic receptors as attachment sites.

We first noticed that both the reported molecular weight and tissue distribution of β-adrenergic receptors matched those of reovirus type 3 receptors. Based on this, more detailed studies were conducted which demonstrated that (i) anti-Id3 specifically immunoprecipitated ligand affinity-purified β-adrenergic receptor and (ii) the two-dimensional gel and partial tryptic digest patterns of these receptors were indistinguishable (21). To determine whether anti-Id3 binds functionally active β-adrenergic receptors, coprecipitation studies were conducted using the labeled β antagonist [^{125}I]iodohydroxybenzylpindolol (IHYP). Approximately 10–20% of cell-bound IHYP was precipitated in detergent lysates of R1.1 cells treated with anti-Id3 (21). In addition, immunoprecipitated purified receptor bound IHYP in a solid-phase radioassay, and this binding was blocked (60–90%) by the addition of the specific β antagonist isoproterenol. The specificity of the binding of β ligands to reovirus attachment proteins

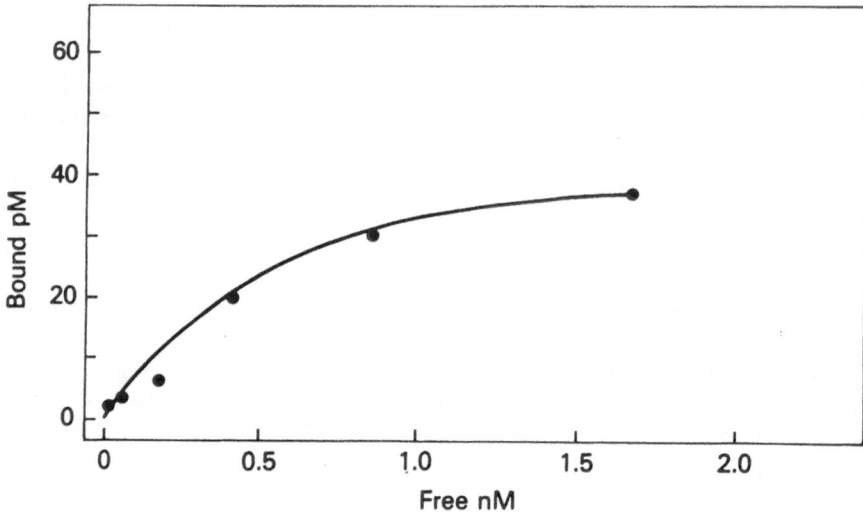

FIGURE 19.1. Binding of [^{125}I] iodocyanopindolol to immunoprecipitated reovirus receptors. Approximately 100 fmol of anti-idiotope-purified receptor was incubated with 20 pM [^{125}I] iodocyanopindolol (20,000 cpm) for 30 min at 30°C. Then bound and unbound ligand were separated on glass fiber filters and the amounts of bound and free label determined.

is demonstrated in Fig. 19.1. In this instance a saturable dose curve is achieved in the binding of [^{125}I]iodocyanopindolol to anti-Id3 precipitated receptors.

Experiments are now in progress to determine whether these purified receptors display the classical rank order of β ligand inhibition seen for β$_2$-adrenergic receptors, as the line R1.1 expresses predominantly this class of β receptor. Studies are also under way to evaluate potential interference of β ligand and reovirus binding and the influence of reovirus or anti-Id3 on β-adrenergic receptor function.

Receptor Modulation and Altered Cell Function

One of the earliest events accompanying reovirus type 3 infection is an inhibition of host-cell DNA synthesis (22). In contrast, type 1 virus shows no such effect. Inhibition is first detected within 12 hours of virus binding, and by 48 hours (for murine L cells) >90% of DNA synthesis is lost. Using viral recombinants such as 1HA3 and 3HA1, in which, respectively, the type 3 HA is placed on a type 1 background or the type 1 HA is placed on a type 3 background, Sharpe and Fields (22) demonstrated that DNA inhibition is linked exclusively to the presence of the HA of type 3 virus. In addition, replication defective particles with an intact HA3 still cause inhibition. Thus, it would appear that this effect is receptor mediated.

To directly address this possibility, we have analyzed the effects of adding anti-receptor antibody (anti-Id3) on L-cell DNA synthesis. This analysis is presented in Fig. 19.2. Treatment with intact anti-Id3 induced a 67% reduction of cellular DNA synthesis, compared to 78% seen with reovirus 3 or the recombinant virus 1HA3. Incubation with type 1 virus or the recombinant 3HA1 had no effect on DNA synthesis. Although not pictured here, the kinetics of anti-Id3 inhibition were indistinguishable from those seen with reovirus type 3. When monovalent fragments (Fab) of anti-Id3 were used, all of this inhibitory effect was lost despite equivalent levels of binding, as determined by fluorescence analysis. These results confirm that inhibition of DNA synthesis by reovirus 3 is mediated through a receptor-driven process which appears partially dependent upon cross-linking or clustering of receptors. While it is tempting to speculate on the potential roles of β-adrenergic receptors in this response, we have no formal data either in support of or against this hypothesis. The addition of β agonists or antagonists to these cells has no effect on DNA synthesis; however, these ligands are all monovalent.

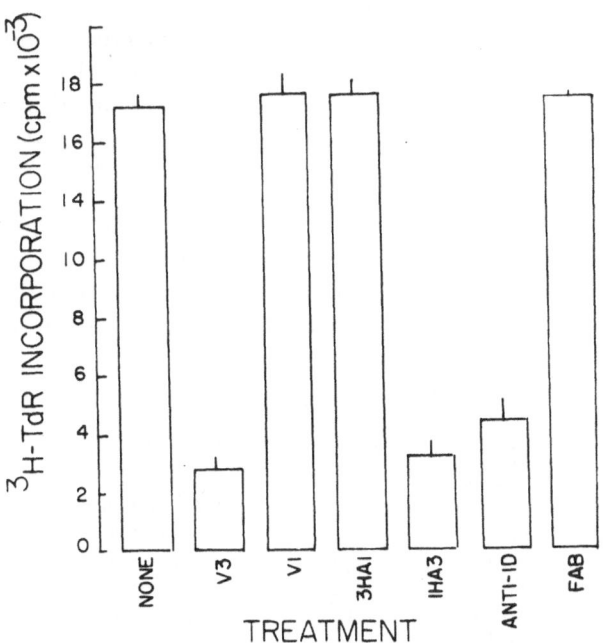

FIGURE 19.2. Inhibition of DNA synthesis by anti-idiotope. L cells were incubated with either reovirus or antibody for 30min at 37°C, then washed and incubated for an additional 16 h prior to a 2-h pulse with [³H] TdR. Type 3, 1, 1HA3, and 3HA1 viruses were used at 20 MOI and antibody at 10 μg/ml.

Use of Anti-Idiotopes as Immunological Mimics

A number of laboratories, including our own, have successfully demonstrated the utility of anti-idiotopes as vaccines (9–12). Immunization of syngeneic mice with purified monoclonal anti-Id3 stimulates type 3 specific cytolytic, delayed type hypersensitivity (DTH) and antibody responses (23,24). Potent and transferrable DTH responses were detected within five days after immunization of mice with 100 μg of anti-Id3 in the absence of adjuvant. The specificity of this response to HA3 is presented in Table 19.2. Animals immunized with anti-Id3 were challenged after five days by subcutaneous footpad immunization with either of the viral serotypes listed and 24 hours later footpad swelling (mononuclear cell invasion) was measured. Results are presented as the percentage of the maximal response of mice immunized with reovirus 3. Significant responses were detected only following challenge with serotypes which bear HA3, thus confirming both the linkage and molecular mimicry of anti-Id3 to the HA3 domain. Similar results were obtained for cytolytic T-cell responses (23).

Our initial studies on the induction of anti-virus antibody responses using anti-Id3 indicated that B-cell-dependent antibody production required more stringent immunization conditions than those required for T-cell activation. Significant responses were induced (in syngeneic mice) only when anti-Id3 was coupled to a carrier protein and when injected in multiple doses in the presence of adjuvant (G.N. Gaulton, submitted for publication). Otherwise, the specificity and relative dose dependence (Fig. 19.3) of antibody responses paralleled those of cellular immune responses. Antibody responses were also elicited in mice of several disparate strains and in three unrelated animal species.

Preliminary results indicate that immunization with anti-Id3 can confer a prophylactic, antipathogenic effect. The pathology of reovirus is characterized by an acute sensitivity of neonatal mice, aged <6 days, to type 3-induced encephalitis and type 1-induced hydrocephalus. Reovirus in-

TABLE 19.2. Specificity of delayed-type hypersensitivity responses elicited by anti-idiotope.

Immunization	Challenge	DTH response
Type 3 virus	Type 3	37.8±2.3
Anti-Id	Type 3	25.5±1.7
Anti-Id	Type 1	8.5±1.5
Anti-Id	1HA3	27.3±2.0
Anti-Id	3HA1	9.5±1.6
Anti-Id	Variant K	11.8±1.7
Saline	Type 3	8.5±1.5

DTH responses are expressed as the mean footpad swelling ± SEM 24 h after challenge, and on day 5 after immunization. Units are mm × 10^{-2}.

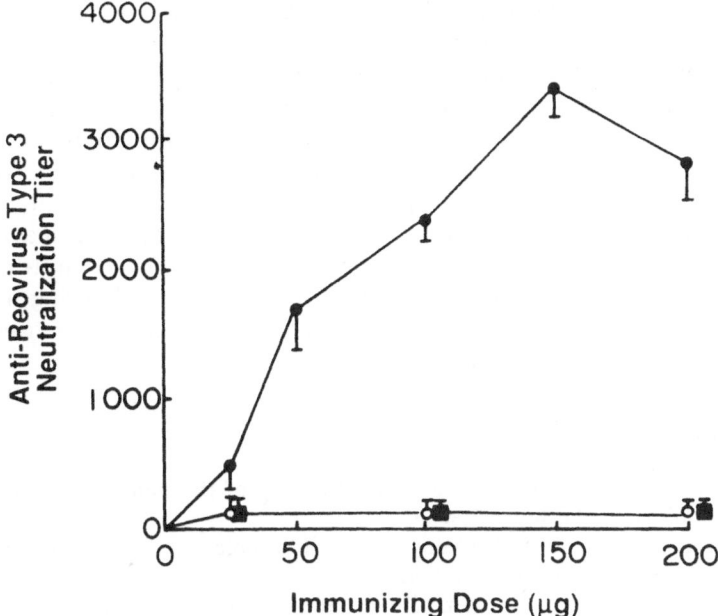

FIGURE 19.3. Dose dependence of neutralizing antibody induction with anti-idiotope. Mice were immunized with varying amounts of antibody in complete Freund adjuvant, as indicated, on day 0, boosted with an equivalent amount of antibody in incomplete Freund adjuvant on day 14, and bled on day 21. Symbols: •, Anti-Id3; ○, isotype-matched control; ■, control anti-Id to Sendai virus.

fectivity thus provides a unique model for studying the effects of maternal immunization with anti-idiotope as a safe fetal or neonatal prophylaxis. Syngeneic female mice were first given a regimen of anti-Id3 immunization which ensured high antiviral titers. Mice were then mated and the offspring given a lethal dose of either reovirus type 3 or 1. Those mice whose mothers had received anti-Id3 therapy were completely resistant to gross disease characteristics and lethality on challenge with type 3 virus. In contrast, littermates of these mice exposed to type 1 virus uniformly developed hydrocephalus.

Conclusion

The results presented here and previously provide substantial evidence that the monoclonal antibody 87.92.6 bears an internal image of the reovirus type 3 hemagglutinin molecule. This unique probe has enabled a variety of biochemical and immunological studies to be conducted to elucidate the structure and function of reovirus attachment proteins on target cells. The most striking observation of these studies is the structural homology between reovirus and β-adrenergic receptors. Whether reovirus utilizes these receptors in unique ways to gain entry and replicate in cells and whether reovirus or anti-Id3 can perturb β-adrenergic receptor function remain obscure. Receptor-mediated inhibition of DNA synthesis by virus and anti-Id3 suggests that interesting relationships will emerge.

The use of anti-idiotopes as potent and specific inducers of antiviral immunity represents a novel approach to vaccine development. This is particularly relevant to situations where immunization with inactivated virus is nonproductive or involves patient risk. The use of anti-Id3 as a maternal vaccine which confers passive protection on offspring from an otherwise lethal reovirus type 3-induced encephalitis demonstrates the practicality of this approach.

References

1. Jerne, N.K. 1974. Towards a network theory of the immune system. Ann. Immunol. (Inst. Pasteur) **125C**:373.
2. Gaulton, G.N., and M.I. Greene. 1986. Idiotypic mimicry of biological receptors. Annu. Rev. Immunol. **4**:253.
3. Gaulton, G.N., M.S. Co, H.D. Royer, and M.I. Greene. 1985. Anti-idiotypic antibodies as probes of cell surface receptors. Mol. Cell. Biochem. **65**:5.
4. Sege, K., and P.A. Peterson. 1978. Anti-idiotypic antibodies against anti-vitamin A transporting protein react with prealbumin. Nature (London) **271**:167.
5. Schechter, Y., R. Maron, D. Elias, and I.R. Cohen. 1982. Autoantibodies to insulin receptor spontaneously develop as anti-idiotypes in mice immunized with insulin. Science **216**:542.
6. Guillet, J.G., S.V. Kaveri, O. Durieu, C. Delavier, J. Hoebeke, and A.D. Strosberg. 1985. Beta-adrenergic agonist activity of a monoclonal anti-idiotypic antibody. Proc. Natl. Acad. Sci. USA **82**:1781.
7. Co., M.S., G.N. Gaulton, B.N. Fields, and M.I. Greene. 1985. Isolation and biochemical characterization of the mammalian reovirus type 3 cell-surface receptor. Proc. Natl. Acad. Sci. USA **82**:1494.
8. Gaulton, G., M.S. Co, and M.I. Greene. 1985. Anti-idiotypic antibody identifies the cellular receptor of reovirus type 3. J. Cell. Biochem. **28**:69.
9. McNamara, M.K., R.E. Ward, and H. Kohler. 1984. Monoclonal idiotope vaccine against *Streptococcus pneumoniae* infection. Science **226**:1325.
10. Reagan, K.J., W.H. Wunner, T.J. Wiktor, and H. Koprowski. 1983. Anti-idiotypic antibodies induce neutralizing antibodies to rabies virus glycoprotein. J. Virol. **48**:660.
11. Sacks, D.L., K.M. Esser, and A. Sher. 1982. Immunization of mice against African trypanosomiasis using anti-idiotypic antibodies. J. Exp. Med. **155**:1108.
12. Uytdehaag, F.G., and A.D. Osterhaus. 1985. Induction of neutralizing antibody in mice against poliovirus type II with monoclonal anti-idiotypic antibody. J. Immunol. **134**:1225.
13. Weiner, H.L., D. Drayna, D.R. Averill, Jr., and B.N. Fields. 1977. Molecular basis of reovirus virulence: role of the S1 gene. Proc. Natl. Acad. Sci. USA **74**:5744.

14. Burstin, S.J., D.R. Spriggs, and B.N. Fields. 1982. Evidence for functional domains on the reovirus type 3 hemagglutinin. Virology 117:146.
15. Fields, B.N., and M.I. Greene. 1982. Genetic and molecular mechanisms of viral pathogenesis: implications for prevention and treatment. Nature (London) 300:19.
16. Nepom, J.T., H.L. Weiner, M.A. Dichter, M. Tardieu, D.R. Spriggs, C.F. Gramm, M.L. Powers, B.N. Fields, and M.I. Greene. 1982. Identification of a hemagglutinin-specific idiotype associated with reovirus recognition shared by lymphoid and neural cells. J. Exp. Med. 155:155.
17. Kauffman, R.S., J.H. Noseworthy, J.T. Nepom, R. Finberg, B.N. Fields, and M.I. Greens. 1983. Cell receptors for the mammalian reovirus. II. Monoclonal anti-idiotypic antibody blocks viral binding to cells. J. Immunol. 131:2539.
18. Noseworthy, J.H., B.N. Fields, M.A. Dichter, C. Sobotka, E. Pizer, L.L. Perry, J.T. Nepom, and M.I. Greene. 1983. Cell receptors for the mammalian reovirus. I. Syngeneic monoclonal anti-idiotypic antibody identifies a cell surface receptor for reovirus. J. Immunol. 131:2533.
19. Inada, T., and C.A. Mims. 1984. Mouse Ia antigens are receptors for lactate dehydrogenase virus. Nature (London) 309:59.
20. Fingeroth, J.D., J.J. Weis, T.F. Tedder, J.L. Stominger, and D.T. Fearon. 1984. Epstein-Barr virus receptor of human B lymphocytes is the C3d receptor CR2. Proc. Natl. Acad. Sci. USA 81:4510.
21. Co, M.S., G.N. Gaulton, A. Tominaga, C.J. Homcy, B. N. Fields, and M.I. Greene. 1985. Structural similarities between the mammalian beta-adrenergic and reovirus type 3 receptors. Proc. Natl. Acad. Sci. USA 82:5315.
22. Sharpe, A.H., and B.N. Fields. 1981. Reovirus inhibition of cellular DNA synthesis: role of the S1 gene. J. Virol. 38:389.
23. Sharpe, A.H., G.N. Gaulton, H.C. Ertl, R.W. Finberg, K.K. McDade, B.N. Fields, and M.I. Greene. 1985. Cell receptors for the mammalian reovirus. IV. Reovirus-specific cytolytic T cell lines that have idiotypic receptors recognize anti-idiotypic B cell hybridomas. J. Immunol. 134:2702.
24. Sharpe, A.H., G.N. Gaulton, K.K. McDade, B.N. Fields, and M.I. Greene. 1984. Syngeneic monoclonal antiidiotype can induce cellular immunity to reovirus. J. Exp. Med. 160:1195.

20
The Production and Characterization of Anti-Idiotypic Antibodies Directed Against a Monoclonal Anti-Bovine Herpesvirus I Antibody

CECELIA A. WHETSTONE, LOREN A. BABIUK, AND
SYLVIA VAN DRUNEN LITTEL-VAN DEN HURK

Introduction

Studies with herpes simplex virus have shown that herpesvirus glycoproteins are incorporated into the plasma membrane of host cell (1,2), eventually make up part of the virion envelope (3,4), and, among other biological and immunological functions, effect attachment and penetration of the virus into the cell (4,5). The cell surface receptors involved, however, have not been identified. It has been reported that bovine herpesvirus 1 (BHV-1) has more than 25 structural polypeptides (6,7), 11 of which are glycosylated, and indirect evidence (8,9) indicates that some of these glycoproteins are involved in viral attachment to the host cell. As with herpes simplex virus, the structure and distribution of the cell surface receptors have not been determined. Recently, anti-idiotypic (anti-Id) antibodies have been utilized to define ligand receptors (10,11) and cell surface receptors for reovirus (12). We have used a similar approach in an attempt to characterize the cell surface receptor for BHV-1.

Methods and Results

Anti-Idiotypic Antibody Preparations

Mouse immunoglobulin (Ig) from an anti-BHV-1 specific monoclonal antibody, 1E11, was HPLC purified on a TSK3000SW size exclusion column. Polyclonal anti-Id antibodies were prepared in four New Zealand white rabbits by inoculating them subcutaneously along the back with 300 μg of either HPLC-purified 1E11 whole Ig emulsified in complete Freund adjuvant (CFA) or Fab fragments (obtained by papain digestion followed by protein A-Sepharose CL-4B purification) of 1E11 in CFA. Two additional intramuscular injections of 100 μg of antigen in phosphate-buffered saline were given to each rabbit on days 7 and 30, and sera were collected 10 days later. Each antiserum was made idiotype specific by repeated passages over Sepharose 4B columns to which mouse Ig was coupled. Monoclonal anti-Id antibodies were made by injecting BALB/c mice intraperitoneally with 7 μg of HPLC-purified 1E11 emulsified in incomplete Freund adjuvant. At 3 weeks, mice were tested for serological response to the antigen. Three days prior to fusion, an additional 15 μg of purified 1E11 was given intravenously. The fusion and hybridoma culturing protocol was done by standard methods as described elsewhere (13). Hybridomas were cloned using a fluorescence-activated cell sorter. Ascites fluids were produced in pristane-primed BALB/c mice and purified on DEAE Affi-Gel blue prior to use in assays.

Characterization of Anti-Idiotypic Antibodies

Primary screening for anti-Id activity was done by dot blot. Rabbit sera, Fab preparations of rabbit sera, hybridoma fluids, and Fab preparations of hybridoma fluids were absorbed onto nitrocellulose paper, blocked with blocking buffer (0.1 M Na_2HPO_4, pH 7.5, 0.5 M NaCl, 0.5% BSA, 1.0% Tween 80, 0.08% NaN_3), and then exposed to an ^{125}I-labeled 1E11 Fab probe. As a control, all samples were also exposed to an Fab preparation of ^{125}I-labeled anti-canine adenovirus (CAV) monoclonal antibody of the same isotype as 1E11. Negative controls included monoclonal antibodies against CAV, pseudorabies virus (PRV), *Leptospira*, and *Chlamydia*.

Iodinated whole and Fab fragments of the original monoclonal antibody, 1E11, were found to bind to the Fc regions of all mouse IgGs tested including monoclonal ascites fluids and hybridoma culture supernatants to CAV, PRV, *Leptospira*, and *Chlamydia*, as well as the specific anti-Id antibodies. When the Fab portion of these monoclonal reagents was tested, however, only the anti-Id antibodies reacted with the 1E11 Fab probe.

Both the whole Ig and Fab fragments of the rabbit and monoclonal anti-1E11 preparations were tested for anti-Id activity in competitive neutralization assays. Serial twofold dilutions (1:4–1:2048) of 1E11 were each mixed with a putative anti-Id sample and incubated for 1 h at 37°C. BHV-1 (50TCID$_{50}$) plus guinea pig complement (GPC') were then added and the mixture was incubated for an additional hour at 37°C. Aliquots (0.1 ml) of each dilution mixture were inoculated onto fresh Madin-Darby bovine kidney (MDBK) cell monolayers in 96-well tissue culture plates which were held for 1 week at 37°C in a humidified 5% CO_2 atmosphere. Endpoint titers were determined on the basis of cytopathic effect (CPE) in the cell cultures. Controls included 1E11 without anti-Id sample, anti-Id samples without 1E11, 1E11 with normal rabbit serum or Fab fragments of normal rabbit serum, 1E11 with anti-CAV monoclonal antibody or Fab fragments from the anti-CAV monoclonal antibody, and BHV-1 titration without antibody. The virus neutralization titer of 1E11 in this test system was consistently 1:2048.

Results of competitive neutralization tests done with the rabbit sera are shown in Table 20.1. All four of the anti-Id rabbit antisera, at dilutions of 1:100 or less, competed with 1E11 and inhibited the neutralization of BHV-1 by that monoclonal antibody. Fab preparations of the same rabbit anti-Id antisera caused inhibition of the neutralization titer when diluted 1:10,000. Undiluted or 1:10 dilutions of normal rabbit serum caused inhibition of neutralization, but the normal serum diluted 1:100 had no effect. None of the rabbit anti-Id preparations neutralized BHV-1. Regression analysis of the neutralization titers showed that both the whole Ig and the Fab preparations differed significantly ($p < 0.01$) from the control group.

Whole, undiluted Ig from 12 monoclonal anti-Id antibody preparations, were used in competitive neutralization tests with 1E11, resulting in nearly complete inhibition of the neutralization of BHV-1 by 1E11. However, 1:10 dilutions of Fab preparations of the same monoclonal antibodies resulted in only moderate inhibition of 1E11 neutralizing activity, similar

TABLE 20.1. Competitive neutralization titers[a] of anti-BHV-1 monoclonal antibody, 1E11, when used in competition with rabbit anti-idiotypic antibodies.

Rabbit serum	Neutralization titers[b] for following rabbit serum dilutions:					
	Undiluted	1:10	1:100	1:1,000	1:5,000	1:10,000
Control[c]						
#1	1:128[d]	1:512	≥1:2,048	≥1:2,048	NT[e]	NT
#2	1:64	1:619	≥1:2,048	≥1:2,048	NT	NT
#3	1:128	1:128	≥1:2,048	≥1:2,048	NT	NT
#4	1:128	1:256	≥1:2,048	≥1:2,048	NT	NT
Anti-Id whole Ig						
#1	1:64	1:159	1:256	1:1,024	NT	NT
#2	1:16	1:90	—[f]	—[f]	NT	NT
#3	1:64	1:112	1:512	1:2,048	NT	NT
#4	1:16	1:90	1:256	1:1,024	NT	NT
Anti-Id Fab preparation						
#1	NT	1:64	1:128	1:64	1:256	1:256
#2	NT	1:512	1:64	1:64	1:256	1:256
#3	NT	1:64	1:64	1:64	1:128	1:256
#4	NT	1:256	1:64	1:16	1:256	1:256

[a]All titers are based on results from two or more tests.
[b]Virus neutralization titer of 1E11 without sample ≥1:2,048.
[c]Control sera are preinoculation sera from the anti-Id rabbits.
[d]Titers were figured from Reed-Muench calculations for 50% end points.
[e]NT, Not tested.
[f]These concentrations of rabbit anti-Id competed with 1E11 at the 1:128, 1:256, and 1:512 dilutions of 1E11, did not compete at the lower dilutions of 1E11, and enhanced neutralization at the higher dilutions of 1E11.

to that caused by whole rabbit anti-Id Ig. The enhanced competition seen with whole Ig could be explained by specific binding of 1E11 to the Fc region of mouse IgG. Control tests using anti-CAV monoclonal samples established the fact that these did not compete with 1E11, and none of the anti-Id monoclonal samples neutralized BHV-1 directly.

The ability of the anti-Id antibodies to block BHV-1 binding to cells was tested in vitro by incubating uninoculated MDBK cell cultures for 1 h at 37°C with 1:10 dilutions of Fab preparations of rabbit or monoclonal anti-Id antibodies. The treated cells were then inoculated with serial 10-fold dilutions of BHV-1. At the end of 1 week, cultures were examined for CPE, and cultures pretreated with anti-Id antibodies were compared to cultures which had received virus alone. The virus titer in cells that had been pretreated with rabbit anti-Id Fab was reduced by 0.5–2.0 logs, whereas the monoclonal anti-Id Fab preparations reduced virus titer by 0.5 log. Fab preparations from normal rabbit and anti-CAV monoclonal antibody did not block BHV-1 binding to cells and had no effect on virus titer.

In order to further test whether the monoclonal anti-Id antibodies were true internal image idiotopes, an indirect fluorescent antibody (IFA) test was used. Rabbit anti-BHV-1 antibody was applied to cytocentrifuge preparations of anti-Id-secreting hybridoma cell cultures (fixed in 95% ethanol with 5% acetic acid) to test whether the anti-BHV-1 antibody would recognize BHV-1-like conformations expressed on the surface Ig of these cells. The cultures were subsequently treated with a fluorescein-conjugated goat anti-rabbit Ig. Of the eight cell cultures tested, two stained well (Fig. 20.1), five stained moderately, and one did not stain at all. Controls consisted of the anti-Id-secreting hybridoma cells exposed to normal rabbit serum and SP2/0 cells not secreting anti-Id antibody exposed to the anti-BHV-1 antiserum. Neither of these controls stained, indicating the absence of nonspecific reactivity in the test system.

Discussion

The results confirm that our anti-1E11 antibodies interfered with specific neutralization and attachment reactions of BHV-1. This indicates that interactions at, or very near, specific viral and host-cell receptor sites have occurred. From these data we conclude that our reagents are anti-idiotypic in nature and that at least a portion of the viral glycoprotein has been mimicked by the anti-1E11 antibodies.

The observation that our anti-Id antibodies competed with 1E11 in neutralization tests but only partly blocked binding of BHV-1 to MDBK cells may be explained by the fact that the anti-Id antibodies were raised against a monoclonal antibody, 1E11, that recognizes a single epitope on the BHV-1 viral glycoprotein (GVP) 6/11a/16. Previous studies (8,9,14,15) have

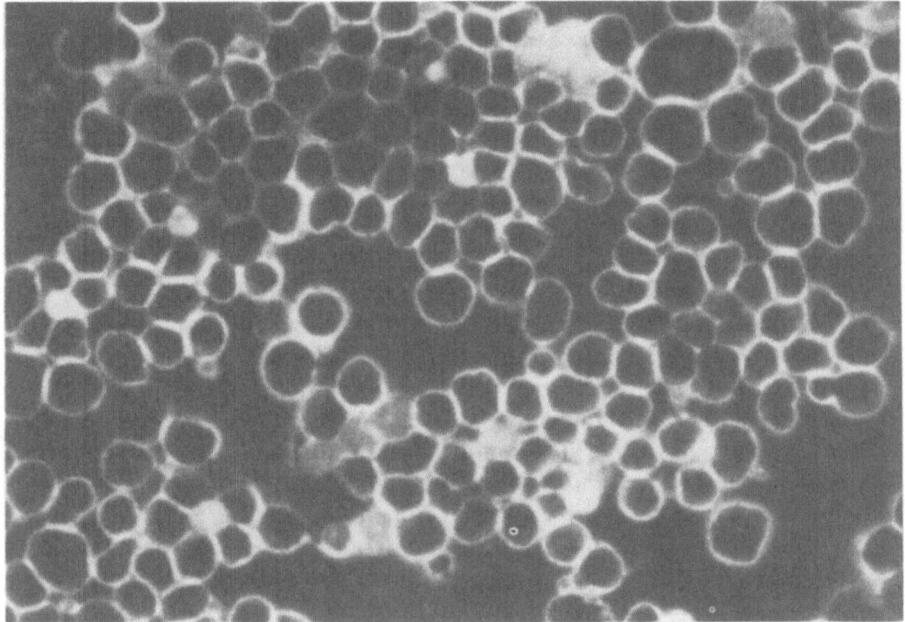

FIGURE 20.1. Hybridoma cells secreting anti-idiotypic antibody, 3C10 E10, stained with fluorescein-conjugated rabbit anti-BHV-1 antibody.

shown that at least three BHV-1 glycoproteins, with 17 epitopes, are involved in either neutralization or complement-mediated lysis of infected cells. Furthermore, two of those glycoproteins, GVP 6/11a/16 and GVP 11b, are believed to contain "receptor determinant" sites involved in virus attachment to host cells. Thus, our anti-Id antibodies, by mimicking the GVP 6/11a/16 viral receptor, would be expected to interfere with site-specific 1E11 neutralization (which it did) but would not be expected to block host cellular receptors for GVP 11b (or others) on cell surfaces. One would, therefore, expect them to be less efficient in blocking virus attachment (which may involve multiple attachment sites) than in interfering with the more specific neutralization interaction with the monoclonal reagent. Anti-Id antibodies against a monoclonal antibody specific for a neutralizing epitope on GVP 11b are now being produced to further investigate this hypothesis.

The high titer ($\leq 1:10,000$) of the inhibition observed with the rabbit anti-Id Fab preparations compared to the low level (undiluted and 1:10) of nonspecific competition seen with the normal (preinoculation) sera is further evidence of the specificity of the anti-Id reagents. The increased efficiency of Fab fragments of rabbit anti-Id sera over the whole anti-Id sera in the competitive neutralization tests was not unexpected. The re-

316   C.A. Whetstone, L.A. Babiuk, & S. vanDrunen Littel-vandenHurk

versal of this situation with the monoclonal antibodies, however, was not anticipated. It is our opinion that the greater efficiency of the whole monoclonal antibodies in competing with 1E11 was due to the demonstrated affinity between the 1E11 antibody and the Fc region of mouse IgG and was not due to a specific anti-Id reaction. The reasons for the apparently specific reactivity of the Fab regions of the 1E11 mouse Ig with the Fc portions of all mouse IgGs tested have not yet been determined. The binding of 1E11 to the Fc region of mouse Ig had a significant effect on the results of this study, and their interpretation, and is a factor to be considered in future studies with anti-Id monoclonal antibodies.

True anti-Id antibodies with conformations that are internal images of the original paratope should mimic antigen in various immunological and virological reactions. We observed this when we reacted rabbit anti-BHV-1 antibody with hybridoma cells that were secreting anti-Id antibodies to our original 1E11 anti-BHV-1 monoclonal antibody. In this reaction, varying degrees of staining were seen with seven of the eight hybridoma cultures tested. This variation may have been due to differences in affinity, or it may indicate that some of the anti-Id antibodies were against an Fab region other than the BHV-1 virus-specific internal image region. The failure of one of the anti-Id-secreting hybridoma cell populations to stain when exposed to rabbit anti-BHV-1 antibody suggests that this anti-Id antibody identified a structural determinant on the 1E11 molecule, and therefore it was not recognized by the rabbit used to produce the BHV-1 antiserum.

Acknowledgments. The authors wish to acknowledge and thank Dr. Michael Hall for the HPLC preparations, Dr. Jerome Sacks for statistical analysis, Dr. Martin Van Der Maaten for critical review of this manuscript, Ms. Donna Bortner for excellent technical assistance, Ms. Ione Peterson for invaluable assistance in the preparation of monoclonal antibodies, and the National Veterinary Services Laboratories, Ames, Iowa, for providing the rabbit anti-BHV-1 antiserum.

References

1. Glorioso, J.C., and J.W. Smith. 1977. Immune interactions with cells infected with herpes simplex virus: antibodies to radioiodinated surface antigens. J. Immunol. 1:114.
2. Roizman, B., and P.G. Spear. 1971. Herpes virus antigens on cell membranes detected by centrifugation of membrane-antibody complexes. Science 171:298.
3. Olshevsky, U., and Y. Becker. 1972. Surface glycoproteins in the envelope of herpes simplex virus. Virology 50:277.
4. Sarmiento, M., M. Haffley, and P.G. Spear. 1979. Membrane proteins specified by herpes simplex viruses. III. Role of glycoprotein VP7(B$_2$) in virion infectivity. J. Virol. 29:1149.

5. Manservigi, R.P., P.G. Spear, and A. Buchan. 1977. Cell fusion induced by herpes simplex virus is promoted and suppressed by different viral glycoproteins. Proc. Natl. Acad. Sci. USA **74**:3913.
6. Bolton, D.C., Y.C. Zee, and A.A. Ardans. 1983. Identification of envelope and nucleocapsid proteins of infectious bovine rhinotracheitis virus by SDS polyacrylamide gel electrophoresis. Vet. Microbiol. **8**:57.
7. Misra, V., R.M. Blumenthal, and L.A. Babiuk. 1981. Proteins specified by bovine herpesvirus 1 (infectious bovine rhinotracheitis virus). J. Virol. **40**:367.
8. van Drunen Littel-van den Hurk, S., and L.A. Babiuk. 1985. Antigenic and immunogenic characteristics of bovine herpesvirus type-1 glycoproteins GVP 3/9 and GVP 6/11a/16, purified by immunosorbent chromatography. Virology **144**:204.
9. van Drunen Littel-van den Hurk, S., J. van den Hurk, J.E. Gilchrist, V. Misra, and L. A. Babiuk. 1984. Interactions of monoclonal antibodies and bovine herpesvirus type-1 (BHV-1) glycoproteins: characterization of their biochemical and immunological properties. Virology **135**:466.
10. Linthicum, D.S., and M.B. Bolger. 1985. Using molecular mimicry to produce anti-receptor antibodies. BioEssays. **3**:213.
11. Strosberg, A.D., J.G. Guillet, S. Chatmat, and J. Hoebeke. 1985. Recognition of physiological receptors by anti-idiotypic antibodies: molecular mimicry of the ligand or cross-reactivity? Curr. Top. Microbiol. Immunol. **119**:91.
12. Noseworthy, J.H., B.N. Fields, M.A. Dichter, C. Sobotka, E. Pizer, L.L. Perry, J.T. Nepom, and M.I. Greene. 1983. Cell receptors for the mammalian reovirus. I. Syngeneic monoclonal anti-idiotypic antibody identifies a cell surface receptor for reovirus. J. Immunol. **131**:2533.
13. VanDeusen, R.A., and C.A. Whetstone. 1981. Practical aspects of producing and using anti-viral monoclonal antibodies as diagnostic reagents. Proc. Am. Assoc. Vet. Lab. Diag. **24**:211.
14. van Drunen Littel-van den Hurk, S., and L.A. Babiuk. 1986. Synthesis and processing of bovine herpesvirus 1 glycoproteins. J. Virol. **59**:401.
15. van Drunen Littel-van den Hurk, S., J. van den Hurk, and L.A. Babiuk. 1985. Topographical analysis of bovine herpesvirus type-1 glycoproteins: use of monoclonal antibodies to identify and characterize functional epitopes. Virology **144**:216.

Index